Statistical Procedures for Analysis of Environmental Monitoring Data and Risk Assessment

ISBN 0-13-675018-4
90000
9 780136 750185

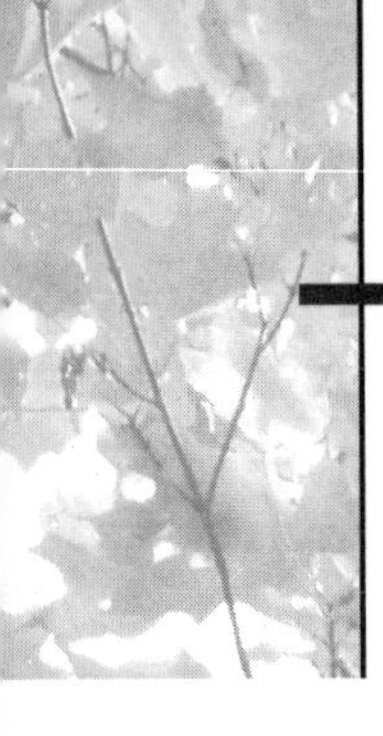

Prentice Hall PTR Environmental Management and Engineering Series

Ritchie and Hayes: ***A Guide to the Implementation of the ISO 14000 Series on Environmental Management***

Louvar and Louvar: ***Health and Environmental Risk Analysis: Fundamentals with Applications***

McBean and Rovers: ***Statistical Procedures for Analysis of Environmental Monitoring Data and Risk Assessment***

Yen: ***Environmental Chemistry: Essentials of Environmental Chemistry for Engineering Practice, Volume 4A***

Yen: ***Environmental Chemistry: Chemical Principles for Environmental Processes, Volume 4B***

Crognale: ***Environmental Management Strategies: The 21st Century Perspective***

Statistical Procedures for Analysis of Environmental Monitoring Data and Risk Assessment

Edward A. McBean
Associate, Conestoga-Rovers and Associates

Frank A. Rovers
President, Conestoga-Rovers and Associates

Prentice Hall PTR, Upper Saddle River, New Jersey 07458
http://www.phptr.com

Editorial/Production Supervision: *Precision Graphics*
Acquisitions Editor: *Bernard Goodwin*
Marketing Manager: *Kaylie Smith*
Cover Design: *Anthony Gemmellaro*
Cover Design Direction: *Jerry Votta*
Manufacturing Manager: *Alan Fischer*

Prentice Hall books are widely used by corporations and government agencies for training, marketing, and resale.

The publisher offers discounts on this book when ordered in bulk quantities.
For more information, contact Corporate Sales Department, phone: 800-382-3419;
fax: 201-236-7141; e-mail: corpsales@prenhall.com or write:
Prentice Hall PTR
Corporate Sales Department
One Lake Street
Upper Saddle River, NJ 07458

Printed in the United States of America

10 9 8 7 6 5 4 3 2 1

ISBN 0-13-675018-4

Prentice-Hall International (UK) Limited, London
Prentice-Hall of Australia Pty. Limited, Sydney
Prentice-Hall Canada Inc., Toronto
Prentice-Hall Hispanoamericana, S.A., Mexico
Prentice-Hall of India Private Limited, New Delhi
Prentice-Hall of Japan, Inc., Tokyo
Simon & Schuster Asia Pte. Ltd., Singapore
Editora Prentice-Hall do Brasil, Ltda., Rio de Janeiro

This book is dedicated to
Matthew, Derek, and Melissa,
and
Wendy, Tim, Chris, and Robin,
for they are the future

CONTENTS

PART IV RISK

CHAPTER 12

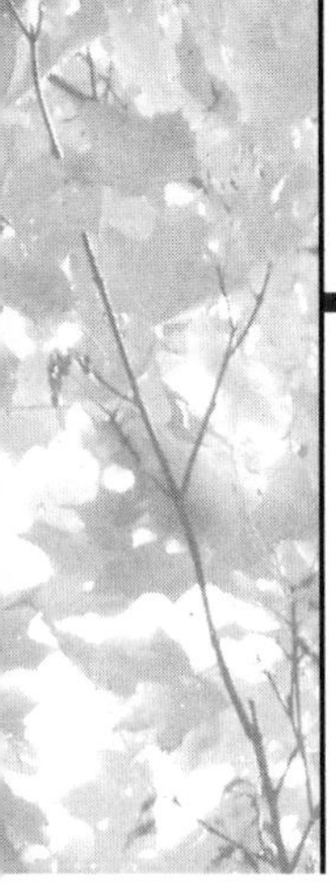

PREFACE

The intent in this book is to carefully explain features of various statistical analysis procedures. Unlike much of the professional literature, however, this text makes special effort to describe statistical techniques in terms comprehensible to the non-statistician. This is accomplished by downplaying mathematical notation, comprehensively explaining the development of equations, and emphasizing example applications. Thus, as successive problems of environmental-monitoring interpretation are developed, the text describes through use of simple examples how each procedure is utilized. References are provided, with particular emphasis on works describing applications reported in the technical literature. Problems included at the end of the chapters stress fundamentals and increase the usefulness of this book as a classroom text.

The collection and laboratory analyses of samples needed to characterize environmental quality are already expensive. Further, as society expresses increasing concern for environmental protection and as instrumentation technology can detect contaminants at ever lower concentrations, expenditures for monitoring environmental quality will rise with time.

As a direct consequence of the rising costs of environmental monitoring, it is essential to use available environmental quality data effectively. Effective utilization involves answering questions such as, "Is the environmental quality acceptable?" and, "Is the environmental quality improving or deteriorating?" Responding to these types of questions requires interpretation of data, and this stage of assessment is beset with difficulties. Some difficulties with interpreting environmental-monitoring results:

(i) Since the data are frequently expensive to accumulate, the data sets being interpreted are usually very modest in size.

(ii) The data may involve a vector of chemical and biological constituent measurements because consideration must typically be given to a range of constituents. Correlation between the constituents may help the infilling of missing data or the identification of outlier data.

(iii) Early detection of any deterioration in environmental quality is highly desirable because early detection may provide the opportunity for controlling the problem at a lower cost before the problem magnifies. Any procedure for identifying early warning signals must not, however, falsely identify a problem of apparent environmental deterioration when one does not actually exist; nor should it fail to identify a problem when one does exist.

(iv) The vagaries of nature introduce significant noise and sources of variability such as seasonality effects. This can make the identification of trends more difficult.

(v) The derivation of quantitative risk assessments is in many ways data dependent. But will the information returned by these risk estimates be worth additional data collection efforts?

The net result of difficulties such as the five mentioned is that making sense of environmental quality data necessarily involves statistical interpretation. Statistical interpretation procedures must be sensitive to small changes in environmental quality and yet recognize the potentially substantial costs of any additional data collection requirement.

The need for the statistical interpretation of environmental quality data is widespread. The range of concerns for each environmental media—air quality, surface water quality, groundwater quality, and soil contamination—are similar in many respects. Yet there is no single statistical analysis procedure universally applicable to the variety of problems associated with environmental quality data. Instead, the practitioner needs to have an array of statistical procedures available. A multitude of statistical analysis tests are available, but each of the tests possess assumptions that may or may not be appropriate for specific circumstances. Computer programs now becoming widely available facilitate use of various procedures. The difficulty remains for the student and the practitioner to learn which conditions dictate a particular procedure and which conditions render it highly inappropriate.

Following the introduction (Chapter 1), the book is organized into four parts as follows:

Part I Chapters 2 through 5 develop the fundamental measures used to describe data and the distributions employed to describe the data.

Part II Chapters 6 and 7 describe procedures commonly utilized to detect changes occurring over time, the detection of outliers and the mathematical procedures for quantifying coincidental behavior in data sets.

Part III Chapters 8 through 11 describe the bases used in hypothesis testing to determine when there are differences in environmental quality at various locations. Problems of censored data are considered as they influence the utilization of alternative tests.

Part IV Chapter 12 focuses on the interrelationships between risk assessment and the data upon which the risk characterization procedures rely. Simulation procedures for risk characterization using sampling methodologies from probability distributions of data are described.

AUTHORS OF THE BOOK

Edward A. McBean (B.A.Sc. from the University of British Columbia and S.M., C.E., and Ph.D. from the Massachusetts Institute of Technology) is an associate of Conestoga-Rovers and Associates and president of CRA Engineering Inc. Dr. McBean's experience includes more than 20 years as a faculty member at the University of Waterloo and the University of California. Much of the focus of Dr. McBean's research and professional work has been on the specific problem of interpretating environmental quality data. He is the author of more than 300 technical articles and has authored or edited eight books.

Frank A. Rovers (B.A.Sc. and M.A.Sc. from the University of Waterloo) is president of Conestoga-Rovers and Associates, an environmental engineering company with more than 850 employees located in 29 offices. Mr. Rovers has been involved for more than 25 years in a very large number of environmental engineering problems dealing with the complete spectrum of environmental quality issues. Frank is the author of numerous technical journal articles dealing with the interpretation of environmental quality data.

Both authors have been heavily involved in the teaching of professional development courses, including those at the University of Wisconsin-Madison, University of Toronto, Nova Scotia Technical College, and UCLA.

ACKNOWLEDGEMENTS

We are under no delusion that the work reported in this book is just *our* work. Clearly the material is the product of many people's efforts. Our intent was to assemble and organize the considerable range of experience and understanding culled from literature about statistical evaluation of environmental quality data.

In addition to the literature, we have drawn upon the experience and efforts of many individuals, and for this assistance we are grateful. During the years preceding publication, the authors worked closely with many colleagues and students, among them:

- The employees of CRA who so generously provided examples. The advice and assistance of many is acknowledged, with special mention of John Donald, Klaus Schmidtke, Darrell O'Donnell, Mark Schwark, and Wes Dyck.
- All the people who examined drafts of the book and whose comments for improvements were valuable. In particular, useful comments by Bill Lennox, Aditya Tyagi, and many students are gratefully acknowledged.
- The secretarial staff at CRA who so obligingly "revised the last revision." In this respect, special acknowledgement must be given to Maria Manoli, who continued to remain cheerful in the face of numerous rewrites.
- Melissa McBean, whose preparation of figures for this book is also gratefully acknowledged.

To all of the above, we owe our sincere thanks for their assistance.

In an undertaking of the magnitude of this text, it is not possible to avoid errors, and for this we apologize in advance. Any corrections, criticisms, or suggestions for improvements will be greatly appreciated by the authors. We would also welcome any additional information and data that would make future editions of the book more complete.

Edward A. McBean
Frank A. Rovers

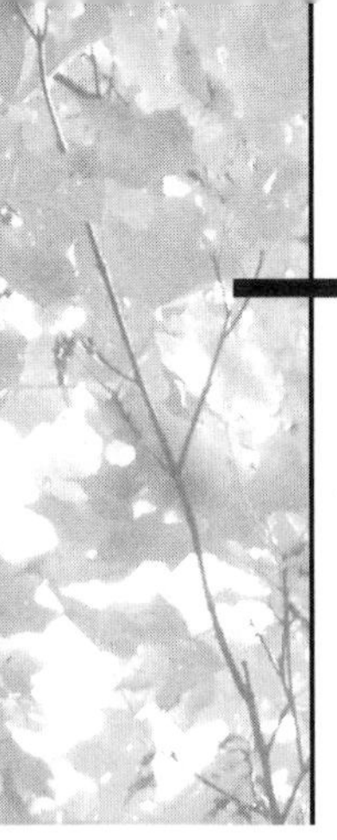

About the Icons

When you see this icon, you will be directed to other volumes in the *Prentice Hall PTR Environmental Management and Engineering Series.* The concept behind the Environmental Management series is to provide professionals and those in training with comprehensive information linking the scientific principles of environmental science and engineering, governmental regulations, and international standards with their practical applications in order to manage the diverse and complex problems facing today's environmental and engineering professionals. For a list of the volumes in this series and their corresponding volume icons, see the series page opposite the title page.

When you see this icon, you will be directed to files on the diskette included with this book.

CHAPTER 1

CHARACTERISTICS OF ENVIRONMENTAL QUALITY DATA

1.1 INTRODUCTION

Concern with the quality of the environment is pervasive. Members of the public, politicians, the media, lawyers, scientists, engineers, and so on are all watchful of current quality levels and the perceived trends in these levels. Much of the concern with environmental quality is real and appropriate, arising from a legacy of inadequate environmental protection, while some of it is only perception. As a result of this environmental awareness, professionals throughout the environmental industry must attend more carefully to environmental quality.

During the past several decades, the public has increasingly demanded that risks associated with exposure to environmental contaminants be reduced. Demands for the protection of human health and ecosystems have been directed in part toward government agencies responsible for soil, water, and air quality. The result has been a dramatic increase in the necessity for professionals to analyze and interpret environmental quality data. The collection of samples from the field, the need for laboratory analyses of these samples, and the time required for interpreting the resulting information can represent enormous expenditures for both governmental agencies and corporations. Consequently, when analyzing the resulting data, diligence must ensure its correct interpretation.

Two among the many reasons for increased focus on environmental quality are perhaps most basic. First, population increases and urban densification have led to locally concentrated pollutant discharges and deteriorated environmental quality. Second, enhanced laboratory technologies have enhanced the ability to measure chemical concentrations at levels not previously quantifiable. With the intensive monitoring efforts now being carried out, instances of deteriorated environmental quality that might otherwise have remained undetected may now be identified.

Current focus on environmental quality has intensified the need to understand the evolution of environmental quality conditions and their assessment.

While partial understanding of environmental change and its assessment is possible through statistical interpretation of environmental quality data, by no means is it sufficient in isolation. The use of statistics is only one of the tools we employ when interpreting environmental quality data. The statistical analyses described in this book must be tempered with numerous other considerations—mathematical modeling of air quality and surface as well as subsurface water and soil quality, for example. Statistical analyses are not an interpretation of the facts, but when properly used, these analyses make the facts easier to see and allow other evidence to enter into judgments about environmental phenomena (Unwin et al., 1985).

As will be seen in the chapters that follow, an extensive theoretical basis exists for statistical methodologies. However, much of the statistics literature is premised on the availability of very lengthy data records, which seldom exist for environmental phenomena. In interpreting the data, we must therefore understand the assumptions and limitations of the statistical procedures and how these affect the interpretation. In addition to brief data sets, other difficulties of dealing with environmental data include the presence of numerous parameters, high degrees of variability in some constituents, and censored data where "censored" indicates that the magnitude of a constituent is known only as being less than some specified magnitude. As well, analyses must account for the frequent occurrence of incompatible data due to different sampling methods, different laboratory analytical methods, and/or different times or spatial intensities of monitoring. Frequently, some of the information available is anecdotal. For example, some data are the result of nonstatistical sampling and grab samples, while other studies may be vague about where, how, and when the samples were collected.

Environmental processes are characterized as conceptually complex, multifaceted, chaotic, and dynamic; in using statistics, however, we try to make sense of the complexity. As a result, we cannot use many of the procedures described by statisticians in the available texts (at minimum we have to be inventive in the application of these procedures). This book addresses these concerns by providing examples and discussions of the advantages and disadvantages of the various statistical analysis procedures and by using statistics to determine which procedure to use. Clear, complete, statistically accurate, and understandable information is essential in making informed decisions. The study of statistics attempts to model order in "disorder," as in the quantification of equation error or measurement error. But one must always temper statistics with a clear understanding of the problem so that spurious information is not introduced nor valid information omitted. Statistical analyses of data are not an interpretation of the facts; statistical analyses are just another way of making the facts easier to see, and therefore interpret (McBean and Rovers, 1990).

1.2 CHARACTERISTICS OF ENVIRONMENTAL QUALITY DATA

The nature of the variability of environmental quality data will greatly influence how the statistical analysis of the data is undertaken. The specifics of statistical analyses will depend on the way the phenomena of interest is defined and sampled. In general, the ability of a sample of environmental quality data to characterize the population from which it is drawn is related to (a) the size of the sample, (b) the degree to which it was selected at random, and (c) the degree of independence among the observation that make up the sample.

1.2.1 Indications of the Sources of Variability in Environmental Data

If extensive monitoring requirements are specified in governmental legislation, this will translate into sizable expenditures in terms of time and dollars. Regardless of time and money spent, however, samples of environmental quality are always just subsets of the populations that are of interest. To clarify this somewhat, some of the features that contribute to the variability and problems of data analysis in ground water quality phenomena are described in Table 1.1. The result is that in many assessments of environmental phenomena, estimates of groundwater quality must be developed from only very brief data records, "brief" both temporally and spatially. The result is that there may be substantial quantities of data, but only a small amount may be usable for specific applications.

These features stress the difficulty with statistical interpretation of environmental data. The analytical procedure must be sensitive to small changes (e.g., early detection of contamination is desirable), and yet a point of diminishing marginal returns also occurs. There may be more data collected than is necessary to make a decision, and thus some of the money spent in data collection may be unnecessary.

The result is that we must be inventive in terms of how statistical analyses proceed. Many standard statistical analysis techniques that are valid for long records have little credibility if only a brief record is available; because the techniques must be modified to correct for biases created due to the brevity of the record. A further complicating factor is that many of the data records are highly variable or "noisy" due to, for example, seasonal phenomena. An additional consideration arising in part with improved instrument technology is that we can measure features that previously were reported only as censored (less than) data. Further, a number of chemicals have maximum concentration levels (MCLs) to which humans and the environment are exposed while not incurring injury, where these MCLs are very close to the technological instrumentation capability. The result is that problems associated with statistical analyses of censored data sets are increasing.

For the variety of reasons previously indicated, there is not a universally appropriate approach to statistical analyses. Instead, what is frequently needed is

TABLE 1.1 Indications of features that contribute to the variability and problems of data analysis in ground water quality

- Sample collection may involve drilling, sampling, and laboratory analysis for many water quality constituents. The expense argues for utilization of brief records.
- Sample collection and laboratory analyses have inherent difficulties resulting in uncertainties in subsequent interpretation of the findings.
- Many important groundwater phenomena take years to evolve, making the available timeframe for sampling programs only statistical "windows" of temporally varying processes.

a series of approaches each of which possesses attributes useful for addressing a particular question.

1.2.2 Independence of Successive Data Values

Time-series analyses are pertinent to the problem of estimating trends and cycles (e.g., seasonal variations). For example, consider the tendency for a ground water monitoring well yielding high chloride concentrations today to yield similar values tomorrow, and for nearby monitoring wells to also yield sampling aliquots with elevated chloride concentrations. These types of sampling results are not necessarily independent, one from another. Similarly, replicate sampling (e.g., the splitting of field or laboratory samples into several samples) does not create independent samples. As a result, there are differing degrees of independence of monitoring results and this continuum must be considered during the statistical analyses of the resulting data.

Many statistical analysis procedures assume independence of data. Dependent samples exhibit less variability and statistics determined from dependent data will therefore have underestimated sample variances. Dependence can severely influence the results of testing any hypothesis. (Example 1.1 demonstrates hypothesis testing.)

Concerns with hypothesis testing include those dealing with independence/dependence of successive samples. One approach to minimizing dependence in samples is to allow sufficient time between sampling times to allow the "real time memory" of the system to be exceeded. For example, the statistical analysis of annual peak flows (the highest flow in an individual year) involves sta-

---

EXAMPLE 1.1—Example of Statistical Hypothesis Testing

Consider the question of whether a landfill is leaking leachate which will contaminate the underlying ground water, as depicted on Fig. 1.1.

One way to consider this question is to monitor the ground water quality both upgradient (point A) and downgradient (point B) from the landfill. We might then compare the quality at B and A to determine if there is a difference. Thus, an hypothesis might be the following:

Hypothesis: There is no statistically significant difference between the quality at the two locations

Outcome 1. If we accept the hypothesis, then we are concluding there is insufficient information to indicate the landfill is leaking.

Outcome 2. If we reject the hypothesis, evidence exists to indicate that the landfill is leaking.

The question of hypothesis testing in environmental phenomena is a recurring one. The details of hypothesis testing will be a recurring question throughout this book, and the quantitative aspects of hypothesis testing will be left to later chapters.

---

tistical analysis of independent events because the peak flow in one year is unlikely to be related to the peak flow in a later year.

For some situations, we can avoid the problem of dependent data. We might "deseasonalize" the data by removing the periodic characteristic(s) associated with the seasons. However, the ability to remove seasonality is typically constrained by the brief length of environmental data which limits our ability to isolate the seasonal variability from the other sources of variability present in the data set.

1.2.3 Uncertainties and Errors in Environmental Quality Data

There are different levels of "observational" data. For example, *proxy* data are observation of one variable that have a high probability of being indicative of levels of another variable. Such data are indirect "observations." Another example is remotely sensed data by which many "indices" are derived from satellite data, including temperature and vegetation. Image classification techniques may allow patterns in "data" to be recognized as signatures of certain environmental features.

In the strictest sense, there are virtually no data that are direct observations of a variable. Data on animal (or human) demography are collected by observing various signs, such as spores, tracks, nests, houses, income tax reports, and so on. Even something as obviously observable as the digital elevation is subject to interpretation according to the methods used to produce the measurements. The validity of these "data" depends on features such as one's confidence in a particular measurement method and the reliability of calibration techniques of the instruments, all of which must then be considered when interpreting the data.

Quantifying observations involves the employment of sensory techniques, indirect measurements of related variables, and various levels of processing. It can

FIGURE 1.1 Schematic depiction of monitoring in the vicinity of a landfill

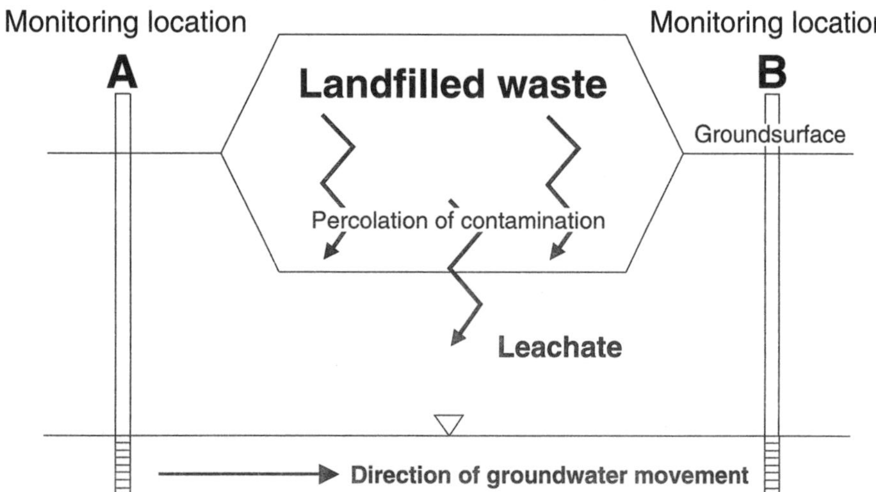

then be argued that data are observed only in the context of the experimental design in which they are produced. It becomes a matter of interpretation as to what degree of processing is appropriate to produce a quantified observation that will then be called "data."

Errors in sampling procedures, inadequate sample storage and preservation techniques, and laboratory analytical errors are examples of errors in environmental data sets. As a demonstration of the multifaceted initiation points for such errors to exist in a data set, further examples of the sources of error in the collection and analysis of groundwater quality data are listed in Table 1.2 and for air quality in Table 1.3.

There is always a degree of uncertainty associated with each discrete measurement of environmental quality. In interpreting data, each discrete measurement is really a range of statistically probable values instead of a single value. There are two subdivisions of reproducibility criteria, namely replication and repeatability. Replication is when two or more results are obtained by the same operator in a given laboratory using the same apparatus of successive determinations on identical test material within a short period of time. Often this is done during quality assurance and quality control testing of a laboratory, to ensure that the lab results are trustworthy. Alternatively, repeatability is a quantitative expression of the random error associated in the long run with a single operator in a given laboratory obtaining successive results with the same apparatus under constant operating conditions on identical test material. Obviously, the requirements for quality assurance and quality control can be substantial.

Many of the statistical analyses described in this book are concerned with sampling errors and the estimation of population characteristics from samples of data. The fact that sampling errors are inherent in random data does not mean, however, that statistical manipulation and sophistication can in any way overcome faulty data. The quality of any statistical analysis is no better than the quality of the data utilized. Furthermore, statistical considerations should not be used

TABLE 1.2 Examples of sources of error in the sampling and analysis of ground water quality data

- Sampling of a nonhomogeneous region in which wells and springs intersect more than one chemical type in water can lead to misinterpretation.
- Piezometers and wells that are inadequately flushed out prior to sampling of groundwater may render a sample of the groundwater unrepresentative of conditions in the adjacent soil environment.
- Sampling stations that are subject to temporal variations of chemical concentrations can exhibit significant sampling error.
- Cross-contamination of a sample may occur at a time of sampling as a result of an unclean container into which the sample is placed.
- An error in the laboratory protocol of the experiment can occur during the laboratory analysis.
- Improper preservation techniques can alter the sample. For example, groundwater samples are often particularly susceptible to changes in the pressure of oxygen and carbon dioxide and improper sample storage. Improper preservation techniques can result in a chemical alteration of the sample as it adjusts to new equilibrium conditions. In the case of pH levels, groundwater samples have been shown to increase as much as 1.0 pH units due to CO_2 escape to the atmosphere during storage.

TABLE 1.3 Examples of possible sources of error in the collection of air quality data

- Instrumentation error due to poor calibration or "drift" of the calibration of the instrument with time
- Channel error incurred during transmission from the monitoring locations to a central data processing unit
- Fluctuations in meteorological conditions or in quantities being released by the emission source resulting in nonrepresentative air quality conditions

to replace judgment and careful thought in analyzing data. Statistics must be regarded as a tool or an aid to understanding, but never as a replacement for careful thought.

1.2.4 ISO Considerations

In response to concerns about quality assurance and quality control (QA/QC), the ISO 14001 standard is a common-sense approach to managing environmental issues that integrate management controls. Ritchie and Hayes (1998) describe a guide for managing environmental programs to provide evidence that an environmental program is comprehensive regarding environmental management. As Ritchie and Hayes indicate, the characterization of physical parameters, chemical constituents, and biological organisms should be a cornerstone of environmental programs and process engineering.

Kuhre, in a series of books (1995, 1996, 1997), describes the background information necessary to understand the ISO 14000 management system, the auditing system, and the labeling/marketing. The need for QA/QC in dealing with all aspects of monitoring and reporting are pervasive and must precede statistical analyses.

1.3 SOME SUMMARY INDICATIONS OF APPROACHES FOR STATISTICAL ANALYSES

The concern with the statistical interpretation of data is widespread. However, the variability encountered in one circumstance is quite possibly very different from that encountered in another circumstance. There is no convenient recipe for an approach that can be universally applied. Instead, statistical analyses of environmental data have become the science of collecting, analyzing, and interpreting data with the findings at each decision point assisting in identifying the next stage of analysis. Statistics is concerned with scientific methods for collecting, organizing, analyzing, summarizing, and presenting data, as well as with drawing valid conclusions and making reasonable decisions. Statistical analyses do not consist of a standard set of rules. Instead, analyses involve successive tests and refinements, with each test an improvement in understanding the data (and its information content). The findings of the tests may well be that additional statistical analyses are needed.

Since there is no convenient recipe, the practitioner needs to have available at his or her disposal a set of approaches, with varying capabilities for a particular application to a class of problems. In selecting the procedure for use in a particular application, there are no absolute rules, only guidelines. To a large extent, the selection of the best procedure involves careful scrutiny of both the characteristics of the problem at hand and the assumptions implicit in the particular statistical interpretation.

An analyst must still understand the basics in terms of both how to characterize a problem and the method by which the results may be interpreted. With the automated calculation procedures available in today's software, it is all too easy to employ a computer package without understanding the basis for the statistical procedures. The intent in this book is to explain the features of the various techniques in terms understandable to the nonstatistician and to provide some order to the available procedures by presenting the advantages and disadvantages, and the limitations of the procedures. The discussion included in this book describes some of the sources of error and how these error sources should be reflected in the statistical analyses and in their interpretation. Proper application of statistical methods by someone who understands the utility and limitations of these methods can be most helpful in revealing the information that the data hold.

The emphasis in the following chapters is on examples to develop and illustrate the equations and relationships and then to demonstrate the usefulness and application of the various procedures. References are provided for those wishing to pursue specific features in greater depth. Much of the theoretical background is omitted, with the focus being on the engineering application to environmental quality data.

The various statistical methods are tools for data analysis which, like any tools, have proper and improper applications. The person using a statistical method is responsible for its proper application because the value of the results obtained depends on it. Even in the best circumstance, however, statistical data analysis can provide only evidence, never proof. Inferential statistics are aimed at distinguishing between random noise in the data and the real effects that are of interest. Only through careful consideration and interpretation of all the evidence can one hope even to begin finding answers to questions about causes for and effects on environmental quality.

Morrison and Henkel indicated "Alas statistical inference is not scientific inference. To have the latter we will have much more than the facade that claims of (statistical) significance provide. There are, of course, no computational formulas for scientific inference: the questions are much more difficult and the answers much less definite than those of statistical inference".

In this context the contents of the book are organized in the following manner. Chapter 2 focuses strictly on the fundamentals of statistical characterization of data. Chapters 3 through 5 examine the attributes of commonly used probability distributions, as appropriate to environmental quality data. Chapter 6 uses these distributions to develop alternative types of control charts and to identify data outliers. Chapter 7 examines the benefits of correlation and regression to better summarize data behavior. Chapters 8 through 11 examine different procedures for hypothesis testing. Chapter 8 considers relatively standard tests, whereas Chap. 9 examines procedures for multiple comparisons, and Chaps. 10 and 11 describe procedures appropriate when the data include numerous censored (less than) data. Finally, Chap. 12 draws the preceding chapters together as statistical interpretation influences risk assessment and data management.

Statistical interpretation of environmental quality data has a major role to play in areas such as qualifying effects, assessing consequences, measuring risks, and interpreting evidence. The intent is that this book will assist the student and the practitioner in all of these areas.

1.4 REFERENCES

KUHRE, W.L. *ISO 14001 Certification, Environmental Management Systems.* Upper Saddle River, NJ: Prentice Hall, 1995.

KUHRE, W.L. *ISO 14020s Environmental Labeling—Marketing.* Upper Saddle River, NJ: Prentice Hall, 1996.

KUHRE, W.L. *ISO 14001 Certification, Environmental Management Systems.* Upper Saddle River, NJ: Prentice Hall, 1997.

McBEAN, E., and F. ROVERS. "Flexible Selection of Statistical Discrimination Tests for Field-Monitored Data," in *Groundwater and Vadose Zone Monitoring,* pp. 256–265, eds. D.M. Nielsen and A.I. Johnson. Philadelphia, PA: American Society for Testing and Materials, 1990.

RITCHIE, I., and W. HAYES. *A Guide to the Implementation of the ISO 14000 Series on Environmental Management.* Upper Saddle River, NJ: Prentice Hall, 1997.

UNWIN, J., R.A. MINER, G. SREVERS, and E. McBEAN. "Groundwater Quality Data Analysis," National Council for Air and Stream Improvement Technical Bulletin, No. 462 (1985).

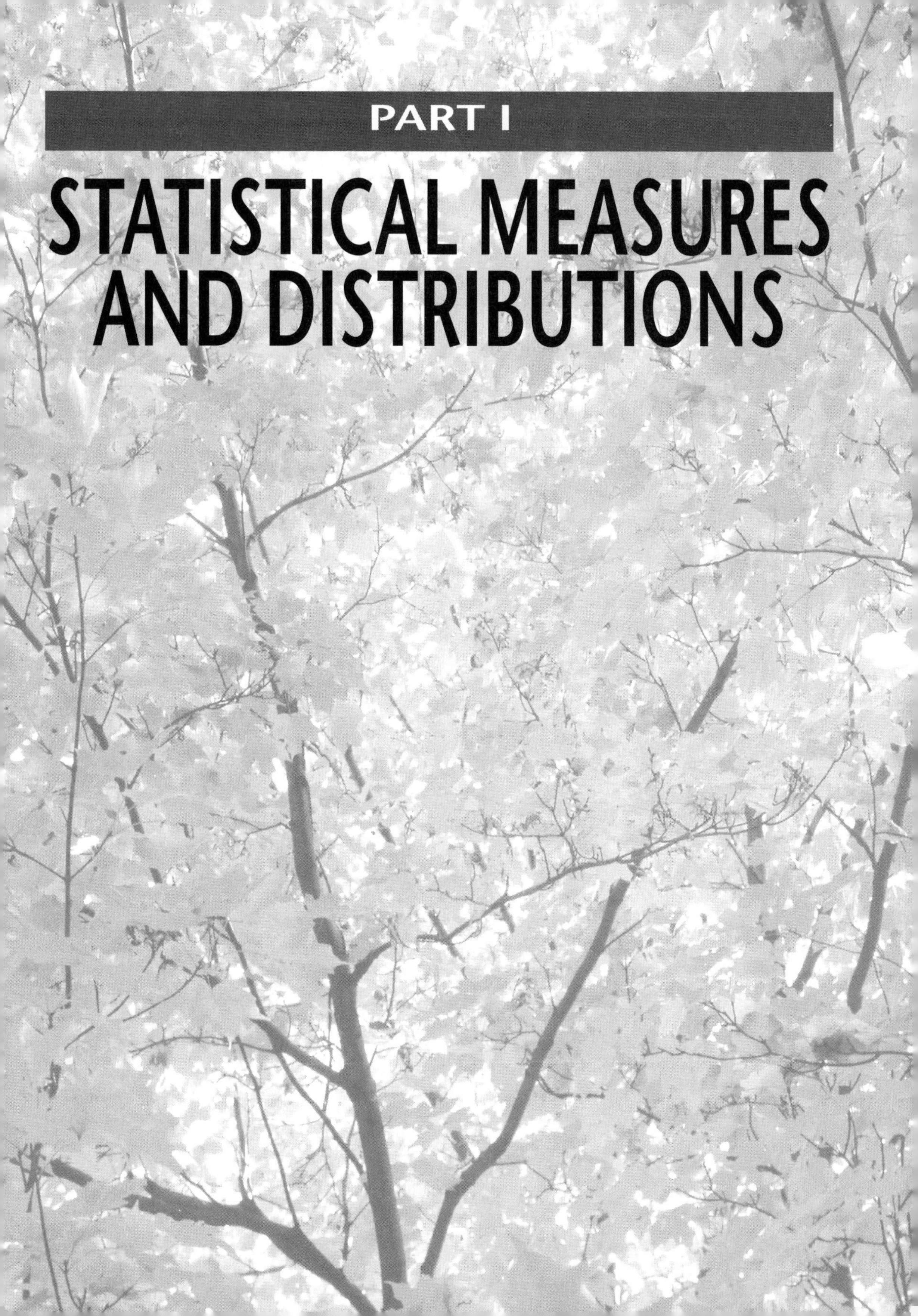

PART I

STATISTICAL MEASURES AND DISTRIBUTIONS

CHAPTER 2

STATISTICAL CHARACTERIZATIONS OF DATA

2.1 INTRODUCTION

As emphasized in Chap. 1, for many environmental quality data sets, only brief data record lengths are available. The brevity of environmental quality data bases is the result of numerous causal factors, for example,

1. the expense of collecting and analyzing field data,
2. limited sampling opportunity because of the lengthy evolution of environmental phenomena, and
3. heterogeneities arising from phenomena such as seasonality, trends, missing data, and the vagaries of nature, all of which limit the record length available for quantification of individual features.

Factors such as these often make a considerable challenge of extracting information from the available environmental data record.

The brevity of most environmental data records means that the statistical analysis procedures described in many books have only limited applicability to the analyses of environmental quality data. Always keep in mind that we have only samples, not population statistics, and carefully consider what the records' brevity implies for the statistical analyses of environmental quality data.

In addressing the issue of 'samples' versus 'populations', this chapter develops summary statistical descriptors of data distribution. In looking for ways to summarize the data set, we move from individual data points to summary descriptors of the data set.

2.2 SAMPLES AND POPULATIONS

Observed data represent only a sample from the population, so statistical procedures must be sensitive to this fact. Theoretically, as the sample size becomes very large, the summary descriptors of the samples approach those of the population values. Generally, however, samples much smaller from the population are collected because it is too expensive and time consuming to collect large numbers of measurements.

Statistical analyses work well only if the sample is typical in a way that enables us to make predictions about the entire population. A sample statistic is a value that expresses an important property of the data set and this information is utilized to estimate specific characteristics of the population. Because of the variability inherent in environmental quality data, a sample statistic will usually differ from the corresponding parameter of the population, but the sample information is the only information currently available. The process of estimating the characteristics of a population through sampling is referred to as *inference,* the limitations of which will be a focus throughout this book.

In mathematical terms, we are dealing with a sample of n individual values, as

$$\{x\} = x_1, x_2, x_3, \ldots x_n \qquad [2.1]$$

where { } will be used throughout the text to indicate a vector of individual values. The n individual observations are a sample from the set of possible outcomes of the experiment (population).

2.3 PROBABILITY AND STATISTICS

Probability and statistics are two different concepts. Probability, a measure of the likelihood or chance that something will happen, is always a number between zero and one, where zero represents the impossible event and one represents the certain event. Alternatively, statistics are aspects that are measured or counted. The interesting part of statistics begins when attempts are made to find the relationship between one statistic and another.

2.4 GRAPHICAL DATA DESCRIPTORS

The first step in data analysis is often a graphical study of the characteristics of the data sample. Graphical data descriptors illustrate the character of the data set, but do not summarize it. If, for example, you want to identify the probability distribution function (e.g., see Chaps. 3 through 5), it is often helpful to develop a frequency histogram from the data as an estimate of the underlying probability distribution.

EXAMPLE 2.1

A frequency histogram of nickel concentration data as monitored in the groundwater is depicted on Fig. 2.1(a). For example, 11 reported values out of the total of 174 values are within the interval between 60 and 70 µg/L.

Alternatively, the relative-frequency-of-occurrence histogram is presented as Fig. 2.1(b). Thus, for example the ordinate frequency value of concentration in the interval between 60 and 70 µg/L is 11/174 or approximately .063. The sum of the ordinates over the range of individual class intervals equals unity.

2.4.1 Histograms of Data

A histogram is a plot that indicates the frequency at which data occur within discrete intervals. Specifically, the relative number of occurrences of an event (e.g., within the class interval or concentration interval 5 to 10 µg/L) is denoted as $P(E_1)$, which is the probability of event E_1, equivalent to the fraction (n_1/n) if the event occurred n_1 times in n sampled results. The frequency distribution is often represented in the form of a histogram, the principle of which is based on the fact that the area of each rectangle represents the proportion of observations falling in that interval.

FIGURE 2.1(a) Frequency-of-occurrence histogram of nickel concentrations

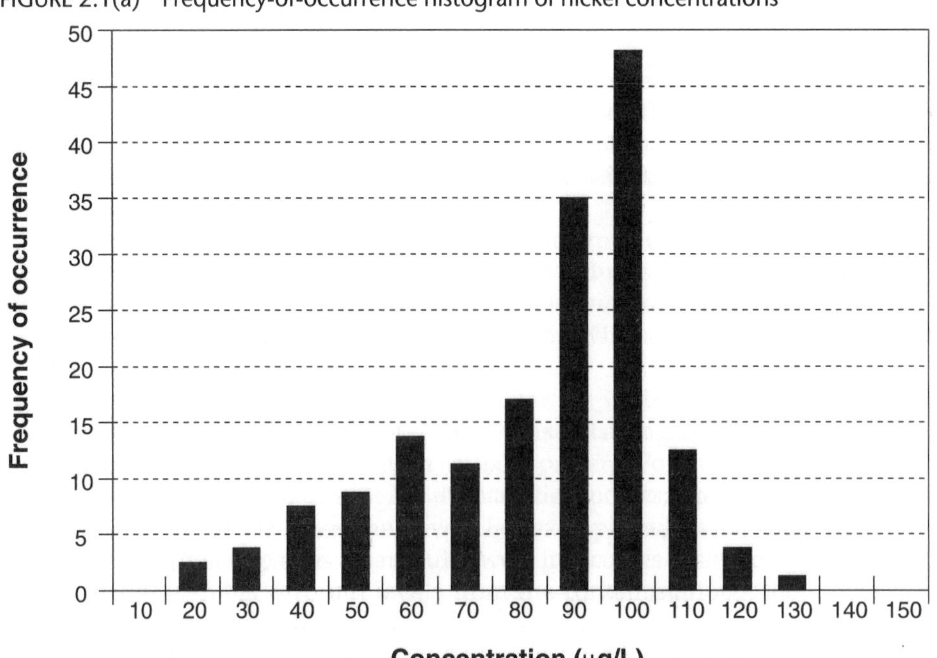

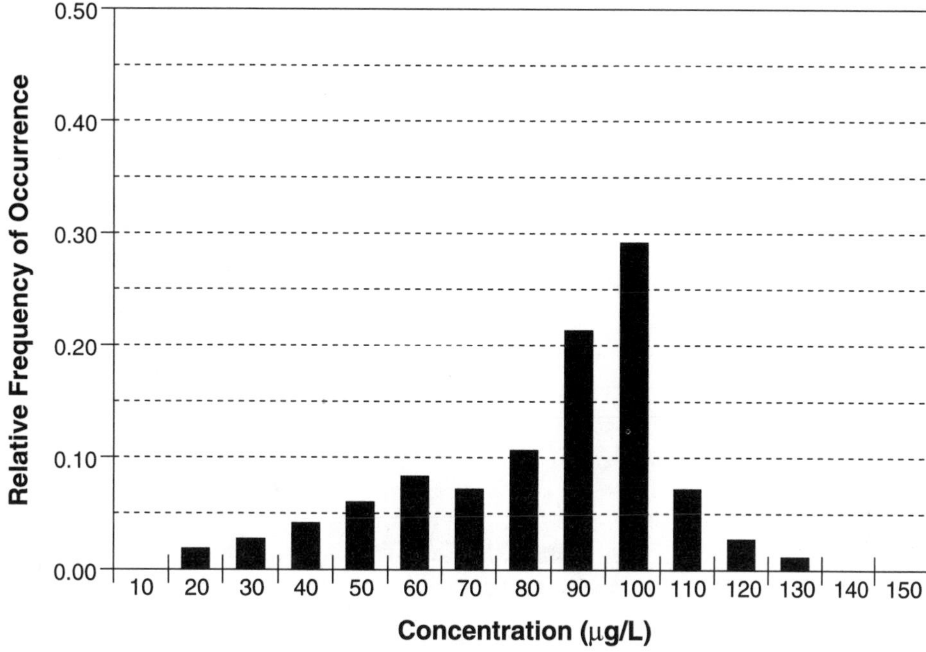

FIGURE 2.1(b) Relative frequency of occurrence of nickel concentrations

A frequency histogram thus consists of a set of rectangles having (a) bases on a horizontal axis (the x-axis) with centers at the designated class marks and lengths equal to the class interval sizes, and (b) areas proportional to the class frequencies. Boundaries of intervals are chosen so that no observation falls on a boundary.

According to Shaw (1964), a class interval is best chosen between one-quarter and one-half the standard deviation of the data (see Sec. 2.5.2 for details on how to calculate the standard deviation). If the class interval selected is too large, the true form of the distribution is masked or blurred. If the class interval selected is too small, then too many gaps appear in the resulting histogram. As the class interval of a histogram decreases for large samples, it becomes easier to pass a smooth curve through the classes. Thus, the effectiveness of a graphical analysis in characterizing the data (and for identifying the probability density function) depends on the sample size and the interval selected to plot the abscissa. For small samples, it is difficult to separate the data into groups that will provide a reliable indication of the frequency of occurrence.

Histograms are a familiar method of displaying numerical information and so provide a rapid visual characterization of the data. Note, for example, how the histograms of concentration data in Figs. 2.2(a) through (c) present three distinct shapes. The data graphed in Fig. 2.2(a) would be referred to as negatively skewed, since the long tail of the histogram is to the left; Fig. 2.2(b) presents data that are

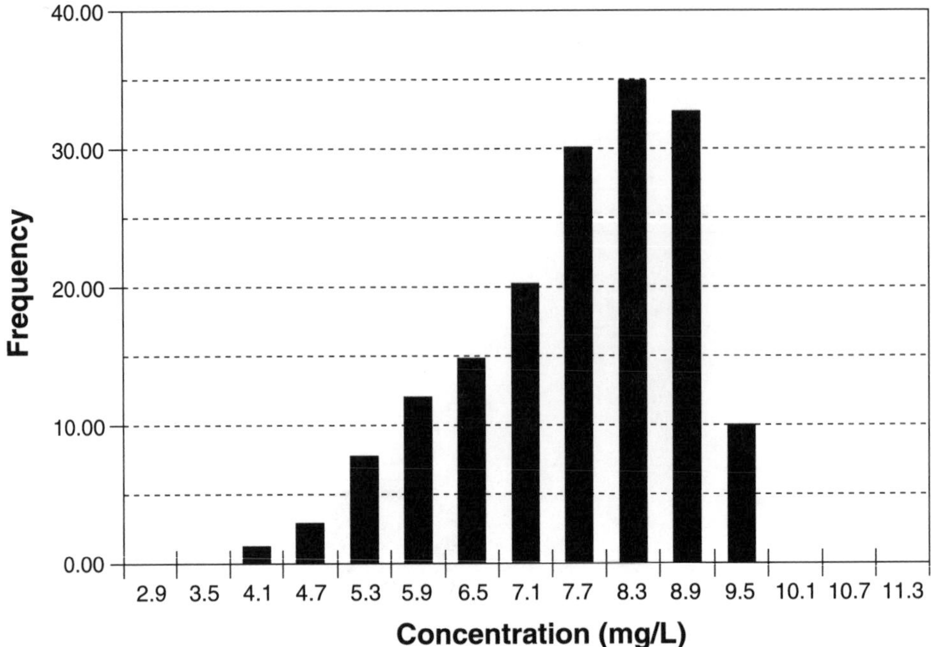

FIGURE 2.2(a) Histogram of dissolved oxygen concentrations as an example of negatively skewed data

FIGURE 2.2(b) Histogram of chloride concentrations as an example of symmetric data

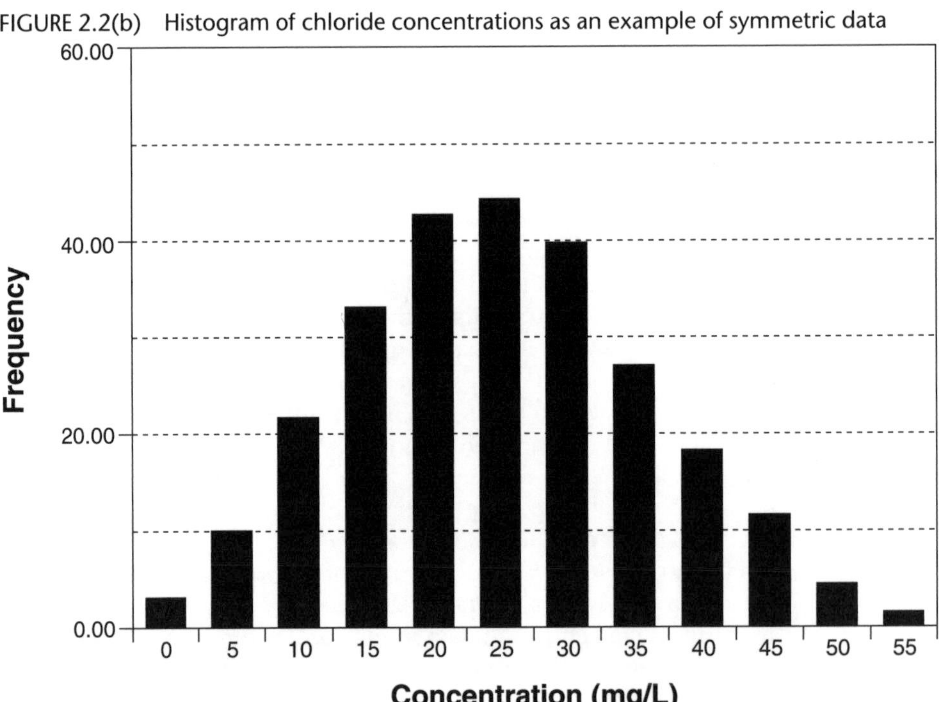

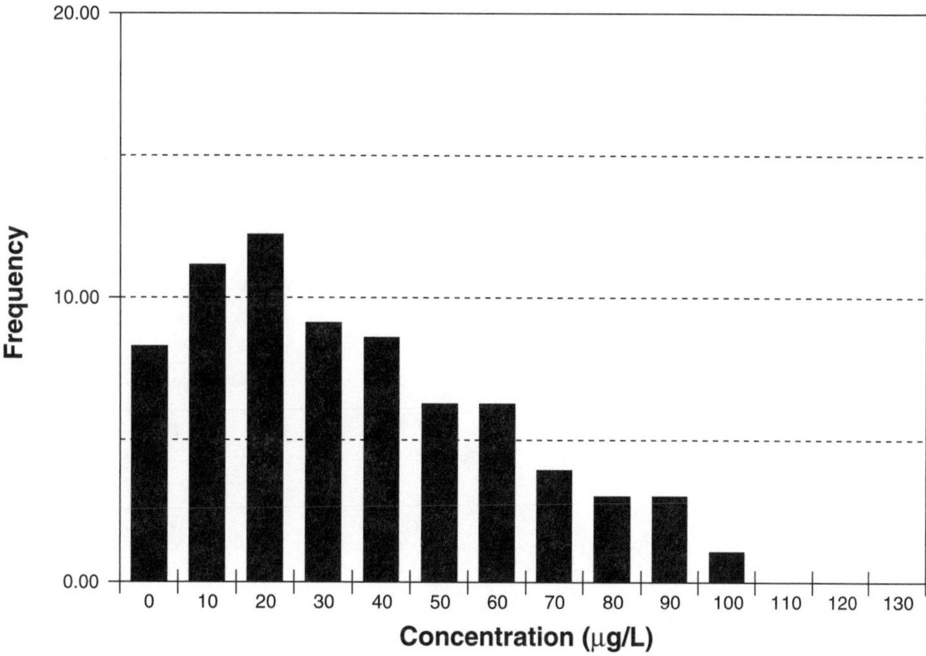

FIGURE 2.2(c) Histogram of lead concentrations as an example of positively skewed data

relatively symmetrical; and Fig. 2.2(c) illustrates data that would be referred to as positively skewed.

To summarize the obvious advantages of histograms as a means of visual representation:

1. The total range of the data in a sample is readily apparent.
2. The most frequently occurring value, the mode, can be easily recognized.
3. The range of greatest abundance can be readily identified.
4. The general shape or character of the density distribution of data is apparent.

2.4.2 Probability Density Functions

As the number of samples n becomes large and the interval discretization small, the histogram will tend to a smooth continuous curve, referred to as the probability density function. Only a very large number of samples will permit a complete delineation of the probability density. A probability distribution is a plot of the probability density (i.e., relative frequency) versus the data values to describe the behavior of a random variable. Often the resulting plot of the probability distribution can be represented as an equation. This mathematical representation is called a probability density function, denoted $f(x)$. Distributions relevant to environmental quality data include the normal distribution, the lognormal distribution, the Poisson distribution, and so on, details of which are examined closely in Chaps. 3

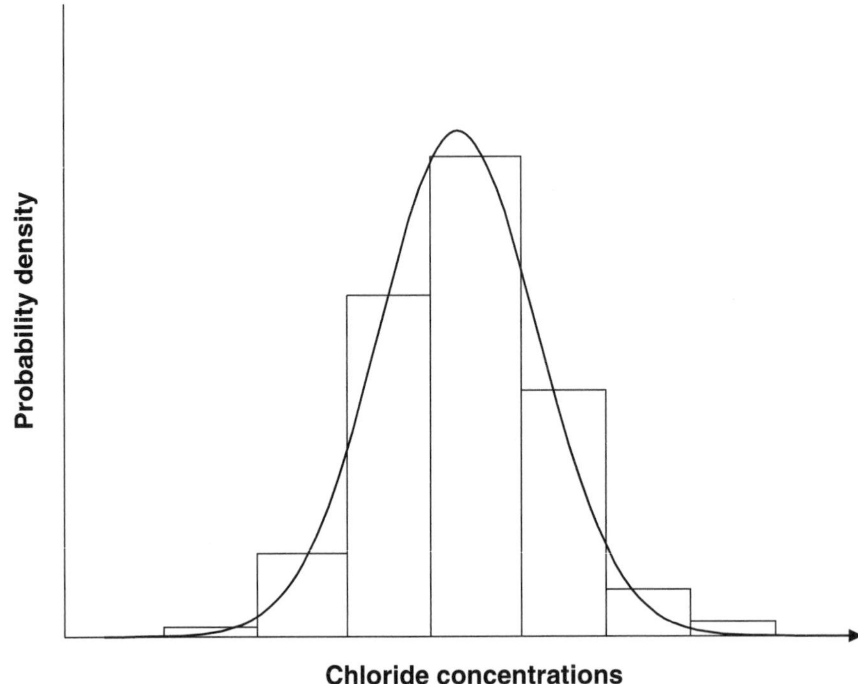

FIGURE 2.3 Probability distribution function fit to histogram of chloride data

through 5. For environmental engineering problems, we frequently assume a distribution and that "fits" the data (Fig. 2.3); that is, the distribution appears to be a reasonable descriptor of the data. There are tests that evaluate the "goodness of fit" of the distribution to the data.

There is no requirement that the probability density be unimodal or single peaked, as it is in Fig. 2.3. Probability density functions of data are generally single peaked, but for particular applications the distribution may be bimodal, such as in Fig. 2.4. Bimodal distributions may, as in Fig. 2.4, be due to seasonal loading patterns where higher wastewater lagoon loadings are discharged to the receiving water body during the winter, as opposed to during the summer. However, for the majority of environmental quality data concerns, the distribution of the data may be assumed to be unimodal and, equally important, most statistical procedures assume unimodal distribution.

We must emphasize that environmental quality monitoring data are not the same thing as any probability density distributions selected to describe them. We do, however, seek a good fit between them. To ensure that the distribution assumed is a reasonable descriptor of the data, one or more of a number of alternative tests will be utilized, as elaborated in Chap. 3.

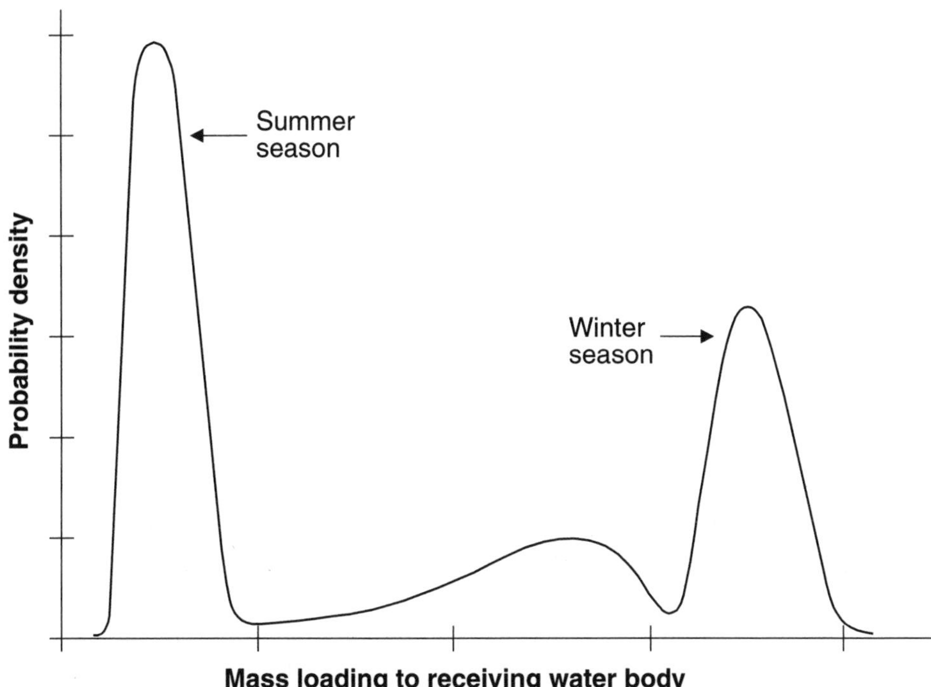

FIGURE 2.4 Example of a bimodal distribution function

2.4.3 Cumulative Distribution Functions

A probability distribution function can also be represented by its cumulative probability distribution function, $F(x)$. The cumulative probability function is obtained by adding (accumulating) the individual increments of the probability distribution function. The cumulative probability distribution function is defined as the probability that any outcome in X is less than or equal to a stated limiting value x. In mathematical terms,

$$F(x) = \text{Prob}\,[X \leq x] = \int_{-\infty}^{x} f(x)dx \qquad [2.2]$$

As x approaches infinity, the area under the curve approaches unity since by definition the area under the probability distribution function is unity.

The probability distribution function is the slope of the cumulative probability function. An example of the cumulative distribution function is depicted in Fig. 2.5, as derived from the data presented in probability distribution format in Fig. 2.3.

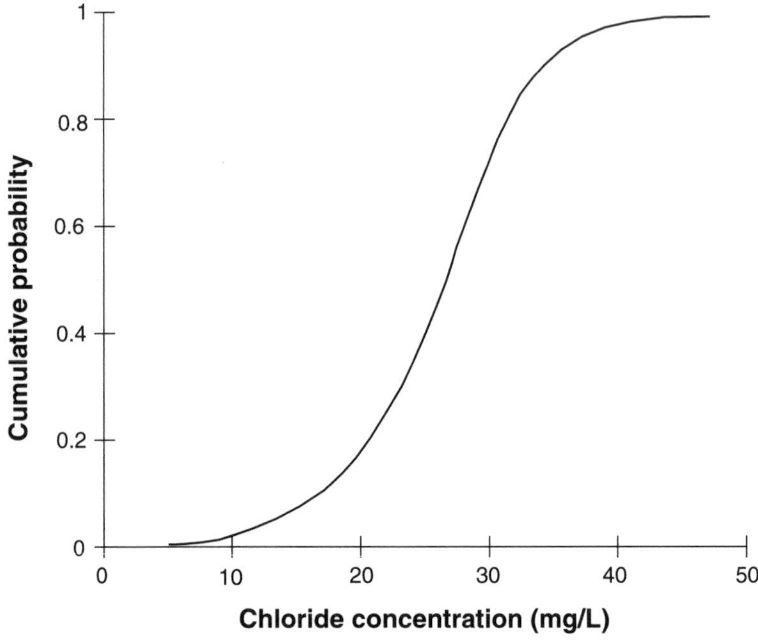

FIGURE 2.5 Cumulative distribution function of chloride concentrations

2.5 SUMMARY MEASURES OF THE DISTRIBUTION OF DATA

There are many mathematical measures that can objectively summarize the nature of a data set. This variety is required for several important reasons including:

1. to help identify an appropriate probability distribution to characterize the data set, and
2. to characterize the data in a manner that allows comparisons between data sets.

For example, we might be interested whether the mean concentration of the river water quality changes as a result of a wastewater discharge. For this problem, we are interested in determining whether the summary statistics of the water quality are different for monitoring results both upstream and downstream of the point of discharge. The parameters utilized for these purposes are generally summary measurements of distribution properties; one parameter is the central tendency of the data and the other is the degree of spread of the distribution.

In keeping with the earlier comments that environmental quality records are normally brief, the emphasis in all of the discussion that follows is on sample statistics, as opposed to population statistics.

2.5.1 MEASURES OF CENTRAL TENDENCY

Measures of central tendency are concerned with identifying the average value of a set of data. However, *average* is a very imprecise term. There are numerous ways to calculate a measure of central tendency, each of which has a different interpretation.

1. **Mean:** The arithmetic mean is the first moment about the origin (sometimes referred to as the data's center of gravity), as calculated by the following equation:

$$\overline{x} = \frac{\sum_{i=1}^{n} x_i}{n} \qquad [2.3]$$

As the number of samples n increases, the sample mean $\overline{x}$ approaches the population mean, μ. The mean is the most common measure of central tendency.

It is important to distinguish carefully between the mean $\overline{x}$ of a set of observations or of a frequency distribution, and the mean μ of the corresponding probability distribution function toward which $\overline{x}$ tends as the number of observations increases. Thus, $\overline{x}$ is the sample mean while μ is the population mean. A measure such as the mean $\overline{x}$, which can be calculated from an observed frequency distribution, is called a *statistic* while the corresponding measure of the probability distribution μ is referred to as a *parameter.*

2. **Median:** The median is the middle value of the set of data or, the median has the property that one-half the values exceed it and one-half are less than the median. If the sample size n is an odd number, the sample median is the middle term when the samples are arranged in increasing order. When n is even, there is no middle term and the median is defined as the average of the term just below the middle and the term just above the middle.

As is apparent from Ex. 2.4, when there are data outliers, the mean may not be a good representation of the central tendency of the data. The median is superior to the mean as a measure of central tendency when the data are skewed.

EXAMPLE 2.2

Values of air quality monitoring results for benzo(a)pyrene were recorded as follows: 5.2, 7.3, 3.4, 4.4, 7.4, 4.3, 2.3, 1.2, 5.6, 4.5, 7.5, 6.1, 4.4 $\mu g/m^3$.

The mean of the monitoring results is:

$$\overline{x} = \frac{63.6}{13} = 4.89 \ \mu g/m^3$$

EXAMPLE 2.3

Determine the median of the air quality measures of benzo(a)pyrene as reported in Ex. 2.2.

First, the values are ranked in order of magnitude as: 7.5, 7.4, 7.3, 6.1, 5.6, 5.2, 4.5, 4.4, 4.4, 4.3, 3.4, 2.3, 1.2. Since the number of samples is odd, the median is 4.5. If the number of samples had been even, then the average (mean) of the middle two numbers is utilized.

EXAMPLE 2.4

Determine the median and the mean of the hydraulic conductivity of a clay soil, given the three measurements provided.

The monitoring values (all in cm/s) are as follows:

3.8×10^{-8} 5.0×10^{-8} 4.8×10^{-6}

The mean value $\bar{x}$ is

$$\bar{x} = 1.63 \times 10^{-6}$$

whereas the median m is

$$m = 5.0 \times 10^{-8}$$

Note the difference in magnitude between the mean and the median values. This occurs because the mean is strongly influenced by the single value, which is two orders of magnitude higher than the second highest value. The result is that, when there are a few, very high values, the mean is not necessarily representative of the central tendency of the data set (since the mean is virtually the largest value divided by the number of samples and is insensitive to the magnitude of the two other values). The relative magnitudes are depicted in Fig. 2.6.

FIGURE 2.6 Comparison of the mean and median values from Ex. 2.4

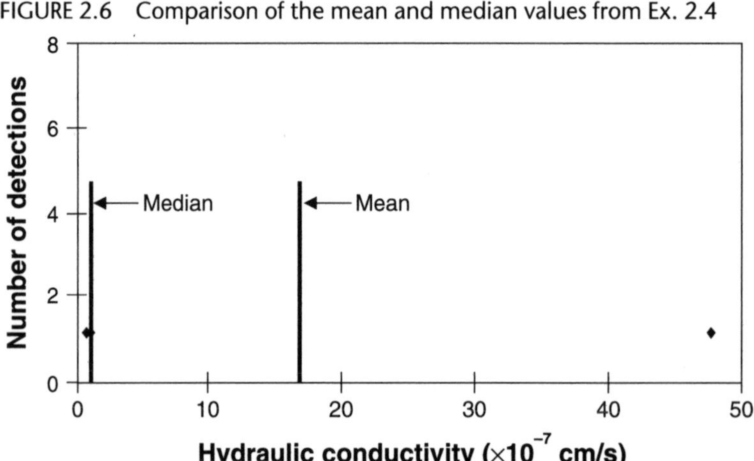

3. **Mode:** The third measure of central tendency is the mode, the most commonly occurring value. The mode is not widely employed because it forms a poor basis for any further arithmetic calculations.

   For a unimodal distribution (i.e., a distribution possessing a single peak) that is moderately skewed (asymmetrical), we have the approximate empirical relation

$$\text{Mean} - \text{Mode} = 3\,(\text{Mean} - \text{Median}) \qquad [2.4]$$

EXAMPLE 2.5

The mode of the data illustrated in Fig. 2.7 is 1.00 mg/L, the median is 2.71 mg/L, and the mean is 4.45 mg/L.

4. **Harmonic Mean:** The harmonic mean H of a set of n numbers $x_1, x_2, x_3, \ldots x_n$ is defined as the reciprocal of the arithmetic mean of the reciprocals of the numbers; that is,

$$H = \frac{1}{1/n \sum_{i=1}^{n} 1/x_i} = \frac{n}{\sum_{i=1}^{n} 1/x_i} \qquad [2.5]$$

FIGURE 2.7 Schematic depiction of three measures of central tendency of data

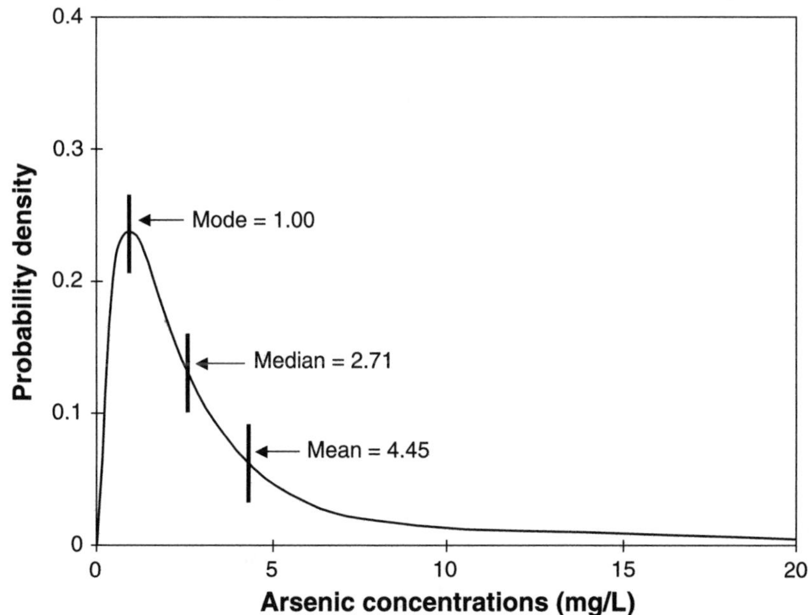

EXAMPLE 2.6

Determine the harmonic mean of the monitoring values of 2, 4, and 8.
Using Eq. [2.5] the harmonic mean is found to be 3.43

5. **Geometric Mean:** The geometric mean G of a set of n positive numbers $x_1, x_2, x_3, \ldots x_n$ is the nth root of the product of numbers

$$G = \sqrt[n]{x_1\, x_2\, x_3 \ldots x_n} \qquad [2.6]$$

The geometric mean of a set of positive numbers $x_1, x_2, x_3, \ldots x_n$ is less than or equal to their arithmetic mean but is greater than or equal to their harmonic mean

$$H \leq G \leq \bar{x} \qquad [2.7]$$

The equality signs in Eq. [2.7] apply only if all the numbers $x_1, x_2, x_3, \ldots x_n$ are identical.

EXAMPLE 2.7

Determine the geometric mean of the numbers 2, 4, and 8. Using Eq. [2.6], the geometric mean is

$$G = \sqrt[3]{2 \times 4 \times 8} = 4$$

EXAMPLE 2.8

The set of numbers 2, 4, 8 has arithmetic mean 4.67, geometric mean 4 and harmonic mean 3.43.

6. **Summary Comments on Measures of Central Tendency:** When the distribution of the data is symmetrical and unimodal, the values of the mean, median, and mode are identical. For example, this is true for the normal distribution. When the distribution of data is skewed, such as the data depicted in Fig. 2.2(a), the relative locations of each of the measures of the central tendency are as schematically depicted on Fig. 2.7.

2.5.2 Measures of the Dispersion of Data; Variance, Standard Deviation, and Range

The most common measures of dispersion or spread of the data around the mean are the variance, standard deviation, and range.

1. **Variance and Standard Deviation:** The more representative of the measures of dispersion of the data (in that they reflect the array of data) are the variance and the standard deviation. The variance and standard deviation are defined as

$$S^2 = \text{sample variance} = \frac{\sum_{i=1}^{n}(x_i - \overline{x})^2}{n - 1} \qquad [2.8]$$

and

$$S = \text{standard deviation of sample} = \sqrt{\frac{\sum_{i=1}^{n}(x_i - \overline{x})^2}{n - 1}} \qquad [2.9]$$

The dimensions for the standard deviation are the same as the dimensions of the mean; this is one of the primary reasons that the standard deviation is utilized so extensively.

As characterized by both Eqs. [2.8] and [2.9], these measures are the sample estimates of the population parameters. The calculation of the population values would proceed in accord with the equations as:

Population Variance

$$V = \sigma^2 = \frac{\sum_{i=1}^{N}(x_i - \mu)^2}{N} \qquad [2.10]$$

and

Population Standard Deviation

$$\sigma = \sqrt{\frac{\sum_{i=1}^{N}(x_i - \mu)^2}{N}} \qquad [2.11]$$

where Eqs. [2.10] and [2.11] apply only when N is the number of elements in the population.

When the distribution of a random variable is not known but a set of observations $(x_1, \ldots x_n)$ is available, the moments of the distribution of n can be estimated with the sample values. However, two quantities are often considered when estimating these summary descriptors of the data—bias and variance. *Bias* measures the difference between the average value of an estimate and the quantity being estimated, whereas the variance measures the spread or width. The bias is especially marked in small samples. A bias correction involving $(n - 1)/n$ is referred to Bessel's correction function. The correction function is a defined approximation intended to reflect the uncertainty associated with including estimated parameters in the calculation of moments. For example, a correction function shows up in the variance and standard deviation calculation (but not in the calculation of the mean) because the calculation sequence for the variance and standard deviation includes the estimated mean, $\bar{x}$. In words, the variance is increased by the ratio $n/(n - 1)$, to reflect the fact that implicit in the calculation sequence is the sample-estimated parameter $\bar{x}$. Generally, with $n > 15$, the sample estimates of the standard deviation and the population value are similar. This is evident by examination of $n/(n - 1)$ when n is large, the ratio of $n/(n - 1)$ is very close to unity.

EXAMPLE 2.9

Estimate the variance and the standard deviation of the lead concentration data listed below.

Lead concentrations (*μg/L*)

37.0	34.9	28.4	27.3	31.8
28.3	26.9	19.9	31.8	32.6

In summary, $n = 10$, $\bar{x} = 29.9$ μg/L, and $S = 4.84$ μg/L

Degrees of Freedom

The term *degrees of freedom* will be referred to throughout the book. The number of degrees of freedom of a statistic is defined as the number of independent observations n in the sample (i.e., the sample size) minus the number k of population parameters, which must be estimated from sample observations. In symbols, the number of degrees of freedom, df, is defined as

$$df = n - k \qquad [2.12]$$

2. **Measures of the Range of Data:** An additional measure of dispersion or spread of data is the range. The range of a data set is simply the difference between the largest and the smallest value. Since only two of the values from the entire data set are utilized, the range is not a very useful parameter because it is sensitive to only these two values.

   A measure similar to the range is the *quartile* and *'10–90' percentile range* (the range which contains 80 percent of the data). Quartiles divide the data into four parts, each containing the same number of values. The difference between the upper quartile and the lower quartile is referred to as the interquartile range. A variation on the range that is not so sensitive to the extreme values is the 10–90 percentile range, defined as:

$$P_{90} - P_{10} \quad [2.13]$$

3. **Mean Deviation of Data:** The mean deviation, or average deviation, of a set of n numbers $x_1, x_2, \ldots x_n$ is

$$\text{Mean deviation} = \frac{\sum_{i=1}^{n} |x_i - \bar{x}|}{n} \quad [2.14]$$

4. **Modified Interquartile Range (MIQR):** The MIQR is the difference between the 75th and 25th percentiles of the data divided by 1.34. The MIQR has a divisor of 1.34 because, for data distributed in accord with the normal distribution (see Chap. 3 for further description of the normal distribution), the standard deviation (SD) = MIQR. If the SD and MIQR are quite different, the distribution of data may be substantially skewed, since data that are skewed have a high standard deviation. The advantage of the MIQR is that it is not influenced by outliers.
5. **Box-Whisker Plots:** Box-Whisker plots are potentially useful ways of summarizing the various measures of the spread or dispersion of the data. Box-Whisker plots are effective for exploratory data analysis as a way to visualize the spread of the data. Specifics of Box-Whisker plots vary from one application to another, but the essence involves depiction of the mean, a measure of the spread (such as the quartile), and the range (highest to lowest). The spread is the "box" and the range is the "whiskers." Fig. 2.8 provides an example of a Box-Whisker plot. The Box-Whisker plots present a useful and quick graphical summary of data from different locations or from one time period to another.

2.5.3 Skewness

Another common descriptor of a data set is the skewness, a measure of the symmetry. The sample estimate of the skewness (a), is determined from

$$a = \frac{n \sum_{i=1}^{n} (x_i - \bar{x})^3}{(n-1)(n-2)} \quad [2.14]$$

The population measure of the skewness, α, is calculated from:

$$\alpha = \frac{\sum_{i=1}^{N} (x_i - \mu)^3}{N} \quad [2.15]$$

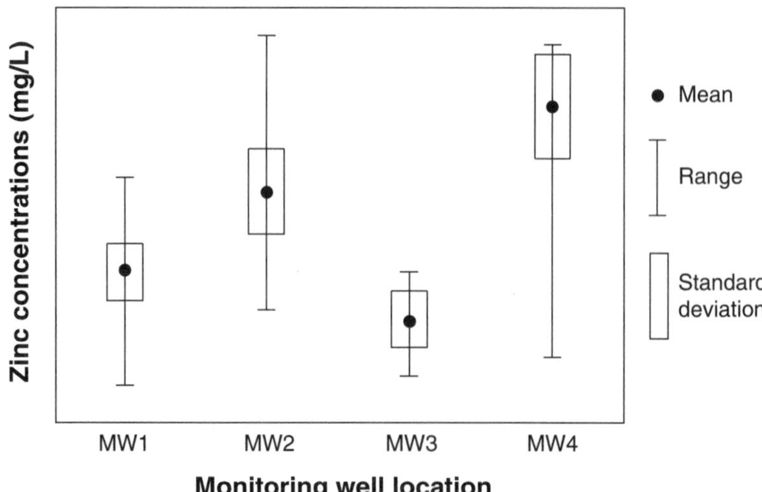

FIGURE 2.8 Box-whisker plots as summaries of environmental quality data

The sample estimate differs from the population estimate as a result of the bias correction; hence the $n/((n-1)(n-2))$ term (rather than $1/N$) in Eq. [2.14].

Because the magnitude of the skewness as determined using Eq. [2.14] depends on the units of the variable employed, a common procedure is to calculate a normalized skewness C_s by dividing Eq. [2.14] by the cube of the standard deviation, or,

$$C_s = \frac{n\sum_{i=1}^{n}(x_i - \overline{x})^3}{(n-1)(n-2)S^3} \qquad [2.17]$$

where S is the standard deviation defined in Eq. [2.9]. C_s is referred to as the coefficient of skewness and is dimensionless. Since Eq. [2.17] is normalized (or unitless) it is effective at comparing environmental quality records across time, space, and/or different constituents.

As most distributions are unimodal (i.e., have a single peak), a primary difference between distributions is their degree of symmetry. Symmetrical distributions of data will have counterbalancing effects from positive and negative deviations from the mean, with the result that the calculation of skewness using Eq. [2.17] will approach zero. Distributions whose right-hand tail is longer than its left-hand tail (e.g., Fig. 2.2(c)) are referred to as skewed to the right (i.e., C_s is positive or is said to have a positive skew); those with a longer left-hand tail (e.g., Fig. 2.2(a)) are skewed to the left (i.e., C_s is negative or is said to have a negative skew). The skewness of a symmetrical distribution, such Fig. 2.2(b), will approach zero.

As a guideline of magnitudes, a distribution is considered highly skewed if the absolute value of the coefficient of skewness is greater than one. Distributions

EXAMPLE 2.10

Calculate the skewness of the data from Ex. 2.9.

Using Eq. [2.14]

$$a = -74.4$$

The negative value of a indicates that the data are skewed to the left. In fact, the single value of 19.9 contributes the majority of the numerator.

The normalized skewness is determined using Eq. [2.17] as

$$C_S = -0.66$$

This magnitude of skewness indicates that the data are moderately skewed.

with C_s from 0.5 to 1 are considered moderately skewed, and those with C_s values from 0 to 0.5 are essentially symmetrical. It is noteworthy that the skewness coefficient value is very sensitive to the number of data points because the calculations involve the cube of deviations from the mean. A robust estimate (i.e., an estimate that is relatively insensitive to individual data values) of the skewness needs approximately 50 data points, but this is just a general guideline and should be assessed before specific applications.

2.5.4 Kurtosis

The fourth moment about the mean is the kurtosis, which is a measure of the "peakiness" of the distribution. The sample estimate of the kurtosis is obtained from the following calculation:

$$K = \frac{n^2 \sum_{i=1}^{n}(x_i - \bar{x})^4}{(n-1)(n-2)(n-3)} \quad [2.18]$$

As with skewness, kurtosis is dimension specific and thus is usually made dimensionless by dividing by the standard deviation to the power 4; that is,

$$C_K = \frac{K}{S^4} \quad [2.19]$$

Usually, an additional, related transformation is to subtract three from the coefficient of kurtosis, as

$$C_{K'} = \frac{K}{S^4} - 3 \quad [2.20]$$

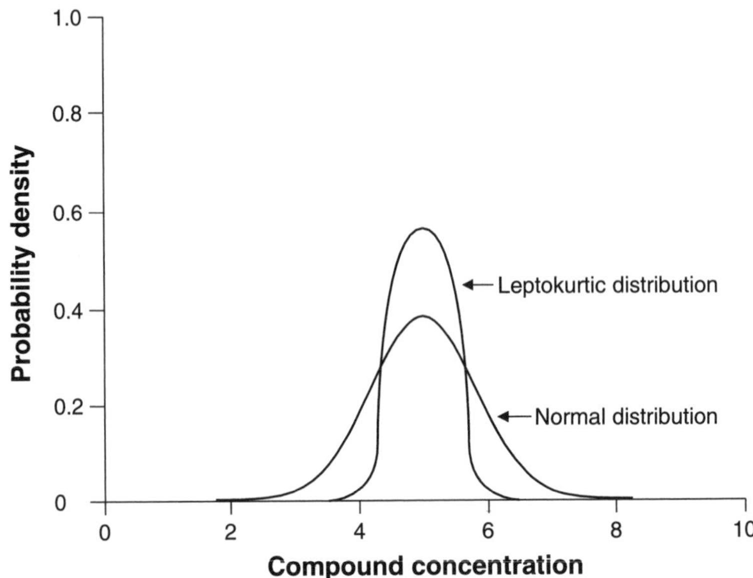

FIGURE 2.9 Leptokurtic distribution

The reason for this transformation is that $C_{K'}$ will equal zero when the data are normally distributed. Normal distribution is described fully in Chap. 3.

If the data being characterized have a relatively greater concentration of probability mass near the mean than does the normal distribution, the kurtosis is greater than three and the distribution is referred to as leptokurtic, as depicted in Fig. 2.9. If the data being characterized has a relatively smaller concentration of

EXAMPLE 2.11

Calculate the kurtosis and the coefficient of kurtosis for the data listed in Ex. 2.9.

The sample estimate of the kurtosis is calculated using Eq. [2.18] as

$$K = 2655$$

The coefficient of kurtosis is calculated using Eq. [2.20] as

$$C_K = 4.85$$

and

$$C_{K'} = 1.85$$

Hence, the data would be described as leptokurtic.

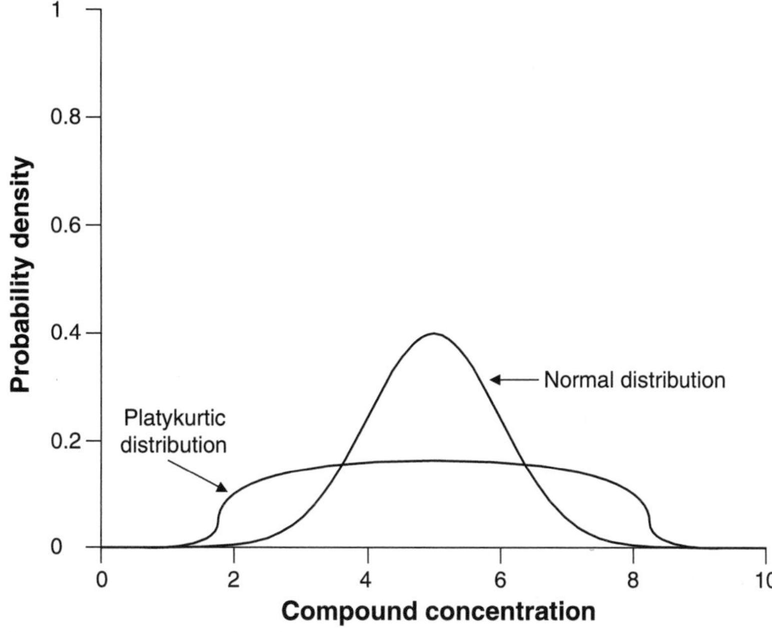

FIGURE 2.10 Platykurtic distribution

probability near the mean than does the normal distribution, the kurtosis will be less than three and the distribution is referred to as platykurtic, as depicted in Fig. 2.10. For leptokurtic distributions, $C_{K'}$ is positive; it is negative for platykurtic and zero for normal (mesokurtic) distributions.

2.5.5 Some Summary Comments

In order to emphasize that sample statistics may differ from population statistics, Table 2.1 summarizes the equations for calculating both.

If only brief data records are available, there is little merit in going beyond the fourth moment (kurtosis) as a means of characterizing the data; consequently, moments higher than kurtosis are infrequent in environmental quality assessments. Since the calculation of the moments involves $(x_i - \bar{x})^m$, where m refers to the moment about the mean, when m is large the resulting moment is very sensitive to individual values.

2.6 FURTHER SUMMARY MEASURES OF THE DISTRIBUTION OF DATA

Section 2.5 developed the equations for the moments of the distribution of data. From these equations, several other useful summary measures exist.

TABLE 2.1 Summary of equations for samples and populations

	Sample equations	Population (N very large)
Mean	$\bar{x} = \dfrac{\sum_{i=1}^{n} x_i}{n}$	$\mu = \dfrac{\sum_{i=1}^{N} x_i}{N}$
Variance	$S^2 = \dfrac{\sum_{i=1}^{n}(x_i - \bar{x})^2}{n - 1}$	$\sigma^2 = \dfrac{\sum_{i=1}^{N}(x_i - \mu)^2}{N}$
Standard deviation	$S = \sqrt{\dfrac{\sum_{i=1}^{n}(x_i - \bar{x})^2}{n - 1}}$	$\sigma = \sqrt{\dfrac{\sum_{i=1}^{N}(x_i - \mu)^2}{N}}$
Skewness	$a = \dfrac{n\sum_{i=1}^{n}(x_i - \bar{x})^3}{(n - 1)(n - 2)}$	$\alpha = \dfrac{\sum_{i=1}^{N}(x_i - \mu)^3}{N}$
Kurtosis	$K = \dfrac{n^2\sum_{i=1}^{n}(x_i - \bar{x})^4}{(n - 1)(n - 2)(n - 3)}$	$k = \dfrac{\sum_{i=1}^{N}(x_i - \mu)^4}{N}$

2.6.1 Coefficient of Variation

The coefficient of variation is often used to describe the relative amount of variation in a population. The sample estimate for the coefficient of variation COV, is defined as:

$$COV = \frac{S}{\bar{x}} \qquad [2.21]$$

The coefficient of variation is dimensionless and thus is effective at characterizing the relative variability of a distribution. The magnitude can be used to compare the extent of variability at alternative sampling locations. The coefficient of variation is a useful measure because the data from different populations or sources, the mean and standard deviation, tend to change together so that the coefficient of variation tends to remain constant. Being dimensionless, it represents a normalizing parameter with which to characterize the scatter of the data.

A small change in Eq. [2.21] yields the coefficient of variation expressed as a percent:

$$COV' = \frac{100\,S}{\bar{x}} \qquad [2.22]$$

EXAMPLE 2.12

Monitoring results at three sampling locations are listed below. Compare the sampling records in terms of the variability

		Sampling Location		
		A	*B*	*C*
		40.0	19.9	37.0
		29.2	24.1	33.4
		18.6	22.1	36.1
		29.3	19.8	40.2
$\overline{x}$	=	29.28	21.48	36.68
S	=	8.74	2.05	2.80
COV	=	0.30	0.10	0.07

In terms of variability (as quantified by the coefficient of variation), the concentrations at A vary the most, followed by B and then C.

The coefficient of variation is also useful and simple as an indicator of skewness. If the data are likely skewed (either positively or negatively), the value of the coefficient of variation will be large (> 1) since the standard deviation is larger than the mean.

2.6.2 Standard Error of the Mean

Statistics like the mean are derived from individual random variables and so are also random in nature; they are random variables. As one of these random variables, the standard error of the mean, $SEM_{\overline{x}}$ is defined as:

$$SEM_{\overline{x}} = \frac{S}{\sqrt{n}} \quad [2.23]$$

The standard deviation and the standard error of the mean ($SEM_{\overline{x}}$) are related to each other and yet are quite different. The $SEM_{\overline{x}}$ is smaller than the standard deviation *S*. The size difference occurs because the $SEM_{\overline{x}}$ is an estimate of the error (or uncertainty) involved in estimating the mean of a sample, and not an estimate of the error (or variability) involved in measuring the data from which the mean is calculated. Variability of the sample mean is much less than that of any single observation, allowing us to better pinpoint the true population mean from the sample average than from any single measurement. The $SEM_{\overline{x}}$ is used when the uncertainty of the estimate of the mean is of concern. The uncertainty decreases as the sample size increases.

In practice the confidence interval *CI* of the mean is expressed as

$$CI = \overline{x} \pm t\frac{S}{\sqrt{n}} \quad [2.24]$$

TABLE 2.2 Table of t-values for different sample sizes

Number of Samples n	$t_{.05,df}$
2	12.71
3	4.30
4	3.18
5	2.78
10	2.26
∞	1.96

where:

$\bar{x}$ = mean
S = standard deviation
n = number of samples
t = t-distribution value

The value of t is both a function of the number of degrees of freedom df (see Sec. 2.5.2) and a level of significance.

For the 95 percent confidence interval, the value of t is as noted in Table 2.2.

The use of t compensates for the tendency of a small number of values to underestimate the uncertainty. A more complete discussion of the basis for the value of t is contained in Chap. 3, Sec. 4. The confidence interval information is presented here only for the sake of completeness and will be dealt with more extensively in Chap. 6.

EXAMPLE 2.12

For the data listed below, compute the 95 percent confidence interval estimate of the mean

25.0, 36.1, 45.2, 19.6, 21.8 mg/L

Mean	$\bar{x} = 31.48$ mg/L
Standard Deviation	$S = 11.44$ mg/L
Number of Samples	$n = 5$

The t-value from Table 2.2 is 2.78. Hence, the 95 percent confidence interval is,

$$\text{Confidence Interval} = 31.48 \pm \frac{2.78\,(11.44)}{\sqrt{5}}$$

$$= 17.26 \text{ to } 45.70 \text{ mg/L}$$ for the 95 percent confidence interval for the mean.

TABLE 2.3 Equations for the standard error of alternative statistics

Sampling distribution	Standard error	Remarks
Means	$\sigma_{\bar{x}} = \frac{\sigma}{\sqrt{n}}$	This is true for large or small samples. The sampling distribution of means is very nearly normal for $n \geq 30$ even when the population is nonnormal.
Standard deviation	$\sigma_S = \frac{\sigma}{\sqrt{2n}}$	For $n \geq 100$, the sampling distribution of S is very nearly normal, σ_S is given by this formula only if the population is normal (or approximately normal)
Median	$\sigma_{med} = \sigma\sqrt{\frac{1}{2n}}$	For $n \geq 30$ the sampling distribution of the median is very nearly normal. The given result holds only if the population is normal (or approximately normal)
Coefficient of variation	$\sigma_v = \frac{v\sqrt{1 + 2v^2}}{\sqrt{2n}}$	Here $v = \sigma/\mu$ is the population coefficient of variation. The given result holds for normal (or nearly normal) populations and $n \geq 100$.

2.6.3 Standard Errors

The standard deviation of a sampling distribution of a statistic is often referred to as its standard error. Summarized in Table 2.3 are the standard errors for a number of statistics of the population. The standard error of the mean, from Sec. 2.5.2, is noted as the first row of Table 2.3; the other rows relate to less frequently used statistics.

2.6.4 Summary Descriptors

The summary equations for the additional descriptors are listed in Table 2.4. These measures are utilized for analyses developed in the chapters that follow.

2.7 SUMMARY

One of the best ways to describe statistically quantitative uncertainty is by using probability distributions. We will use the summary statistics noted in this chapter to fit the distributions to the data. This will be the focus of discussion in Chaps. 3 through 5.

TABLE 2.4 Summary of equations for additional descriptors of data

Coefficient of variation	$COV = \frac{S}{\bar{x}}$
Standard error of the mean	$SEM_{\bar{x}} = \frac{S}{\sqrt{n}}$
Standard error of the standard deviation	$SE_S = \frac{S}{\sqrt{2n}}$

2.8 REFERENCES

SHAW, D.M., "Interpretation Geochimique des elements en trace dans les roches cristallines," Paris: Masisson et Cie, 1964.

2.9 PROBLEMS

2.1. Find the arithmetic mean, median, mode, harmonic mean, and geometric mean for the following vapor concentration (VC) data for benzene in:

Location:	**VC1**	**VC2**	**VC3**	**VC4**	**VC5**	**VC6**	**VC7**	**VC8**	**VC9**
Conc. (ppb)	5	3	6	5	4	2	8	10	2

2.2. The Ontario Ministry of the Environment is conducting an investigation as to the effects of vehicle exhaust on vegetation bordering regional highways. While studying the impact of lead, the following pattern was recorded in 100 samples:

Concentration (μg/L)	**Frequency**
4	20
5	40
6	30
7	10

What is the arithmetic mean, the mode, and the median for the data?

2.3. Mercury concentrations were measured at six different locations downstream from a nine tailings collection area. What is the median concentration for the following data: 0.0, 0.1, 1.2, 0.6, 0.05 and 0.1 μg/L?

2.4. Prove that the sum of the deviations of $x_1, x_2, x_3 \ldots x_n$ from their mean $\overline{x}$ is equal to zero.

2.5. If $z_1 = x_1 + y_1, z_2 = x_2 + y_2, \ldots z_n = x_n + y_n$, prove that $\overline{z} = \overline{x} + \overline{y}$

2.6. The data located in Data set A in the diskette included with this book represents historical results from a different monitoring well. Using these data,

a. plot a frequency histogram of data;

b. compute the sample moments;

c. convert the frequency diagram in (a) to a probability histogram;

d. construct two probability histograms for $n = 50$ using the first and second sets of 50 values. Then construct five probability histograms for $n = 20$ using the five sets of two rows. Discuss the change in shape with sample size.

2.7. What is the difference between the population and a sample?

2.8. The following air quality measurements were taken downwind of a coal burning electrical generating station. What is the probability that the sulphur concentration x will fall within the following ranges?

$$x > 6, 3 < x < 10, x < 5, 1 < x < 9$$

Date/Time	6:00 a.m.	12:00 p.m.	6:00 p.m.	12:00 a.m.
July 4, 1990	34	35	43	22
July 5, 1990	18	9	4	30
July 6, 1990	27	23	11	18
July 7, 1990	13	10	11	17
July 8, 1990	21	16	9	23
July 9, 1990	8	9	14	16
July 10, 1990	3	2	1	15

2.9. For the following data gathered from a stream running alongside an industrial landfill, determine the probability that the concentration of phenols in the creek will exceed, 2, 4, 6, and 8 ppb.

SW1	SW2	SW3	SW4	SW5
7	5	7	4	7
5	7	6	6	5
4	8	4	8	3
0.5	2	7	3	5
6	1	5	6	4

2.10. Calculate the mean, variance, skew, and kurtosis for the following DDT concentrations measured in agricultural land runoff:

a. 0.5, 0.6, 7.0, 1.0, 1.2, 0.01, 0.3, 0.9, 2.4, 0.8 μg/L.

b. 0.3, 4.0, 1.0, 1.5, 3.5, 2.0, 2.4, 1.7, 2.1, 1.9 μg/L.

c. 0.3, 0.4, 6.3, 6.2, 5.3, 4.7, 5.8, 7.0, 6.6, 5.8 μg/L.

2.11. Show that the following two equations for computing the variance of a set of observations are identical when n is large:

Equation 1

$$\frac{1}{n-1} \times \sum_{i=1}^{n} (x_i - \overline{x})^2$$

Equation 2

$$\frac{1}{n-1}\left[\sum_{i=1}^{n} x_i^2 - \frac{1}{n}\left(\sum_{i=1}^{n} x_i\right)^2\right]$$

2.12. For each of the following sets of data:

a. determine the median and the semiquartile value;

b. calculate the arithmetic and geometric mean for each of the four data sets; and

c. calculate the coefficients of variation for each of the four data sets.

Data set 1	Data set 2	Data set 3	Data set 4
0.13	0.14	0.15	0.16
1.50	1.85	6.17	1.18
17.41	16.96	7.32	2.84
19.33	32.67	16.51	16.06
30.40	38.61	21.46	18.37
35.03	44.53	21.96	33.45
47.99	54.31	27.74	33.58
51.35	54.41	28.36	52.93
56.36	55.90	37.13	16.06
58.50	59.97	52.35	54.77
71.05	73.85	54.02	60.23
74.66	80.66	65.85	62.79
80.87	86.03	71.73	63.52
82.28	89.16	71.95	73.74
85.89	89.98	83.06	82.89
89.60	98.13	86.16	83.94

2.13. The tabulated data below represent hydraulic conductivity measurements from the Keele Valley Landfill in Vaughan, Ontario. Using the data, quantify the arithmetic and geometric mean of the hydraulic conductivity. Which of the arithmetic and geometric mean is more representative of the data?

- 1.01×10^{-8} cm/s;
- 5.65×10^{-9} cm/s;
- 7.24×10^{-9} cm/s;
- 6.66×10^{-9} cm/s;
- 4.15×10^{-7} cm/s; or
- 7.82×10^{-8} cm/s.

2.14. During the course of a groundwater quality monitoring program, the arsenic concentrations at four locations were measured as detailed below:

MW1	MW2	MW3	MW4
2.14	1.97	1.92	1.87
2.12	2.14	2.14	1.74
2.21	2.16	1.85	1.98
2.11	1.87	1.92	1.89
1.99	2.17	2.17	2.01
2.05	2.04	1.97	2.03
2.04	2.01	1.89	1.95
2.00	2.06	2.05	1.86
1.97	2.19		
2.08	2.03		

a. calculate the mean and standard deviation for each of the monitoring wells and for the entire set; and

b. using the data, calculate the standard error of the mean for each ground and for the entire data set.

2.15. The data listed below was measured at a sewage treatment outfall. Compute the mean, variance and standard deviation for the following three circumstances:

a. the data as is;

b. add two to each data value; and

c. multiply each data value by two.

Day 1	Day 2	Day 3	Day 4
0.01	0.87	4.94	0.83
2.82	4.29	2.23	3.32
0.97	3.55	0.60	2.25
4.04	2.57	0.02	1.76
2.93	1.52	0.04	0.29
2.40	0.07	1.89	3.04
1.75	0.46	2.66	3.92
4.48	1.82	2.86	4.01
4.11	0.74	3.01	2.60
3.73	0.83	3.04	1.51

CHAPTER 3

THE NORMAL OR GAUSSIAN DISTRIBUTION

"The probable is usually what happens."

Aristotle

3.1 INTRODUCTION

One of the objectives of data analysis is to describe the problem under scrutiny by using available sample observations to identify the most appropriate population distribution function. This is important because, if we can select a probability distribution function to describe the distribution of the data, then we can make inferences based on the known statistical properties of the distribution selected. From many possible population distribution functions (e.g., normal, lognormal, beta, gamma, etc.) we are interested in selecting one that can be used to make probability statements.

Although in theory a number of alternative distributions may be necessary to describe the wide variety of features that exist in environmental quality concerns, in practice a majority of environmental quality variables exhibit frequency distribution shapes that can be approximated by a relatively small number of distributions. In fact, two distributions, the normal distribution and the lognormal distribution, are applicable to a widespread set of concerns. There is an important supporting rationale for the widespread utility of the normal and lognormal distributions as described in this and the following chapter. Specifically, Chap. 3 focuses on the normal distribution or, as it is also frequently referred to, the Gaussian distribution; Chap. 4 focuses on the lognormal distribution.

The terms 'Gaussian' or 'normal' refer to precisely the same distribution and the two terms are frequently used interchangeably. As reported by Bulmer (1965), the normal distribution was discovered in 1733 by Huguenot refugee Abraham de Moivre, although the distribution is more frequently associated with Gauss who derived it in 1809. The term "normal" is used because so many outcomes may be described by the distribution. In this book the distribution will be referred to by the term *normal distribution.*

The normal distribution applies to many applications, including those within the environmental quality field, in large part because of a mathematical property known as the Central Limit Theorem. One of the most significant theorems in probability, the Central Limit Theorem pertains to the limiting distribution of a sum of random variables. The theorem states that "under very general conditions," as the number of variables in a sum becomes large, none of which is dominant, the distribution of the sum of random variables approaches the normal distribution, almost regardless of the distribution of the individual variables. Thus, any variable that can be regarded as the sum of a large number of small independent contributions is likely to follow the normal distribution. Therefore, if a physical or chemical process in environmental phenomena is the result of the totality of a large number of individual effects, then according to the Central Limit Theorem, the sum of the individual effects will tend to be normally distributed; the existence of some of the individual effects on the high side are counterbalanced by an (approximately) equivalent number of individual effects on the low side such that the resulting shape is symmetrical, as depicted in Fig. 3.1.

As will be apparent in the chapters to follow, the normal distribution is assumed as the underlying distribution for numerous statistical tests. The pervasive importance of the normal distribution will be dealt with extensively in this chapter. The normal distribution is assumed for tolerance intervals and confidence intervals (Chap. 6), the *t*-test (Chap. 8), and parametric ANOVA (Chap. 10), among other statistical procedures. The failure of data to approximate a normal distribution precludes subsequent tests for fear of incorrect conclusions. Much of this chapter focuses on various procedures to determine how well environmental quality data can be characterized by the normal distribution.

FIGURE 3.1 Schematic of the normal or Gaussian distribution

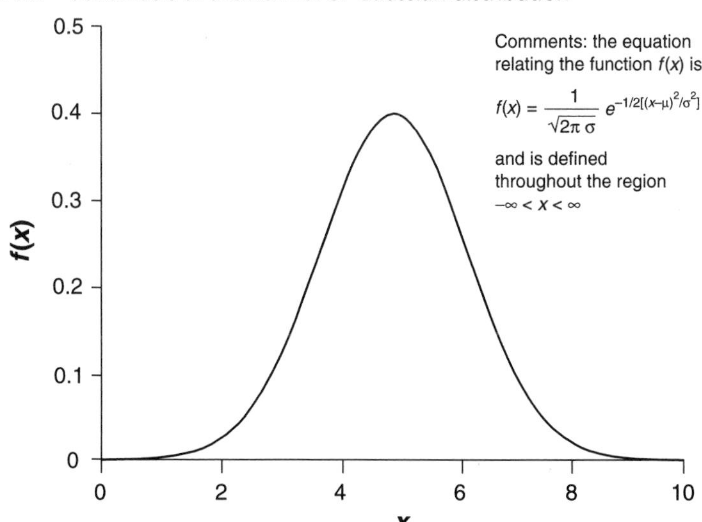

3.2 THE MATHEMATICS OF THE NORMAL DISTRIBUTION

A primary purpose for examining the features of a distribution such as the normal distribution is to characterize a set of data. The first step of data analysis is to identify the probability distribution function that best describes the random nature of a particular variable. The intent is to fit the individual data to a distribution. If a probability distribution function possesses the general shape of the phenomena of interest (e.g., bell-shaped as per Fig. 3.1), the individual values from the data record are used to fit the probability distribution as a means of estimating properties of the data such as exceedance probabilities of specified events. If the data record was very lengthy, one could use the data record itself to assign probability exceedances, but an appropriate probability distribution function lessens the need for a lengthy data record.

As indicated in Sec. 3.1, the normal distribution is the most common descriptor of data. Mathematically, the normal distribution is defined as

$$f(x) = \frac{1}{\sqrt{2\pi}\,\sigma} e^{-1/2[(x-\mu)/\sigma]^2} \qquad [3.1]$$

where Eq. [3.1] is defined for the region $-\infty < x < \infty$, and where μ = the mean (a location parameter), σ = the standard deviation (a scale parameter) and σ^2 is the variance. A summary notational form for this is $N(\mu, \sigma^2)$ where N refers to the normal distribution. For example, $N(0, 1)$ refers to a normal distribution with a mean of zero and a variance (and standard deviation) of unity.

In mathematical terms, Eq. [3.1] gives the height of the normal probability distribution at any point along the scale of x as depicted by the $f(x)$ ordinate in Fig. 3.2. Equation [3.1] extends to infinity in both directions as already indicated above by the defined region. The existence of nonzero values over the range from $-\infty$ to $+\infty$ is one of the major problems associated with applying the normal distribution to environmental quality issues; it allows, for example, nonzero probabilities for negative values, which means there is a possibility of having negative concentrations.

It is significant that the probability distribution described by Eq. [3.1] is in fact a probability density. To estimate the probability of occurrence, the probability density function (Eq. [3.1]) must be integrated over a range. For example, the probability that a sample value x is less than or equal to a specified value X is evaluated by integrating from $-\infty$ up to the level X, using

$$\text{Prob}\,[x \le X] = \int_{-\infty}^{X} \frac{1}{\sqrt{2\pi}\sigma} e^{-1/2[(x-\mu)/\sigma]^2} dx \qquad [3.2]$$

This corresponds to the area under the curve as depicted in Fig. 3.3. A property of the normal distribution (and all probability distributions) is that the total area bounded by the curve and the x-axis is unity. Hence the area under the curve

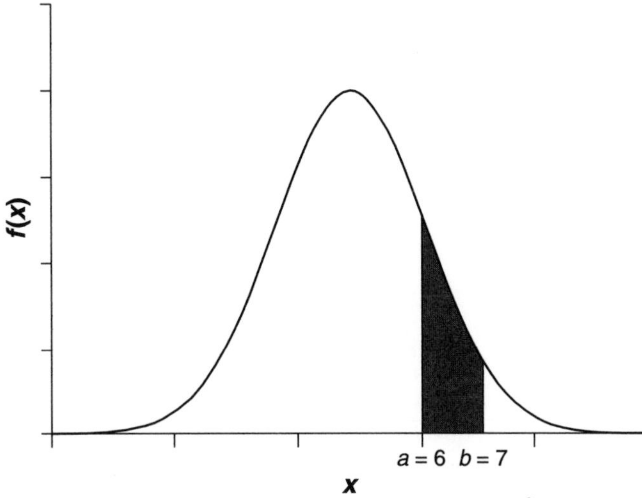

FIGURE 3.2 Probability density of normal distribution

between two ordinates $x = a$ and $x = b$ where $a < b$ represents the probability that x lies between a and b. This relationship is shown in Fig. 3.2; that is,

$$\text{Prob}\ \{a \leq x \leq b\} = \int_a^b \frac{1}{\sqrt{2\pi}\sigma} e^{-1/2[(x-\mu)/\sigma]^2}\,dx \qquad [3.3]$$

FIGURE 3.3 Cumulative density function for normal distribution

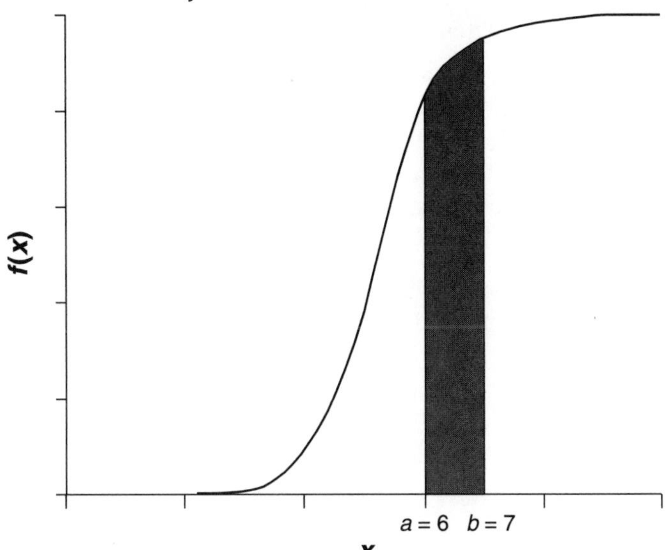

Because of the extensive use of the normal distribution and the infinite number of possible values of μ and σ, probabilities for the normal distribution are usually evaluated using the so-called standard normal distribution. The standard normal distribution is derived by transforming the random variable x to the random variable z using $z = (x - \mu)/\sigma$. When the variable x is expressed in terms of standard units, Eq. [3.1] becomes

$$f(z) = \frac{1}{\sqrt{2\pi}} e^{-(z^2/2)} \qquad [3.4]$$

Equation [3.3] is the probability density function for the standard normal curve and is the form of the information provided in the normal distribution tables. In this case we say that z is normally distributed with mean of zero and variance of unity, or $N(0, 1)$.

As can be seen from Eq, [3.1], the normal distribution is a two-parameter distribution. In other words, all aspects of the distribution are known once the mean μ, and variance, σ^2 are specified. Both of these parameters characterize the central tendencies of the distribution (the mean being the centroid or center of gravity and the variance being the dispersion or spread of the distribution) around the centroid. Therefore, when an issue relates to environmental quality near the "tails" of the distribution (on the very low and/or the very high end of the data record), the normal distribution assumption is not necessarily the best distribution to employ.

The area under any probability distribution curve is equal to unity by definition (i.e., the integration from negative to positive infinity). However, for issues of environmental quality, the full range of the normal distribution is seldom of interest. Of greater interest are features such as how often one would expect values to be within various proximities to the mean. Three commonly employed guidelines are indicated in Table 3.1 and shown graphically in Fig. 3.4. In words, the results from Table 3.1 indicate that, given data that can be described by the normal distribution,

1. you would expect 68 percent of monitoring values to lie within plus and minus one standard deviation from the mean. The normal distribution is symmetrical, and therefore 34 percent of the values should lie between the mean and the mean plus one standard deviation, and 34 percent of the values should be between the mean and the mean minus one standard deviation;

TABLE 3.1 Indications of percentages of monitoring values expected to be in the vicinity of the mean value, assuming the normal distribution

Number of standard deviations from the mean	Percentage of values in vicinity of mean
±1	68.26%
±2	95.45%
±3	99.73%

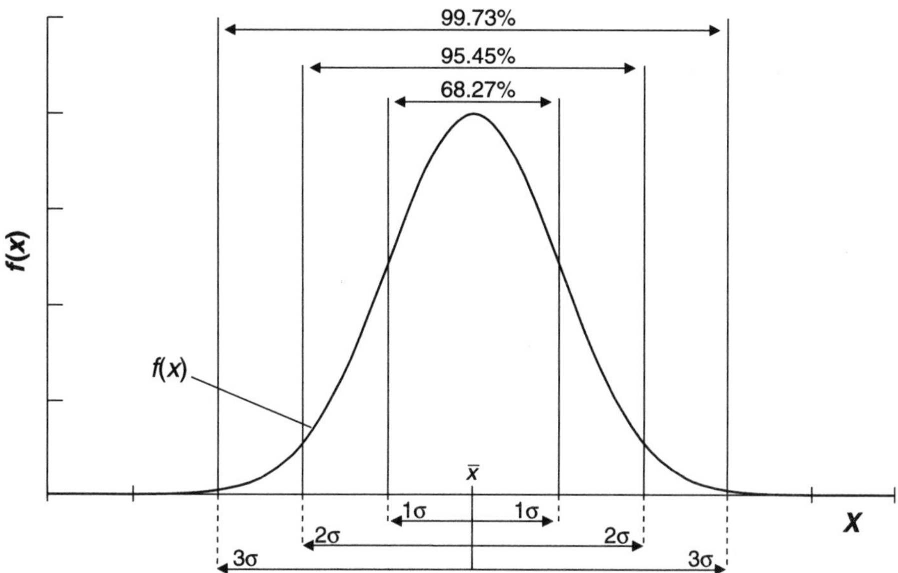

FIGURE 3.4 Probability density of normal distribution and areas under curve for different distances from the mean

2. you would expect 95 percent of the monitoring values to lie within plus and minus two standard deviations of the mean. Because the distribution is symmetrical about the mean, you would expect (approximately) 2.5 percent above the mean plus two standard deviations, and 2.5 percent below the mean minus two standard deviations; and
3. you would expect 99.73 percent of the monitoring values to lie within plus and minus three standard deviations of the mean.

The normal distribution function (from Fig. 3.1 and from the values listed in Table 3.1) illustrates that random fluctuations around the mean (or average) are to be expected but that fluctuations of more than two or three standard deviations are very improbable. The curve is practically at zero for locations beyond three standard deviations from the mean.

Henceforth in this book, sample values, not population statistics, will be utilized in the equations, since in practical applications dealing with environmental quality concerns we never have the population statistics.

3.3 TESTS FOR NORMALITY

As indicated in Sec. 3.2, we utilize the data to fit the normal distribution to allow us to quantify in turn various aspects of the probabilities. Nagging doubts about the implicit assumption that the normal distribution is a reasonable descriptor of the data, as well as suspicions that other distributions may be more appropriate for specific circumstances, beg for criteria to establish the adequacy of the normal distribution. In other words, data must be subjected to goodness-of-fit tests to determine the appropriateness of different distributions. A series of alternative tests of varying complexity embodying several approaches are available to determine whether a probability distribution function adequately describes the random nature of the random variable. A number of these tests will be considered in the sections to follow.

3.3.1 Coefficient of Variation Test for Normality

The coefficient of variation (COV) test is the simplest and least rigorous of the tests available to determine whether the data can be characterized by a normal distribution. Specifically,

$$COV = S/\bar{x} \qquad [3.5]$$

Given that the normal distribution is a symmetrical distribution and that environmental quality data cannot be negative, it is not possible for the data to be described effectively by the normal distribution if S is very large in relation to the mean; in this situation, the data distribution must be highly skewed. For environmental quality data (where individual concentration measurements cannot be less than zero), the failure of the data to be described by the normal distribution is sometimes acknowledged when the coefficient of variation is greater than 0.5 and/or 1.0. If *COV* exceeds 1.0 there is strong evidence that the data are not normally distributed. If the data are from a normal distribution, less than ten percent of all sample coefficients of variation will exceed 1.0 by random chance. If the data are nonnormal, then the sample standard deviation will be large relative to the sample mean, indicating a high probability of a sample coefficient exceeding 1.0.

EXAMPLE 3.1

Results from monitoring chloride levels in the groundwater are summarized in Table 3.2. Utilize the coefficient of variation test to determine if the data can be characterized by the normal distribution.

Solution

From the data in Table 3.2, the statistical parameters are determined as

Mean	$\bar{x}$ =	18.3 mg/L
Standard Deviation	S =	4.06 mg/L
Coefficient of Variation	COV =	.222

In summary, the *COV* is less than unity, which means that (at least by this test) the data can be described by the normal distribution. The general, bell-shaped distribution of the data depicted in the histogram of chloride concentrations on Fig. 3.5(a) confirms that the data are distributed in a relatively symmetrical fashion.

EXAMPLE 3.2

The results from monitoring benzene concentrations in water in a roadside ditch are listed in Table 3.2, Column II. Utilize the coefficient of variation test to determine if the data may be characterized by the normal distribution.

SOLUTION

From the data in Table 3.2, the statistical parameters were determined; specifically,

Mean	$\bar{x} = 4.47$ mg/L
Standard Deviation	$S = 4.98$ mg/L
Coefficient of Variation	$COV = 1.11$

In summary, the *COV* is greater than unity, which indicates that the benzene concentration data are not well described by a symmetrical distribution. The histogram of data as illustrated in Fig. 3.5(b) confirms that the data are not well described by the normal distribution.

The result is that the coefficient of variation test is an easy but approximate measure or indicator of skewness (see Ex. 3.3 on p. 49).

TABLE 3.2 Chemical concentrations in monitoring records

	Column I	Column II
Number of Samples	**Chloride Concentrations (mg/L)**	**Benzene Concentrations (µg/L)**
1	25.25	1.1
2	13.32	1.4
3	15.78	1.8
4	22.63	2.05
5	20.36	2.11
6	21.43	2.15
7	11.74	2.23
8	22.27	2.54
9	18.11	2.58
10	27.50	2.61
11	18.12	2.85
12	16.16	3.15
13	16.39	3.50
14	16.66	4.80
15	18.71	5.12
16	14.90	7.40
17	14.09	10.15
18	15.64	5.35
19	16.09	2.94
20	20.47	23.5

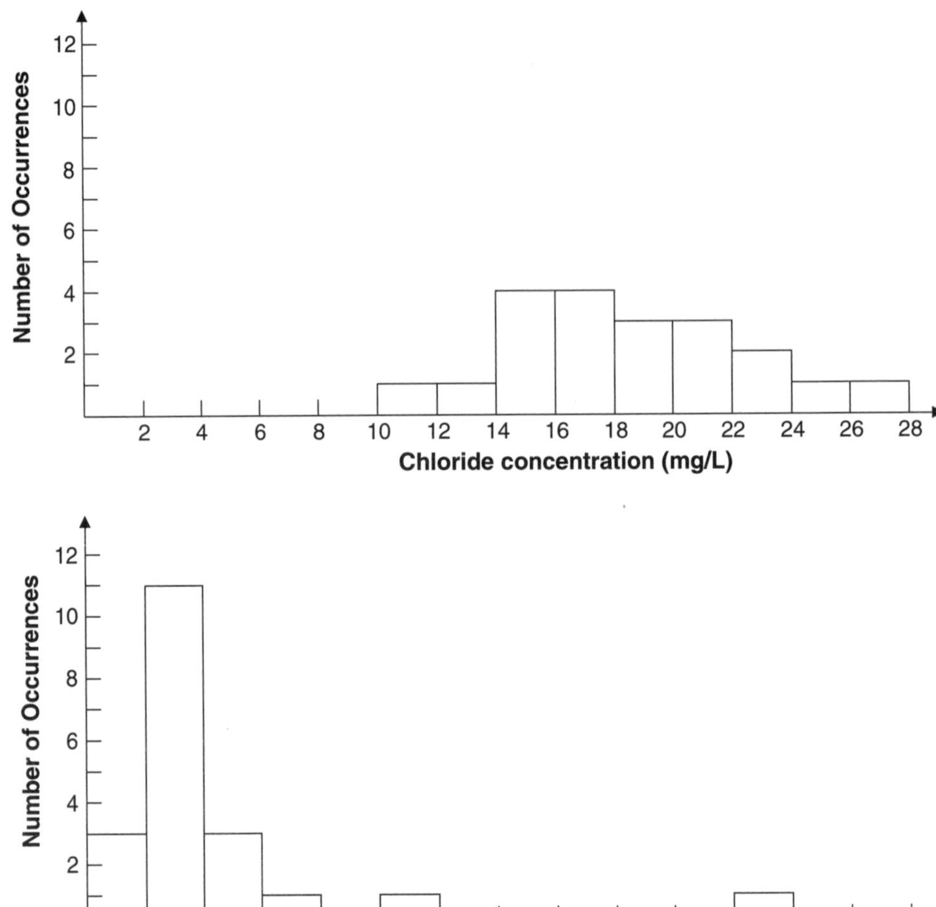

FIGURE 3.5 Histograms of chloride and benzene concentrations

3.3.2 Skewness and Kurtosis Coefficient Tests for Normality

As described in Secs. 2.5.3 and 2.5.4, the normal distribution has zero skewness and a kurtosis coefficient value of three. Some deviations from these magnitudes for samples (as opposed to populations) must be expected. These calculation procedures are computationally simple, utilizing spreadsheet procedures. The skewness coefficient, C_s, for the data should approach zero since the normal distribution is a symmetrical distribution, and the positive deviations should cancel the negative deviations. Although not a firm criterion, when the absolute value of $C_s > 1$, the distribution is not likely a normal distribution. Note that this test using skewness is not robust when the data set is small, because the cube of devia-

EXAMPLE 3.3

Monitoring results at two sampling locations are listed below. Comment on whether the data are symmetrically distributed and/or skewed, using the coefficient of variation test.

	SAMPLING LOCATION	
	A	B
	37.0	2.56
	34.9	1.00
	28.4	.25
	27.3	.60
	31.8	.40
	32.6	.85
	28.3	.38
	26.9	2.25
	19.9	.41
	31.8	1.75
$\bar{x}$ =	29.9	1.05
S =	4.84	.84
COV =	.16	.80

SOLUTION

The coefficient of variation test indicates that the monitoring results at Location A are relatively uniform, whereas those at B are skewed.

tions makes the calculations very sensitive to the individual members of the data set. In summary, for large data sets, the skewness coefficient must be small to support the assumption that the data are adequately described by the normal distribution. For small data sets, the utility of the skewness coefficient as an indicator that the data may be adequately described by the normal distribution is less certain because it is highly sensitive to individual values within the data set. While there are no strict definitions of large and small, five to ten samples is definitely small, and more than 50 is probably large.

Note that just because a distribution is symmetrical does not indicate normality. For example, it will be shown in Sec. 3.5 that the student's t-distribution is symmetrical but not normal (except as the sample size becomes large). The t-distribution has fatter tails; hence a greater proportion of the random values from the t-distribution are farther from the mean than occur with normal distribution. If the absolute value of the skewness coefficient is less than unity, the data do not show evidence of significant skewness (i.e., they may be accepted as normally distributed).

The kurtosis coefficient test is another way to consider whether a data set is normally distributed.

EXAMPLE 3.4

Using the chloride concentration data listed in Ex. 3.1, the skewness coefficient is $C_s = 0.65$. As characterized by the skewness test, the distribution of data is moderately skewed. The chloride concentrations are plotted in histogram form on Fig. 3.5(a).

EXAMPLE 3.5

Using the benzene concentration data from Ex. 3.2, the coefficient of skewness of data is 3.27. This magnitude of skewness coefficient indicates highly skewed data. The benzene concentrations are plotted in histogram form in Fig. 3.5(b) and indicate the right skew of the data. Calculating kurtosis as presented in Sec. 2.5.4, the coefficient of kurtosis = 12.0 and the transformed coefficient for the $C_K' = 12 - 3 = 9$

3.3.3 Probability Plots

Another procedure useful in assessing the fit of a particular distribution for a particular data test utilizes probability graph paper. Probability plots are procedures routinely used for making quantitative and subjective statistical analyses of data sets by plotting the magnitude versus a cumulative distribution function. If the sample of data is from the distribution function that was used to scale the probability paper, the data will appear as a straight line when plotted. The use of probability plots requires only a general understanding of simple statistical concepts.

Cumulative probability plots were first developed by Hazen (1914) as a means of simplifying the interpretation of reservoir storage data. The method has since found widespread application in graphical studies of numeric data in many fields.

The horizontal axis along the top of probability paper characterizes the probability that the value of the random variable will be exceeded (i.e., the exceedance probability). The horizontal axis on the bottom provides the nonexceedance probability as depicted in Fig. 3.6. The abscissa scale of normal probability paper is an arithmetic scale. The ordinates are percentages, descending from 99.99 to 0.01, as indicated in Fig. 3.6. The ordinate scale is stretched at both ends; the distance from 50 to 60 is smaller than that from 60 to 70, and so on. The probability (cumulative percentage) scale is arranged such that a cumulative normal density distribution will plot as a straight line.

Probability plots are available that transform a curvilinear distribution into a straight line. Utilizing results plotted on probability paper, the probability of the concentration exceeding some $x\ \mu g/m^3$ is found by entering the value of the random variable, moving horizontally to the frequency curve, and then moving vertically to the probability scale. Probability paper is specially designed graph paper, of which normal probability paper (Fig. 3.6), is a particular type; other forms of probability paper exist. In all cases the objective of attempts to identify the

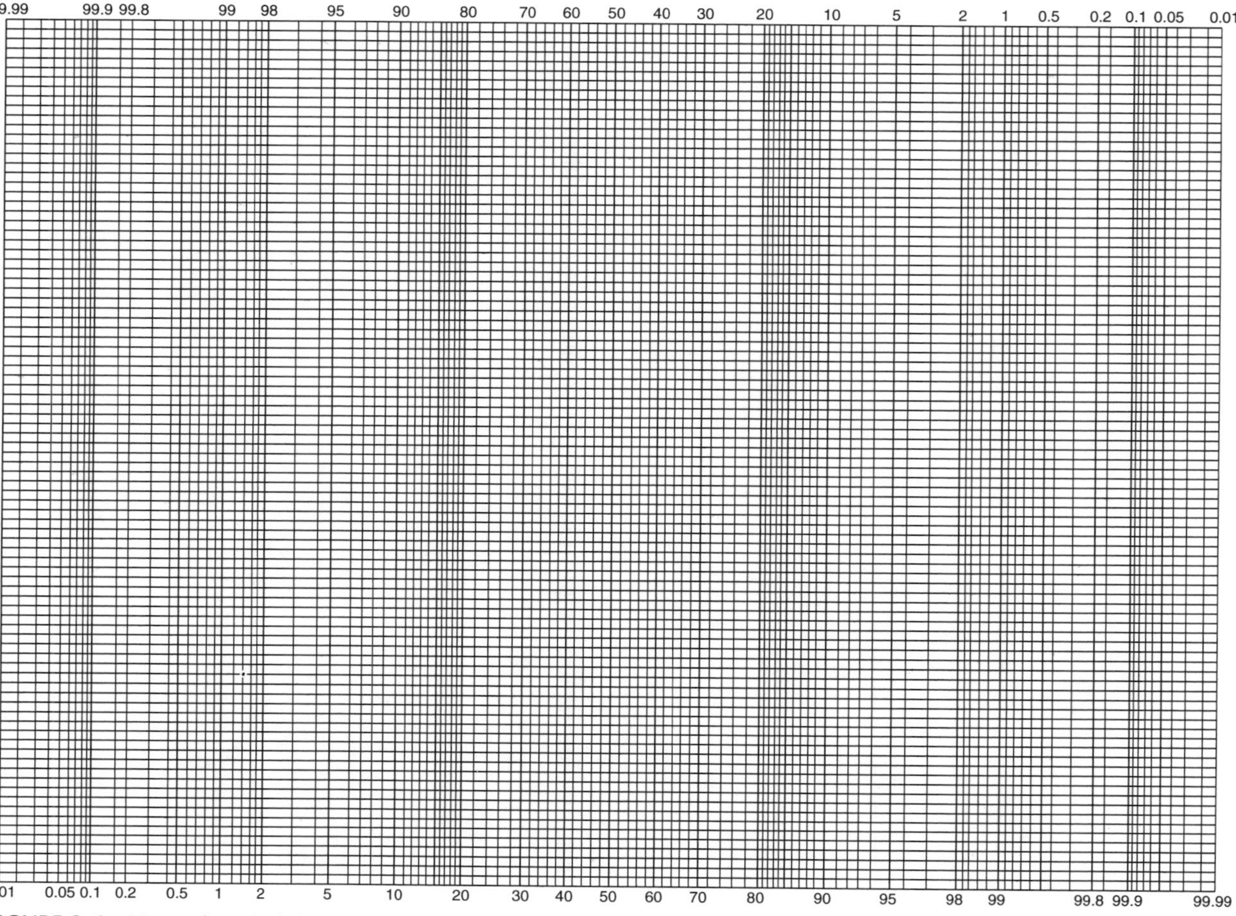

FIGURE 3.6 Normal probability paper

TABLE 3.3 Examples of plotting position formulas

Formula Name	Formula
California	$\frac{m}{n}$
Hazen	$\frac{2m-1}{2n}$
Beard	$1-(0.5)^{1/n}$
Weibull	$\frac{m}{n+1}$
Chegadayev	$\frac{m-0.3}{n+0.4}$
Blom	$\frac{m-3/8}{n+1/4}$
Tukey	$\frac{3m-1}{3n+1}$

Source: Adapted from W. Viessman, G.L. Lewis, and J.W. Knapp, *Introduction to Hydrology*, 3rd ed. (New York, NY: Harper and Row, 1989), p. 722. Reprinted with permission.

underlying population for a data set is to obtain a model that can be used to make probability statements about the occurrence of values of a random variable. By plotting the data and determining if the data plot as a straight line, we can determine the appropriateness of the distribution as a descriptor of the data.

The plotting of the individual data points requires the determination of "plotting position" for the probability. When successive values of the data set are independent, the exceedance probability can be estimated from one of a number of formulae, examples of which are listed in Table 3.3, where m refers to the rank of values, with the largest value equal to one, and n is the total number of values.

Kimball (1946, 1960) reviewed the problem of choosing a plotting position for the ranked observations in a series of n independent identically distributed random variables (nine alternatives). Although the nine different formulae in Kimball (1960) give similar plotting positions near the middle of the data ($i = n/2$), they can lead to quite different plotting positions at the extreme values, near $i = 1$ and $i = n$. In general, the plotting formulae have been described by Blom (1958) as:

$$P_m = \frac{m-c}{n+1-2c} \qquad [3.6]$$

where P_m is the plotting position of the mth point, c is a function of the record length n and the distribution being sampled, and $0 \leq c \leq 1$.

The probability, P_m, is an estimate of the probability of values being equal to or less than the ranked value $p(x \leq X)$. Thus, the probability of a concentration being exceeded can be obtained using a plotting position formula. For example, the expected number of exceedances x for a given number of future measurements N using the Weibull plotting formula is

$$x = N\frac{m}{n + 1} \qquad [3.7]$$

where m and n are as defined previously.

To calculate the recurrence interval R (the number of measurements before an exceedance is expected to occur) a value of $\bar{x} = 1$ can be used in Eq. [3.7]. Rearranging, the recurrence interval or return period can be estimated using

$$R = \frac{n + 1}{m} \qquad [3.8]$$

Thus, the recurrence interval based on the Weibull plotting formula of a concentration to be equaled or exceeded is equal to the number of measurements plus one, divided by the rank of that magnitude of observation.

Assuming that a straight line represents a reasonable characterization of the data, the values of the mean and standard deviation may be read directly from the graph. The mean is simply the 50th percentile point. The standard deviation of the distribution is determined from the slope of the fitted straight line. Because each of the populations is normal (a straight line plot on normal probability paper) the values of the mean, plus and minus one standard deviation, can be estimated from the 16th and 84th percentiles.

The implications of different means but similar standard deviations for different monitoring locations are depicted in Fig. 3.7(a). Alternatively, the implications of similar means but different standard deviations are depicted in Fig. 3.7(b). (See Ex. 3.6 on p. 55.)

The linearity, or lack of linearity, of a set of sample data is used as a basis for determining whether the distribution of the underlying population is the same as that of the probability paper (Benson, 1962). The disadvantages of probability plots are that there is no commonly accepted rule for plotting the data and drawing the lines by eye; nor is there a simple or universally accepted way to judge how well the data points conform to the straight lines (Mage, 1982).

Owing to sampling variations, there will be some departure from a straight line even when normally distributed data are plotted. We are concerned with the need to look for systematic deviations of the type indicating a long-tailed distribution or those indicating a skewed distribution. A situation in which the normal distribution fits the data rather poorly is provided in Ex. 3.7 on p. 57.

A problem with this rather informal approach is that different people will draw different lines when independently analyzing the same probability plot. Hahn and Shapiro (1967) instruct the reader, "If a straight line appears to fit the data, draw such a line on the graph by eye." And, of course, if the data set is small, it is difficult to determine whether a straight line fit is adequate.

Statistical goodness-of-fit tests exist for quantifying whether the data set may be characterized by the normal distribution, as will be examined in Secs. 3.3.4 and 3.3.5. However, a problem with statistical goodness-of-fit hypothesis testing is that few tests are useful with small sample sizes. The negative aspect of the plotting method utilizing the visual comparison of a line with data points is the subjectivity associated with the drawing of the lines.

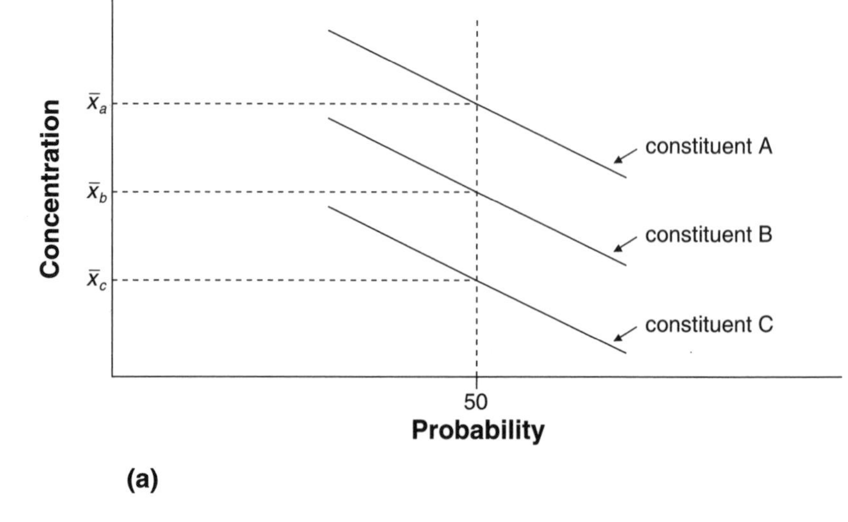

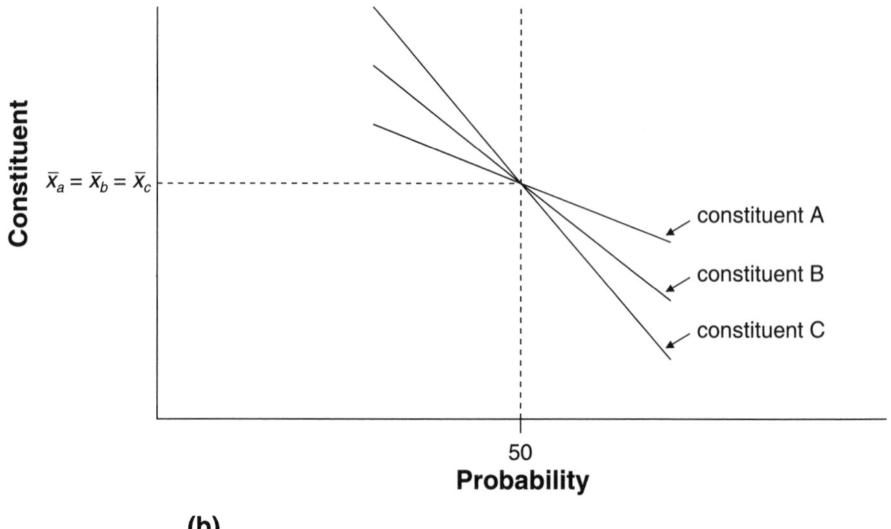

FIGURE 3.7 (a) Schematic of probability plots for three constituents A, B, and C, with different means but same standard deviation. (b) Schematic of probability plots for three constituents A, B, and C, with identical means but different standard deviations.

Probability plots are particularly useful for spotting irregularities within the data. It is easy to determine whether departures from normality are occurring more or less in the middle ranges of the data or in the extreme tails. Probability plots can also indicate the presence of possible outlier values that do not follow the basic pattern of the remainder of the data.

EXAMPLE 3.6

Consider the chloride concentration data listed in Table 3.2. Use the Weibull plotting position formula to plot the data on normal probability paper.

SOLUTION

The columns in Table 3.4 indicate the details of the calculation sequence. The resulting data are plotted in Fig. 3.8. The mean value estimate can be read as the ordinate value corresponding to the 50th percentile or 18 mg/L. The standard deviation can be estimated fairly precisely by ordinate values that correspond to the 16th and 84th percentiles respectively:

$$S = C_{50} - C_{84} = 18.0 - 13.8 = 4.2 \text{ mg/L}$$

These results compare favorably with the values computed from data, as summarized below:

	from the graph	from the data
$\bar{x}$	18.0 mg/L	18.3 mg/L
S	4.2 mg/L	4.1 mg/L

As is apparent from Fig. 3.8, the data points plot as a relatively straight line, confirming that the data are adequately described by the normal distribution.

TABLE 3.4 Calculation sequence for probability graphing of chloride data

Number of samples	Chloride concentrations (mg/L)	Rank	Rank-ordered concentration data	Weibull plotting position[a]
1	25.25	1	27.5	.048
2	13.32	2	25.25	.095
3	15.78	3	22.63	.143
4	22.63	4	22.27	.190
5	20.36	5	21.43	.238
6	21.43	6	20.47	.286
7	11.74	7	20.36	.333
8	22.27	8	18.71	.381
9	18.11	9	18.12	.429
10	27.50	10	18.11	.476
11	18.12	11	16.66	.524
12	16.16	12	16.39	.571
13	16.39	13	16.16	.619
14	16.66	14	16.09	.667
15	18.71	15	15.78	.714
16	14.90	16	15.64	.762
17	14.09	17	14.90	.810
18	15.64	18	14.09	.857
19	16.09	19	13.32	.905
20	20.47	20	11.74	.952

[a]Computed from $m/(n+1)$ where m is the rank and $n = 20$ for the total number of samples.

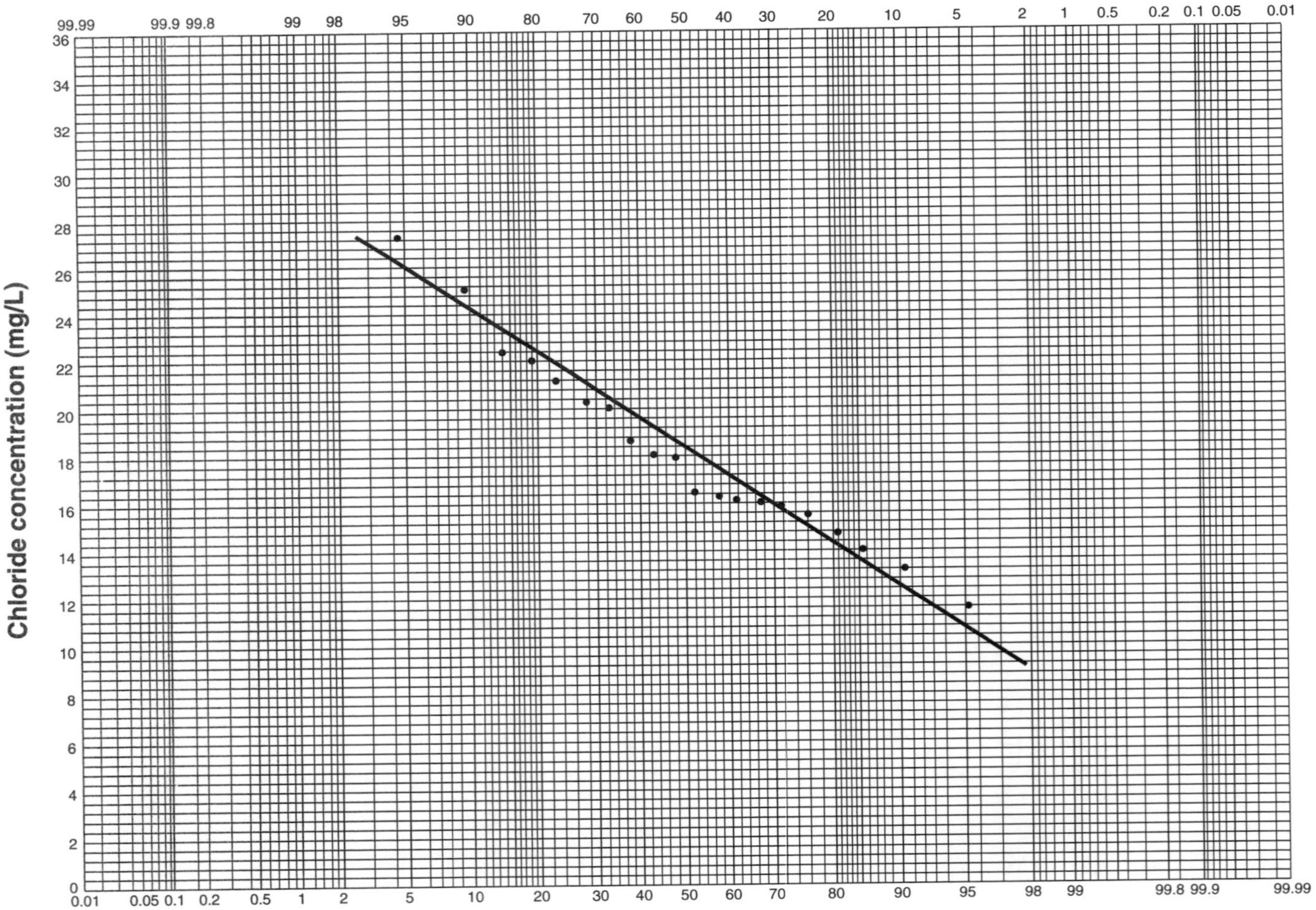

FIGURE 3.8 Normal probability plot for chloride concentrations

EXAMPLE 3.7

Consider the benzene concentration data listed in Table 3.2. Use the Weibull plotting position formula to plot the data on normal probability paper.

SOLUTION

The columns in Table 3.5 summarize the details of the calculation sequence. The resulting data are plotted in Fig. 3.9. As is apparent from the plotted data, which do not plot on a straight line, the normal distribution is not a good descriptor of the data.

Another type of probability paper in common use is lognormal probability paper. Lognormal and normal probability papers have identical probability scales; the difference is that the lognormal probability paper has a logarithmic scale for the ordinate whereas the latter has an arithmetic scale. Further discussion of lognormal paper is provided in Chap. 4.

A given data set may be plotted directly or, if the data do not plot as a straight line, a data transformation which produces a function that is normally distributed may be investigated. Alternative data transformations are infinite in number, but examples include square-root, square, and inverse operations.

TABLE 3.5 Calculation sequence for probability graphing of benzene data

Number of samples	Benzene concentrations (mg/L)	Rank	Rank-ordered concentration data	Weibull plotting position[a]
1	1.1	1	23.5	.048
2	1.4	2	10.15	.095
3	1.8	3	7.40	.143
4	2.05	4	5.35	.190
5	2.11	5	5.12	.238
6	2.15	6	4.80	.286
7	2.23	7	3.50	.333
8	2.54	8	3.15	.381
9	2.58	9	2.94	.429
10	2.61	10	2.85	.476
11	2.85	11	2.61	.524
12	3.15	12	2.58	.571
13	3.50	13	2.54	.619
14	4.80	14	2.23	.667
15	5.12	15	2.15	.714
16	7.40	16	2.11	.762
17	10.15	17	2.05	.810
18	5.35	18	1.8	.857
19	2.94	19	1.4	.905
20	23.5	20	1.1	.952

[a]Computed from $m/(n+1)$ where m is the rank and $n = 20$ for the total number of samples.

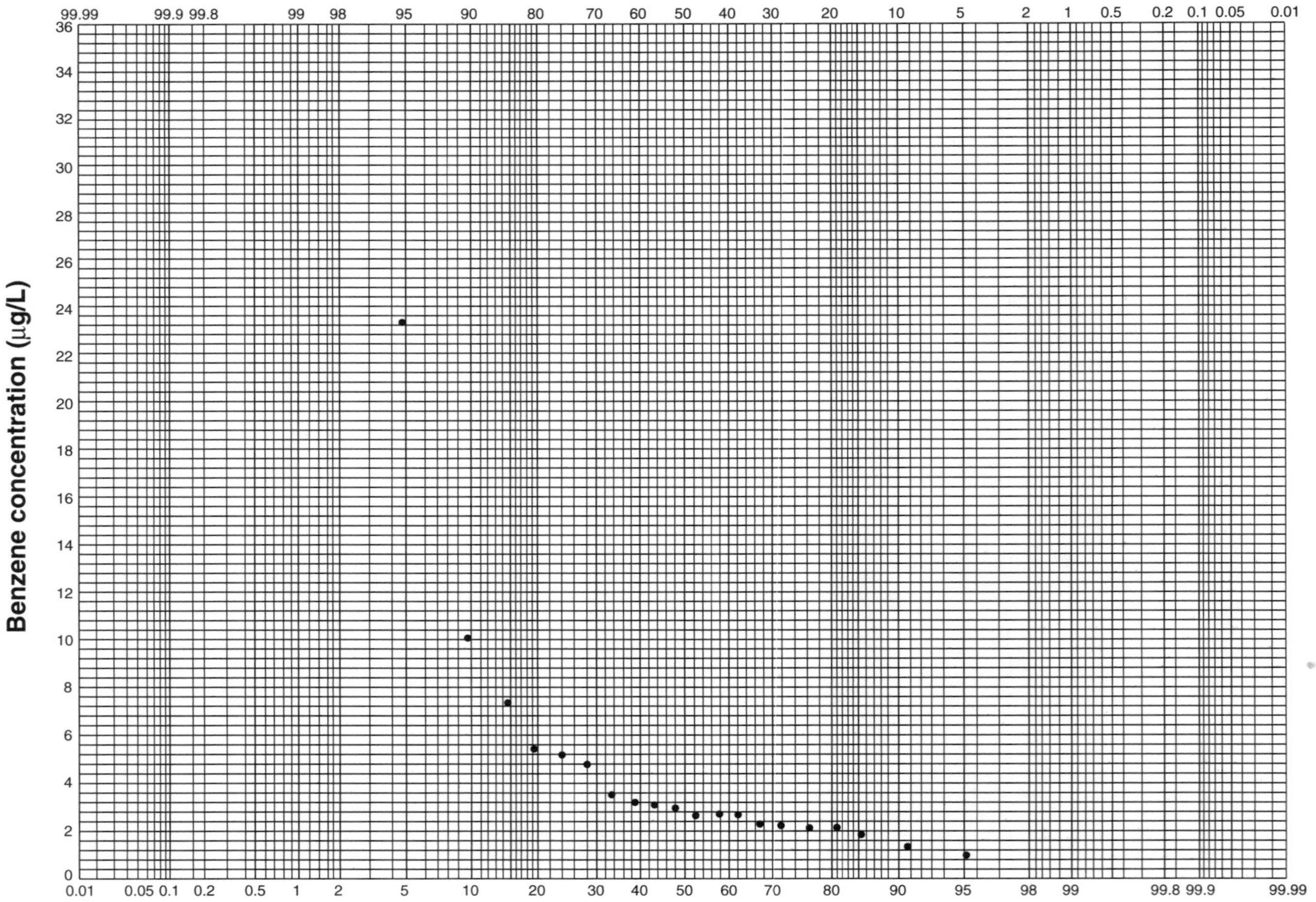

FIGURE 3.9 Normal probability plot of benzene concentrations

3.3.4 The Chi-Square Goodness-of-Fit Test

The chi-square (χ^2) test is one of the most versatile tests in statistical theory. The intent in this test is to evaluate whether the observed frequencies in a distribution differ significantly from the frequencies that might be expected according to some assumed hypothesis.

The first step is to formulate a null hypothesis and an alternative hypothesis, namely:

H_0: The data distribution is normal, with mean $\bar{x}$ and standard deviation S (where $\bar{x}$ and S are specific values for a particular application), and,

H_A: The data distribution is not normal.

The chi-square goodness-of-fit test is always a one-tailed test because the hypotheses are unidirectional, since the random variable is either distributed as specified, or it is not.

Although an acceptable goodness-of-fit test, the chi-square test is not considered the most sensitive or powerful (Gan and Koehler, 1990).

The chi-square test for normality is accomplished by segmenting the range of the variable into a series of intervals or cells, representing distinct, nonoverlapping ranges of data values. In each interval, an expected value is computed based on the number of data points that would be found in each of the intervals if the distribution being evaluated (in this case, the normal distribution) compares to the observed frequencies of data within each of the intervals. The observed data are used to form a histogram, which shows the observed frequencies in a series of k cells. The chi-square cells are usually selected such that the width for each is the same (although this is not a requirement of the test). The test is demonstrated in Ex. 3.9.

Differences from normality tend to show up more clearly in the tails of the distribution (as expected, since the normal distribution is a central-fitting distribution using only the mean and the standard deviation); it is in the tails that the chi-square goodness-of-fit test does poorly because of the need to combine cells to avoid small frequencies. The result is that the chi-square goodness-of-fit test is not a very useful test of normality with small data sets (although a correction factor is possible if $n < 5$). The validity of the chi-square test is best if the number of observations in each group or interval is consistently greater than five. Thus, the chi-square test requires a fairly large sample size.

If the chi-square test indicates that the data are not normally distributed, it may not be clear what ranges of the data most violate the normality assumption. Most tests, such as the ANOVA (see Chap. 9), are not overly influenced by deviations from the normal distribution in the central portion of the data. If the extreme tails are approximately normal in shape even if the middle part of the density is not, these parametric tests will still tend to produce valid results. For chi-square, departures from normality in the middle intervals are given nearly the same weight as departures from normality from the extreme tail intervals. As a result, as will be demonstrated later, the chi-square test is not as powerful for detecting departures from normality in the extreme tails of the data, the areas most crucial to the validity of parametric tests like the t-test or ANOVA (Miller, 1986)

EXAMPLE 3.8

For the chloride concentration data from Ex. 3.5, use the chi-square test to determine whether the data follow the normal distribution at the five percent level of significance.

SOLUTION

Assuming an interval of 2.50 mg/L, the chloride data are used to develop the observed frequencies (O_i). The expected frequencies (e_i) for each cell are computed using the probability distribution function associated with the intended population (in this case, the distribution being considered is the normal distribution). To compute the expected frequencies, the expected probability for each cell is determined and multiplied by the sample size n. The expected probability for cell i, p_i is the area under the probability distribution function between the cell bounds for that cell. The sum of the expected frequencies then equals the total sample size n.

The specifics in relation to the application of the chi-square test are indicated in Table 3.6.

The χ^2 distribution works with $k - 1 - p$ degrees of freedom, where k is the number of intervals or classes into which the observations are divided and p is the number of parameters that have been independently estimated from the data.

A large value of χ^2 indicates there is a large difference between the observed and expected frequencies. The occurrence of large χ^2 would indicate that the null hypothesis can be rejected. To do this, one must check to see whether the calculated value of χ^2 is greater or less than the critical value, χ^2_c given in the appropriate table (see Table A.12). To do this, one must know the degrees of freedom. For this example, the number of degrees of freedom is 8 – 3 = 5. One degree of freedom was lost because the total number of observations is used to compute the expected frequencies, while two were lost for the mean and standard deviation, which were estimated from the sample of 20 observations.

For these data, the degrees of freedom df is $df = 8 - 3 = 5$, and at $\alpha = 0.05$ (i.e., 5% significance level) $\chi^2_c = 11.07 > 3.32$.

The null hypothesis is not rejected and the data may be assumed to be normally distributed at a five percent level of significance.

3.3.5 THE KOLMOGOROV-SMIRNOV GOODNESS-OF-FIT TEST

The Kolmogorov-Smirnov test is another test of the goodness-of-fit of data. The Kolmogorov-Smirnov (e.g., see Lindgren, 1976) is used to test the hypothesis that data come from a specific (i.e., completely specified) distribution. A minimum sample size of 50 is recommended for use of this test. The Kolmogorov-Smirnov test can take one of two forms. A one-sample test can be performed to examine how closely observed probabilities correspond to theoretical probabilities. To accomplish this, the observed cumulative distribution of sample values are compared to a specified continuous distribution function (which, for purposes of this chapter, is the normal distribution). Alternatively, a two-sample test can be performed to test the agreement between two observed cumulative distributions.

TABLE 3.6 Chi-square goodness-of-fit test for the chloride data

Group number	Group interval	Observation[a] (n_i)	Theoretical frequency[b]	$\frac{(n_i - e_i)^2}{e_i}$ [c]
1	< 12.50	1	1.90	.43
2	12.50 –14.99	3	2.54	.08
3	15.00 –17.49	6	3.80	1.27
4	17.50 –19.99	3	4.22	.35
5	20.00–22.49	4	3.60	.04
6	22.50 –24.99	1	2.26	.70
7	25.00 –27.49	1	1.10	.01
8	> 27.50	1	0.52	.44
		$\Sigma = 20$		$\chi^2 = \Sigma = 3.32$

Notes:

[a]Indicates the number of observations within each group interval

[b]For purposes of demonstration, consider Group 3: From the chloride data, the mean $\bar{x} = 18.49$ and the standard deviation $S = 4.63$. The two ends of the group interval in terms of the standardized normal deviate are

1. for the lower end $Z_l = \frac{18.49 - 15.00}{4.63} = .754$

The corresponding area under the standardized normal distribution is .225.

2. for the upper end, $Z_u = \frac{18.49 - 17.49}{4.63} = .216$

The corresponding area under the standardized normal distribution is .415. The difference between the two areas indicates the theoretical probability of a chloride concentration lying within the interval if the data are from a normal distribution (with mean 18.49 and standard deviation = 4.63). Given 20 observations, one would expect (20)(.415 – .225) = 3.80 observations in that interval.

[c]Using the values in the third and fourth column in the formula, compute the chi-square statistic where χ^2 is the symbol for chi-square.

This application tests the null hypothesis, H_0, that two independent samples come from identical continuous distributions. It is sensitive to population differences with respect to location, dispersion, and skewness.

The Kolmogorov-Smirnov statistic is simply the maximum absolute difference between the theoretical and observed cumulative probability distribution, and this distribution is written as D.

Algebraically, the Kolmogorov-Smirnov test considers the deviations between an hypothesized cumulative distribution function such as the normal distribution function, $F(x)$, and the observed cumulative histogram as:

$$F^*(x^{(i)}) = i/n \quad [3.9]$$

in which $x^{(i)}$ is the ith largest observed value in the random sample of size n. The sample statistic D is calculated as

$$D = \max_{i = 1,n} \left\{ \left| (F^*(x^i) - F_x(x^{(i)}) \right| \right\} \quad [3.10]$$

Of interest then is whether the difference D is larger than can reasonably be expected. The test involves consideration of the following hypotheses:

H_0: that x has a specified distribution (the null hypothesis for the Kolmogorov-Smirnov goodness-of-fit test is that the sample data that produced the observed probability distribution have been drawn from a population of the specified theoretical distribution).

H_A: that the distribution of x is other than specified.

If indeed the sample is from the underlying distribution (in this case, the normal distribution) then the Kolmogorov-Smirnov statistic D has a distribution that is independent of the underlying distribution. Since the sampling distribution of D under the null hypothesis is known, values of the critical statistic, $d_{n,\alpha}$ as a function of sample size n and level of significance α, are contained in many standard references (e.g., Benjamin and Cornell, 1970, p. 667). Tables of critical values of D are available (see Table A.13) and are a function of sample size n and level of significance α. Therefore, the hypothesis H_0 is accepted if

$$D \leq d_{n,\alpha} \qquad [3.11]$$

If the calculated value of D is greater than the table of critical values at a specified significance level, the null hypothesis can be rejected at that level.

The degrees of freedom for the Kolmogorov-Smirnov goodness-of-fit test are the number of items in the observed frequency distribution. Harter (1980) provides modified asymptotic formulas for the statistic $d_{n,\alpha}$.

Lilliefors (1967) modified the Kolmogorov-Smirnov statistic tables to test whether a set of observations is from a normal population when the mean and variance are not specified but must be estimated.

The Kolmogorov-Smirnov test has the following advantages over the chi-square test (Benjamin and Cornell, 1970):

1. For continuous distributions, it does not lump data and compare discrete categories, but rather compares all data in an unaltered form;
2. the value of the test statistic D is usually computed more easily than the chi-square statistic; and
3. unlike the chi-square test, the Kolmogorov-Smirnov test is an exact test for all sample sizes. However, as a general statement, the departures from normality have to be large to be detected in sample sizes less than approximately 25.

The Kolmogorov-Smirnov test is generally more efficient than the chi-square test for goodness-of-fit for small samples, and it can be used for very small samples where the chi-square test does not apply. Note as well that the test as presented above was applied to compare the data with the normal distribution; however, there is no need to restrict the test to the normal distribution since the theory applies to any continuous distribution.

3.3.6 The Shapiro-Wilk Test

The Shapiro-Wilk W test (Shapiro and Wilk, 1965) is another statistical goodness-of-fit test that performs well on small sample sizes and tests the null hypothesis that the data values are random samples from a normal distribution against an unspecified alternative distribution. This test is considered one of the best numerical tests of normality and is particularly useful for detecting departures from normality in the tails of a sample distribution. It can be used in conjunction with a probability plot to measure how well the plotted quantiles are following a straight line (i.e., how well the sample values are correlated with normal quantiles).

The test can be performed on any sample size from 3 to 50. As would be expected, the power of the test increases as the sample size gets larger.

Steps in the calculation involve:

1. Order the sample data.
2. Compute a weighted sum (b) of the differences between the most extreme observations.
3. Divide the weighted sum by a multiple of the standard deviation, and square the result to get the Shapiro-Wilk statistic W:

$$W = \left\{\frac{b}{S\sqrt{n-1}}\right\}^2 \qquad [3.12]$$

where the numerator is computed as

$$b = \sum_{i=1}^{k} a_{n-i+1}\left(x_{n-i+1} - x_{(i)}\right) = \sum_{i=1}^{k} b_i \qquad [3.13]$$

where $x_{(i)}$ represents the smallest ordered value in the sample, and coefficients a_i depend on the sample size n. The coefficients can be found for any sample size from 3 to 50 in Table A.7. The value of k can be found as the greatest integer less than or equal to $n/2$.

Normality of the data should be rejected if the Shapiro-Wilk statistic is too low when compared to the critical values provided in Table A.6. Otherwise, one can assume the data are approximately normal for purposes of further statistical analyses.

The Shapiro-Wilk test statistic W will tend to be large when a probability plot of the data indicates a nearly straight line. Only when the plotted data show significant bends or curves will the test statistic be small (Miller, 1986; Madansky, 1988). See Ex. 3.9 and 3.10.

EXAMPLE 3.9

Use the chloride data of Ex. 3.1 to compute the Shapiro-Wilk test of normality.

SOLUTION

Step 1: Order the data from smallest to largest, as per the table below. Also list the data in reverse order alongside the ordered data.

Step 2: Compute the difference $x_{(n-i+1)} - x_{(i)}$ in Column IV of the table by subtracting Column II from Column III.

I i	II $x_{(i)}$	III $x_{(n-i+1)}$	IV $x_{(n-i+1)} - x_{(i)}$	V a_{n-i+1}	VI b_i
1	11.74	27.50	15.76	.4734	7.461
2	13.32	25.25	11.93	.3211	3.831
3	14.09	22.63	8.54	.2565	2.191
4	14.90	22.27	7.37	.2085	1.537
5	15.64	21.43	5.79	.1686	.976
6	15.78	20.47	4.69	.1334	.626
7	16.09	20.36	4.27	.1013	.433
8	16.16	18.71	2.55	.0711	.181
9	16.39	18.12	1.73	.0422	.073
10	16.66	18.11	1.45	.0140	.020
11	18.11	16.66	−1.45		$b = 17.329$
12	18.12	16.39	−1.73		
13	18.71	16.16	−2.55		
14	20.36	16.09	−4.27		
15	20.47	15.78	−4.69		
16	21.43	15.64	−5.79		
17	22.27	14.90	−7.37		
18	22.63	14.09	−8.54		
19	25.25	13.32	−11.93		
20	27.50	11.74	−15.76		

Step 3: Compute k as the greatest integer less than or equal to $n/2$. Since $n = 20$, $k = 10$.

Step 4: Look up the coefficients a_{n-i+1} from Table A.6 and place in Column V. Multiply the differences in Column III by the coefficients in Column IV and add the first k products to get quantity b. In this case, $b = 17.329$.

Step 5: Compute the standard deviation of the sample, $S = 4.05$. Then

$$W = \left[\frac{17.329}{4.05\sqrt{19}}\right]^2 = 0.964$$

Step 6: Compare the computed value of $W = 0.964$ to the five percent critical value for sample size 20 in Table A.7, namely $W_{.05,20} = 0.905$. Since $W > 0.905$ the chloride data do not show significant evidence of nonnormality.

3.3.7 The Shapiro-Francia Test

When there are more than 50 observations, the Shapiro-Francia test involves a slight modification of the Shapiro-Wilk test. This test has the same advantages as the Shapiro-Wilk test and, like the Shapiro-Wilk test (Shapiro and Francia, 1972), the test statistic of Shapiro-Francia W' will tend to be large when the data follow the normal distribution. This test statistic is defined as

$$W' = \frac{(\Sigma\, m_i x_i)^2}{\Sigma\, (x_i - \overline{x})^2} \qquad [3.14]$$

EXAMPLE 3.10

Use the benzene data of Ex. 3.2 to compute the Shapiro-Wilk test of normality.

SOLUTION

Paralleling the analyses in Ex. 3.10, the tabular array becomes

I i	II $x_{(i)}$	III $x_{(n-i+1)}$	IV $x_{(n-i+1)} - x_{(i)}$	V a_{n-i+1}	VI b_i
1	1.1	23.5	22.4	.4734	10.60
2	1.4	10.15	8.75	.3211	2.81
3	1.8	7.40	5.60	.2565	1.44
4	2.05	5.35	3.30	.2085	.69
5	2.11	5.12	3.01	.1686	.51
6	2.15	4.80	2.65	.1334	.35
7	2.23	3.50	1.27	.1013	.13
8	2.54	3.15	.61	.0711	.04
9	2.58	2.94	.36	.0422	.02
10	2.61	2.85	.24	.0140	.00
11.	2.85	2.61	–.24		$b = 16.59$
12	2.94	2.58	–.36		
13	3.15	2.54	–.61		
14	3.50	2.23	1.27		
15	4.80	2.15	–2.65		
16	5.12	2.11	–3.01		
17	5.35	2.05	–3.30		
18	7.40	1.8	–5.60		
19	10.15	1.4	–8.75		
20	23.5	1.1	–22.4		

Step 2: for $n = 20$, $k = 10$
Step 3: $b = 16.59$ and standard deviation = 4.99
Step 4:

$$W = \left[\frac{16.59}{4.99\sqrt{19}}\right]^2 = 0.582$$

Step 5: Since $W_{.05,20} = 0.905$ and $W < 0.905$, the data show significant evidence of nonnormality by the Shapiro-Wilk test.

As a result, the data should be transformed using natural logs and rechecked using the Shapiro-Wilk test. This alternative is described in Chap. 4, Ex. 4.4.

Steps in the calculation:

1. Order the sample data.
2. Compute a weighted sum of the observations:

$$\text{Weighted sum} = \sum m_i x_i$$

The values for m_i can be approximately computed as

$$m_i = \phi^{-1}\,[i/(n + 1)] \qquad [3.15]$$

where ϕ^{-1} denotes the inverse of the standard normal distribution with zero mean and unit variance.

3. divide the square of the weighted sum by a multiple of the standard deviation to obtain the Shapiro-Francia statistic W'; and

$$W' = \frac{(\Sigma\, m_i\, x_i)^2}{(n - 1)\, S^2\, \Sigma\, m_i^2} \qquad [3.16]$$

4. the normality of the data is rejected if W' is too low when compared to the tabulated critical value.

Normality of the data should be rejected if the Shapiro-Francia statistic is too low in comparison with the critical values provided in Table A.8 of Appendix A. Otherwise one can assume that the data are approximately distributed for purposes of further statistical analyses.

3.3.8 Data Transformations

The preceding seven subsections have all been directed toward alternative tests for assessing whether a data set is adequately described by the normal distribution. Obviously, any and or all of the tests may reject the hypothesis that the data follow a normal distribution. In this circumstance, one possible approach is to apply a data transformation so that the normal distribution is appropriate. In fact there always exists a transformation (or a series of transformations) such that the resulting distribution is normal (although in many applications it may be difficult to determine the one or more transformations that are required). Some of the more commonly used transformations are indicated in Table 3.7.

In some circumstances, a situation exists in which the data contain seasonal variability in addition to random sampling and analytical errors. For this situa-

TABLE 3.7 Common data transformations

Data Transformation	**Calculation Procedure**
Arithmetic	$x' = x + C$
Reciprocal	$x' = \frac{1}{x}$
Arcsine	$x' = \text{arcsine}\,\sqrt{x}$
Logarithmic (base 10)	$x' = \log x$
Logarithmic (base e)	$x' = \ln x$
Square root	$x' = \sqrt{x}$
Cubic root	$x' = \sqrt[3]{x}$

tion, if the data record is sufficiently lengthy, a sinusoidally varying term may be fit to the data and this component thus removed, leaving only the residuals. Such removal of seasonal variability is extensively done in surface water hydrology. Nevertheless, for many environmental quality problems, the ability to "fit" a sinusoidally varying term to a brief record is highly suspect.

A negative aspect of these data transformation approaches is that they have a degree of arbitrariness with respect to the transformation to be employed for a specific application. However, without the data transformation, the characteristics of the data may preclude the utilization of specific tests. The lognormal distribution is a good alternative distribution and will be the focus of analysis in Chap. 4.

3.3.9 Summary of Goodness-of-Fit Tests

With large groups of data, we are able to develop an accurate characterization of the parameters of the population (e.g., kurtosis and skewness) and from these parameters determine if the population is normal (with a specified level of confidence). Alternatively, we can use a chi-square goodness-of-fit test for testing normality. However, when this data set is small (e.g., $n < 25$, these measures or tests (kurtosis, skewness, and chi-square goodness-of-fit) are not accurate indicators of normality.

Unfortunately, all of the available tests for normality do at best a fair job of rejecting nonnormal data when the sample size is small (e.g., $n < 20$ to 30). For example, small samples of untransformed lognormal data can be accepted by a test of normality even though the skewness of the data may lead to poor statistical conclusions later. Most environmental data sets are like this; therefore, utilization of the normal or lognormal distribution as adequate is likely. Plotting tests and coefficient of variation tests give a reasonable basis for judgment with respect to applicability of specific distributions.

3.4 THE *t*-DISTRIBUTION

All of the preceding sections in this chapter have been specifically focused on the normal distribution. However, the normal distribution is actually a special case or subset of the *t*-distribution. The *t*-distribution is an unbounded distribution having a mean of zero and a variance that depends on the scale parameter *df*, the degrees of freedom. As the degrees of freedom approach infinity the variance approaches unity, and the *t*-distribution becomes the standard normal distribution. In general, the standard normal distribution can be used instead of the *t*-distribution for sample sizes greater than 30.

As a way of introducing the *t*-distribution, it is useful to compare the structure of the *t*-table with the table of the standard normal distribution. Recall that the standard normal distribution is not a function of the degrees of freedom; as the number of degrees of freedom increases, the *t*-distribution approaches the normal distribution. The *t*-table in Table 3.8 provides *t*-values when the probability specified at the top of the table is in only one tail (or direction) of the distribution.

TABLE 3.8 Representative values of the t-distribution indicating the relationship to the normal distribution

	Points of the t-distribution[a]		
Degrees of freedom	**.05**	**.01**	**.001**
1	6.314	31.82	318.1
5	2.015	3.365	5.893
10	1.812	2.764	4.144
15	1.753	2.602	3.733
20	1.725	2.528	3.552
25	1.708	2.485	3.450
40	1.684	2.423	3.307
60	1.671	2.390	3.232
100	1.658	2.358	3.160
∞	1.645	2.326	3.090

[a]This table gives the values of t which are exceeded with probability P.
Source: Beyer, W.H. ed., *CRC Handbook of Tables for Probability and Statistics* (Cleveland, OH: Chemical Rubber Company, 1966). Reprinted with permission.

3.5 EXTENT OF USE OF THE NORMAL DISTRIBUTION

As indicated in Sec. 3.1, the normal distribution is widely applicable (even far beyond those concerns relevant to environmental quality data). In later chapters in this book, the normal distribution will be seen to be fundamentally important to many statistical techniques, such as ANOVA (to be discussed in Chap. 9), and the testing of various hypotheses relies on the assumption of the normal distribution, a subject that will be addressed in Chaps. 8 through 10.

It is also noteworthy that while many statistical methods invoke the assumption of normality, the statistical methods remain approximately valid even when there are moderate departures from normality. As such, these procedures are considered robust (not impacted by moderate departures from normality). Robustness (i.e., deviations from normality are acceptable) will be featured at various points in the chapters to follow.

3.6 SUMMARY COMMENTS

"Normal distribution" is so named because it frequently describes phenomena. As ensuing chapters will demonstrate, many subsequent tests are premised on the data being (at least approximately) distributed normally. Some of the tests reviewed need large samples and some can function with very few. Some tests are specific to test the appropriateness of the normal distribution while some are goodness-of-fit tests applicable to any distribution (e.g., Kolmogorov-Smirnov).

3.7 REFERENCES

BENJAMIN, J.R., and C.A. CORNELL, *Probability, Statistics, and Decisions for Civil Engineers.* New York, NY: McGraw-Hill, 1970.

BENSON, M.A., "Plotting Positions and Economics of Engineering Planning," *Proceedings of ASCE Journal of the Hydraulic Division,* 88 (November, 1962), 57–71.

BLOM, G., *Statistical Estimates and Transformed Beta Variables.* New York, NY: John Wiley and Sons, 1958.

BULMER, M.G., *Principles of Statistics.* Cambridge, MA: MIT Press, 1965.

GAN, F.F., and K.J. KOEHLER, "Goodness-of-Fit Tests Based on P-P Probability Plots," *Technometrics,* 32, no. 3 (1990), 289–303.

HAHN, G., and S. SHAPIRO, *Statistical Models in Engineering.* New York, NY: Wiley, Interscience, 1967.

HARTER, H.L., "Modified Asymptotic Formula Critical Values of the Kolmogorov Test Statistic," *The American Statistician,* 34 (1980), 110–111.

HAZEN, A., "Storage to be Provided in Impounding Reservoirs for Municipal Water Supply," *Proceedings of ASCE,* 39 (1914), 1943–2044.

KIMBALL, B.F., "Sufficient Statistical Estimation Functions for the Parameters of the Distribution of Maximum Values," *Annals of Mathematical Statistics,* 17, no. 3 (1946), 299–309.

KIMBALL, B.F., "On the Choice of Plotting Positions on Probability Paper," *Journal of American Statistical Association,* 55 (1960), 546–560.

LILLIEFORS, H.W., "On the Kolmogorov-Smirnov Normality with Mean and Variance Unknown," *Journal of the American Statistical Association,* 62 (1967), 399–404.

LINDGREN, B.W., *Statistical Theory,* 3rd ed. New York, NY: MacMillan Press, 1976.

MADANSKY, A., *Prescriptions for Working Statisticians.* New York, NY: Springer-Verlag, 1988.

MAGE, D.T., "An Objective Graphical Method for Testing Normal Distributional Assumptions Using Probability Plots," *The American Statistician,* 36, no. 2 (1982), 116–120.

MILLER, R.G., Jr., *Beyond ANOVA Basics of Applied Statistics,* New York, NY: John Wiley and Sons, 1986.

SHAPIRO, S.S., and R.S. FRANCIA, "An Approximate Analysis of Variance Test for Normality," *Journal of the American Statistical Association,* 67, no. 337 (1972), 215–216.

SHAPIRO, S.S., and M.B. WILK, "An Analysis of Variance Test for Normality (Complete Samples)," *Biometrika,* 52 (1965), 591–611.

VIESSMAN, W., G.L. LEWIS, and J.W. KNAPP, *Introduction to Hydrology,* 3rd ed. New York, NY: Harper and Row, 1989.

3.8 PROBLEMS

3.1. A random variable has a mean of 0.7 and a standard deviation of 0.1. Assuming the data follow a normal distribution, find the probability that the value of a sample point is:

- **a.** less than 0.55;
- **b.** greater than 0.82; and
- **c.** between 0.61 and 0.88.

3.2. A concentration of a parameter is Gaussian, has a mean value of 25 mg/L, and has a standard deviation of 5 mg/L. Find the value of the concentration such that

- **a.** 25 percent of the area lies above it; and
- **b.** 7 percent of the area lies below it.

3.3. The average annual precipitation for Los Angeles is 350 mm, with a standard deviation of 36 mm. If the amounts can be assumed to follow a normal distribution, find the probability that the precipitation in any one year is

- **a.** greater than 400 mm;
- **b.** less than 250 mm; and
- **c.** either less than 250 mm or greater than 400 mm.

3.4. Using the 20 values in the first two rows of the data from file NORM on the diskette provided, perform a frequency analysis for a normal distribution. Fit the frequency curve using the method of moments, and use the Weibull plotting position formula to plot the data.

3.5. Demonstrate the Central Limit Theorem. Generate summation of random variables from a uniform random number generator; sum them and plot the results for 5, 25, 100, and 250 numbers. A good approximation to $N(0,1)$ is where x_i is from uniform random number from zero to one.

3.6. Measurements of concentration from 200 samples showed a mean of 0.825 mg/L and a standard deviation of 0.042 mg/L. Find the 95 percent confidence limits for the mean concentration.

CHAPTER 4

The Lognormal Distribution

4.1 INTRODUCTION

Discussion in Chap. 3 described in detail the features of the normal distribution, a very general and important distribution. However, two attributes of the normal distribution limit its applicability to environmental quality concerns. Specifically,

1. the normal distribution is symmetrical about the mean and yet many distributions of environmental quality data are not symmetrical (for example, the skewed water quality distribution in Fig. 4.1); and
2. the tails of the normal distribution extend to infinity in both the negative and positive directions. Since negative concentrations are not feasible, the normal distribution may be a poor characterization of environmental quality data.

FIGURE 4.1 Example of concentration data skewed to the right

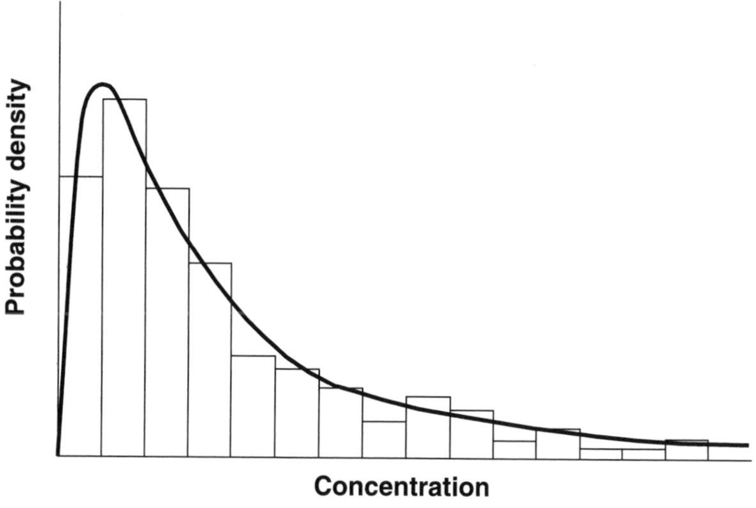

As a result, situations arise where it is inappropriate to characterize the distribution of an environmental quality constituent as Gaussian or normal. An important alternative that is applicable to a wide array of environmental concerns is the lognormal distribution. Specifically, the lognormal distribution precludes values less than zero, and the distribution has no upper bound. Consequently, the lognormal distribution's shape has considerable potential to describe environmental quality data (McBean and Rovers, 1992).

4.2 IMPORTANT FEATURES OF THE LOGNORMAL DISTRIBUTION

4.2.1 The Central Limit Theorem

To better understand why the lognormal distribution is often appropriate to describe environmental quality data, it is useful to establish the parallel to the normal distribution. To do this, consider a variable that is the product of a large number of independent factors, *m*, none of which dominates the resulting product; that is,

$$y_i = x_{i1}\, x_{i2}\, x_{i3}\, x_{i4} \ldots x_{im} \qquad [4.1]$$

where *m* is a large number. In other words, the variable of interest y_i is expressed as the product of a large number of subvariables, each of which is in itself difficult to study and describe. If we take the logarithm of both sides, Eq. [4.1] becomes

$$z_i = \ln y_i = \text{n}\, x_{i1} + \ln x_{i2} + \ln x_{i3} + \cdots + \ln x_{im} \qquad [4.2]$$

which is of the form of the relationship described in Sec. 3.1; the variable z_i is the sum of a large number of *m* variables. Thus, z_i is distributed as per the Gaussian or normal distribution. Therefore, the arguments made in Chap. 3 that related to the Central Limit Theorem and the normal distribution as appropriate to support the concept of the normal distribution can, in a similar way, be used to support the concept of the lognormal distribution to describe data. In its simplest conceptual form, then, the lognormal distribution is viewed as a normal distribution of the logarithms of a set of data. Thus, the lognormal distribution gets its name because the logarithms of random values are normally distributed.

For simple problems such as the summation of dice rolls, it is possible to support the idea that the sum of the dice rolls may be characterized using the Central Limit Theorem, as a normal distribution. In reality, however, it is difficult (if not impossible) to identify the large number of contributing variables that contribute in an additive and/or multiplicative way in environmental quality problems. The water quality in a stream is the consequence of many features—these include the runoff from farmers' fields, drainage from urban streets, wastewater discharges, as well as scour and deposition from the stream bottom, to name only a few. It is difficult to envision explicitly that the combinations from these individual sources are either additive or multiplicative; it is simply necessary to recognize that the concept of the Central Limit Theorem basically supports the idea that environmental phenomena and the consequences of innumerable individual features combine in such a way that their combinatorial effect is not dominated by a single feature but by a combination. The results may be described by a normal and/or lognormal distribution.

4.2.2 The Mathematics of the Lognormal Distribution

If the variables $y_1, y_2, y_3, \ldots, y_n$ of the original data, are lognormally distributed, then the transformed variables $z_1, z_2, z_3, \ldots, z_n$ where $z_i = \ln y_i$, for $i = i, \ldots, n$, are normally distributed. The z_i, $i = 1, \ldots, n$ may then be described by the normal distribution equation (Eq. [3.1]) as per the mathematical details presented in Chap. 3.

The population parameters μ_z and σ_z^2 may be estimated by $\bar{z}$ and S_z^2 in the manner described in Chap. 3. This is done first by transforming all of y_i to z_i by taking logarithms ($z_i = \ln y_i$). The $\bar{z}$ and S_z^2 are then calculated as

$$\bar{z} = \sum_{i=1}^{n} z_i / n \qquad [4.3]$$

and

$$S_z^2 = \frac{\sum_{i=1}^{n} (z_i - \bar{z})^2}{n - 1} \qquad [4.4]$$

See Ex. 4.1 on p. 74.

Chow (1964), in a comprehensive investigation of the lognormal distribution, presents the following relationships for calculating $\bar{z}$ and S_z^2, without taking the logarithms of all of the data, as

$$\bar{z} = 1/2 \ln [\bar{y} / \{C_v^2 + 1\}] \qquad [4.5]$$

and

$$S_z^2 = \ln [C_v^2 + 1] \qquad [4.6]$$

where C_v is the coefficient of variation of the original data ($C_v = S_y / \bar{y}$).

The mean and variance of the lognormal distribution are

$$\mu_y = E[y] = \exp\left(\mu_z + \frac{\sigma_z^2}{2}\right) \qquad [4.7]$$

and

$$\sigma_y^2 = \text{Var}[y] = \mu_y^2 \left[\exp\left(\frac{\sigma_z^2}{2}\right) - 1\right] \qquad [4.8]$$

where $E[\ ]$ is the expectation operator (numerous reference sources discuss the expectation operator; see, for example, Chatfield, 1978).

The coefficient of variation of the y values is

$$C_v = \left[\exp\left(\frac{\sigma^2}{2} - 1\right)\right]^{.5} \quad [4.9]$$

The coefficient of skew of the y values is

$$y = 3\,C_v + C_v^{\,3} \quad [4.10]$$

EXAMPLE 4.1

The data listed below represent measurements of the hydraulic conductivity of the soils utilized in developing the liner at the Keele Valley landfill.

Sample	Hydraulic Conductivity (cm/s)
1	1.08×10^{-8}
2	8.76×10^{-9}
3	1.01×10^{-8}
4	8.07×10^{-9}
5	5.63×10^{-9}
6	4.60×10^{-9}
7	7.42×10^{-9}
8	1.49×10^{-8}
9	3.48×10^{-9}
10	6.40×10^{-8}

a. Calculate the arithmetic mean and the geometric mean. Which of the two measures of central tendency, the arithmetic mean or the geometric mean, is a better indication of the central tendency of the data?

$$\text{Arithmetic mean} = 1.38 \times 10^{-8}\ \text{cm/s}$$

$$\text{Geometric mean} = 9.31 \times 10^{-9}\ \text{cm/s}$$

A histogram of the data is illustrated in Fig. 4.2. The geometric mean is more representative of the bulk of the data.

b. Is the logarithm of the (arithmetic) mean the same as the mean of the logarithm-transformed data?

$$\text{Logarithm of the mean} = \ln(1.38 \times 10^{-8}) = -18.1$$

$$\text{Mean of the logarithms} = -16.19$$

Note that the mean of the original data is not the same as the logarithm of the mean of the original data. With the original data, the mean is very much related only to the high values.

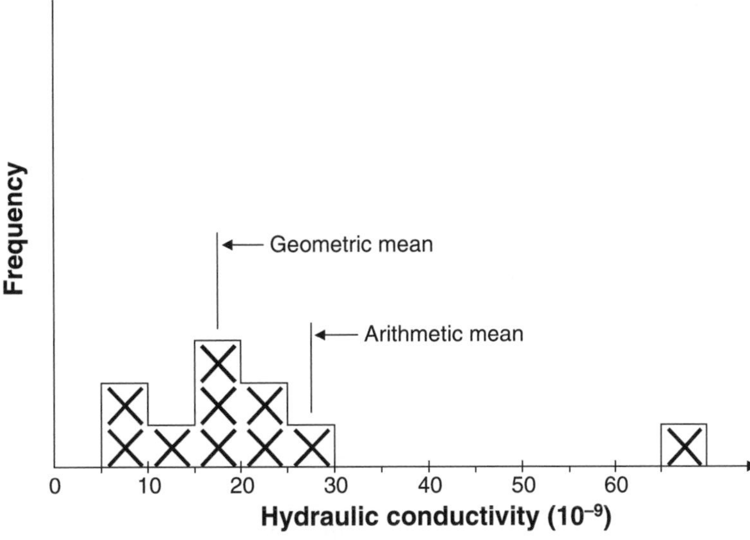

FIGURE 4.2 Histogram of hydraulic conductivity

With the widespread availability of spreadsheet calculation programs, the need for Eqs. [4.4] through [4.10] is considerably less than when the equations were first developed. As well, these equations are premised on the values being the population parameters (as opposed to sample statistics).

There are several very important features associated with the lognormal distribution. First, the lognormal distribution is a two-parameter distribution, which means that to completely characterize the distribution only two parameters need to be quantified. Therefore, as with the normal distribution, the lognormal distribution is a central-fitting distribution, not a fitting to the tails of the distribution. As a result, the lognormal distribution may not provide acceptable accuracy in describing the extreme values of the data.

Because there are just the two fitting parameters implicit in the description of the lognormal distribution and yet the distribution shape is not symmetrical, there is a built-in skewness associated with the distribution. Once the mean and variance are specified, the value of the skewness is fixed. If it is desirable to change the skewness (and thereby to better reflect the tails of the data distribution), one needs to utilize more sophisticated distributions, the subject of Chap. 5.

Transformation to the logarithmic scale is not done to make large numbers look smaller. Performing a logarithmic or other monotonic transformation preserves the basic ordering within a data set so that the data are rescaled with a different set of units. The changes occur because the logarithms of lognormally distributed data are more nearly normal or Gaussian in character, thus satisfying a key assumption of many statistical procedures.

4.2.3 Probability Paper

The primary discussion on probability plotting is contained in Sec. 3.3.3 and should be the first point of reference; discussion herein simply follows along from the discussion in Sec. 3.3.3. The same probability plot technique outlined in Chap. 3 may be used to investigate whether a set of data or residuals follows the lognormal distribution. The procedure is the same except that each observation is replaced by its logarithm. Hence, probability paper for the lognormal distribution is identical to normal distribution paper except that the ordinate is a logarithmic scale. Thus, lognormal probability paper is similar to the normal probability paper in terms of its effectiveness in providing the opportunity to characterize visually whether the lognormal distribution is an appropriate fit to a data set.

EXAMPLE 4.2

Use the Weibull plotting position formula and probability plotting paper to indicate the appropriateness of the following data to be characterized by the lognormal distribution.

Solution

Mercury concentration (μg/L) is 2.2, 1.5, 1.0, 2.2, 3.8, .15, .45, 1.3, 6.4, .85.

We can demonstrate the data are skewed, from the coefficient of variation test.

$$\bar{x} = 3.30$$

$$S_x = 3.70$$

Hence, the coefficient of variation is 3.70/3.30 = 1.12 and greater than unity. The data distribution must be skewed.

It is useful as well, to plot the data on probability paper since deviations of the plotted data from a straight line indicate the data are not well characterized by the normal distribution. The results using normal probability paper are shown in Fig. 4.3 (a) and for lognormal paper in Fig. 4.3(b).

Note that the ordinate on the probability paper may be equally expressed on the logarithmic axis or, equivalently, in arithmetic terms.

In a manner comparable to the discussion in Sec. 3.3.3, both the mean and the standard deviation may be estimated from the plotted lines on the probability paper.

A number of researchers have reported that the effective hydraulic conductivities of soil are not adequately described by their arithmetic mean values (e.g., Dagan, 1979; Donald and McBean, 1994). It is now generally accepted that the hydraulic conductivity of a compacted clay liner follows a lognormal distribution (e.g., see Donald and McBean, 1994) and hence the geometric mean is a better descriptor of the effective hydraulic conductivity of soil.

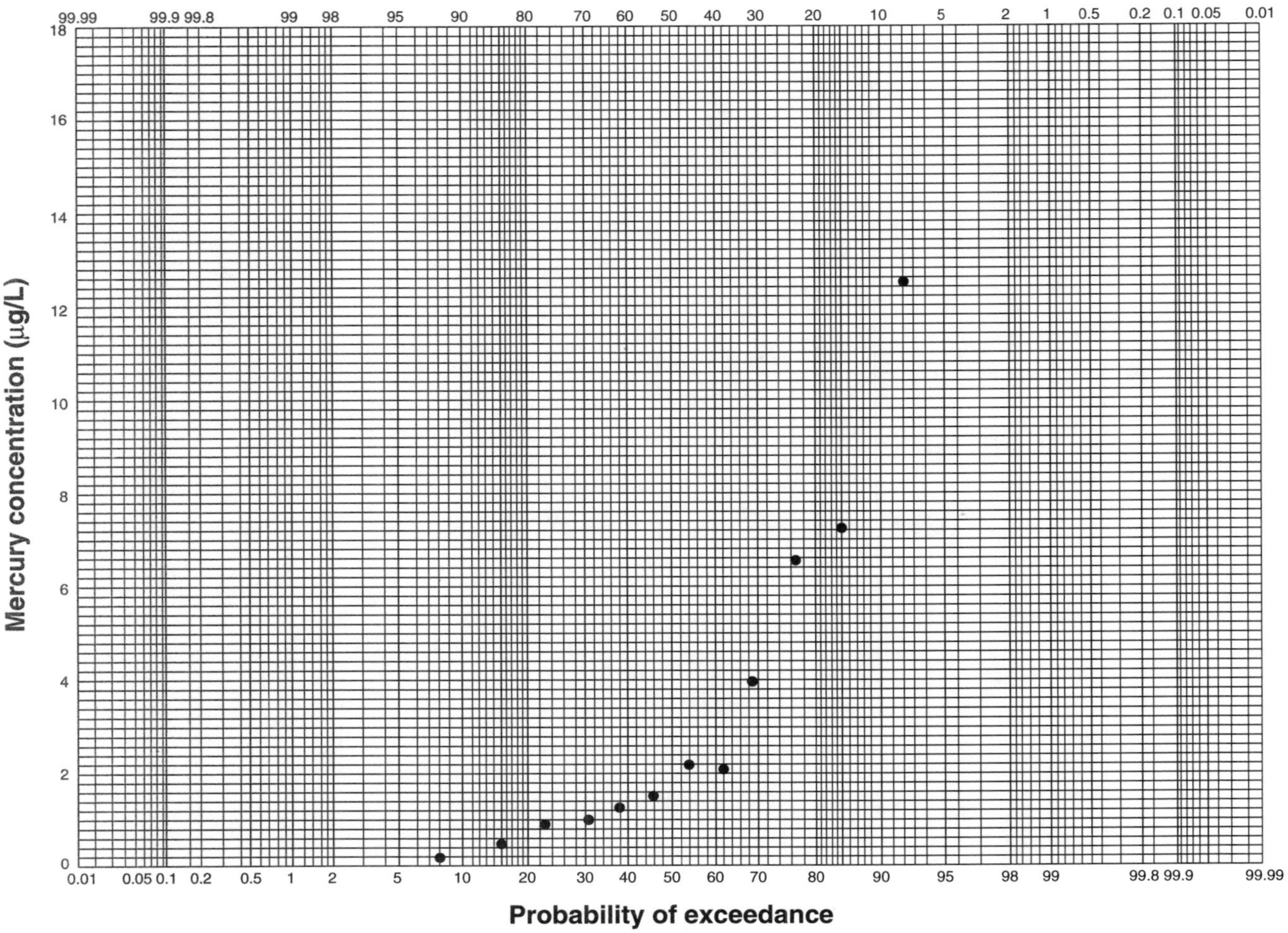

Mercury concentration (μg/L)

Probability of exceedance

FIGURE 4.3(a) Probability plot of mercury data, normal probability paper

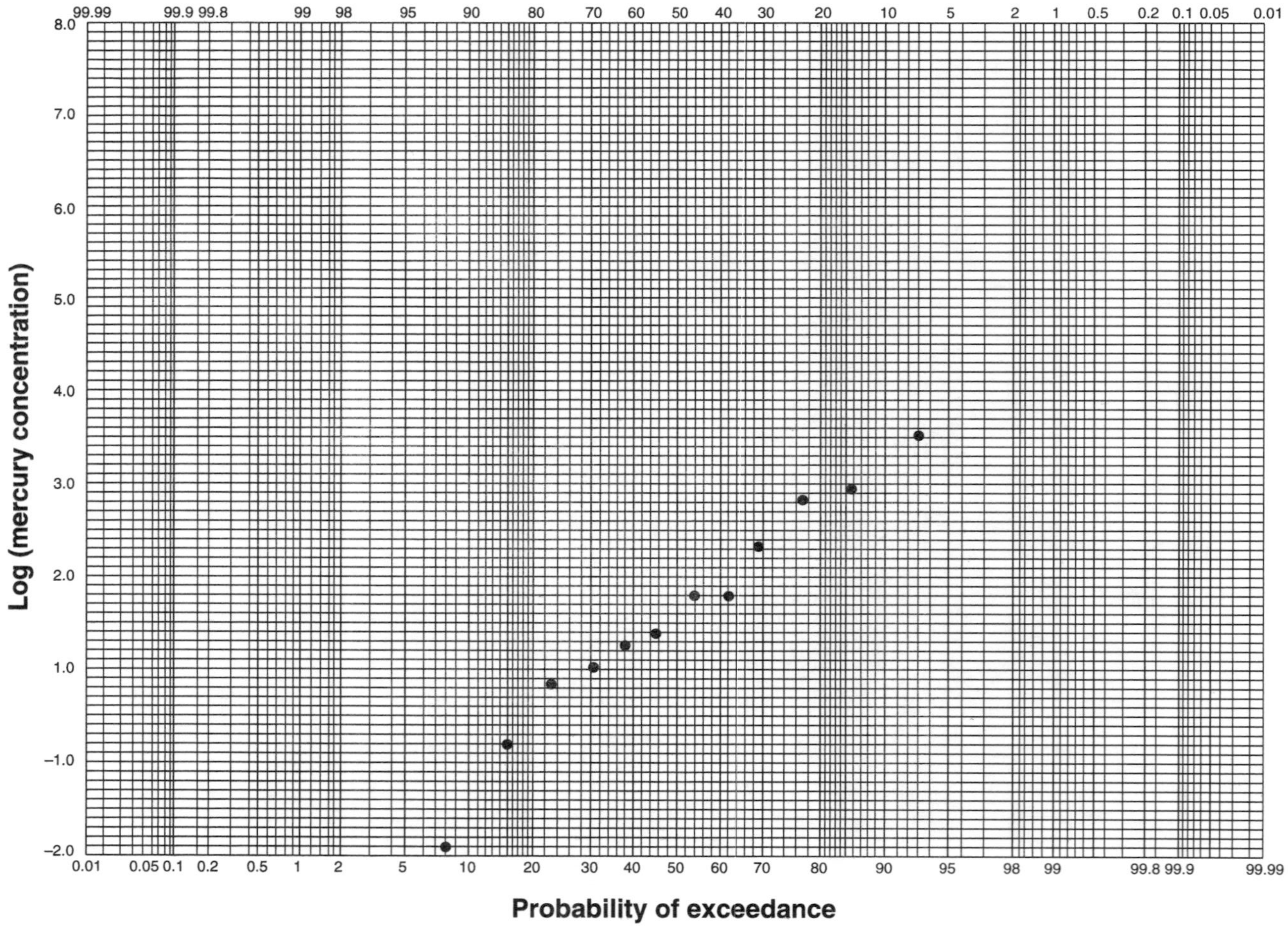

Log (mercury concentration)

8.0 7.0 6.0 5.0 4.0 3.0 2.0 1.0 −1.0 −2.0

99.99 99.9 99.8 99 98 95 90 80 70 60 50 40 30 20 10 5 2 1 0.5 0.2 0.1 0.05 0.01

0.01 0.05 0.1 0.2 0.5 1 2 5 10 20 30 40 50 60 70 80 90 95 98 99 99.8 99.9 99.99

Probability of exceedance

FIGURE 4.3(b) Probability plot of mercury data, lognormal probability paper

EXAMPLE 4.3

Of all soil physical properties, Warrick and Neilsen (1980) illustrate that hydraulic conductivity exhibits the greatest degree of variability. This implies that samples collected from a field site will not all yield the same hydraulic conductivity but rather will be discrete values from an underlying statistical distribution. In addition, given that the behavior of a porous medium is not necessarily best described by its mean value, the mean value of the hydraulic conductivity's statistical distribution does not represent the overall effective or bulk conductivity. Donald and McBean (1994) showed a plot of the logarithms of hydraulic conductivity for samples from an engineered natural soil liner for Metropolitan Toronto's Keele Valley landfill versus the standard normal deviate. The results are illustrated in Fig. 4.4 and include a straight line estimate of the lognormally distributed conductivity measurements. A reasonable characterization of the hydraulic conductivity data by the log-normal distribution is apparent, since the data are reasonably characterized by the straight line. Note that z indicates the number of standard deviations from the mean.

4.3 TESTS FOR LOGNORMALITY

To determine whether the lognormal distribution is an adequate descriptor of a data set, mathematical procedures similar to those described in Secs. 3.3.1 through 3.3.7 may be implemented by applying the mathematical procedures to the logarithms of the data. For this reason, the tests for goodness of fit to the distribution will not be re-examined.

FIGURE 4.4 Probability plot of hydraulic conductivity of an engineered natural soil liner

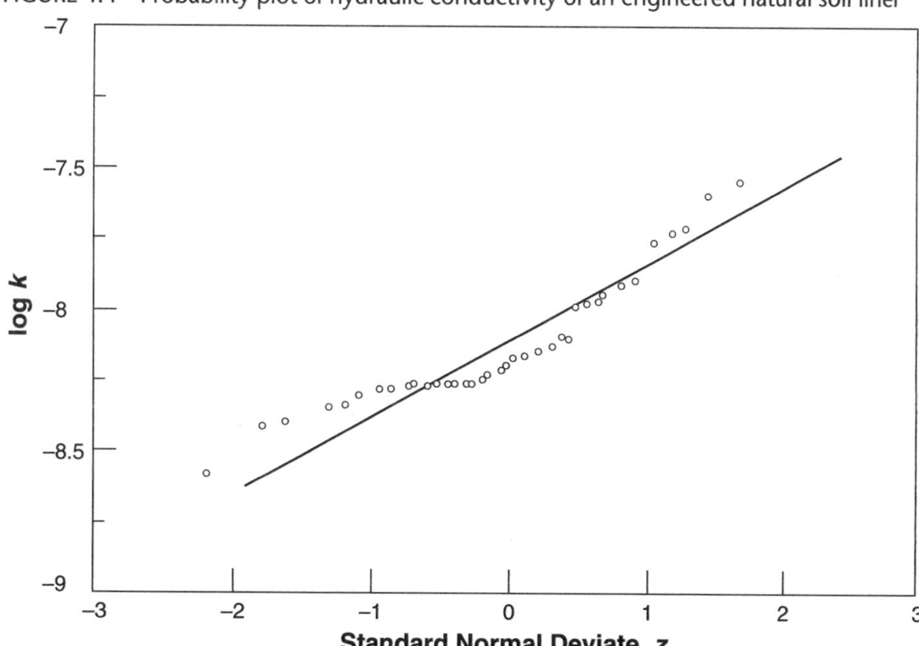

EXAMPLE 4.4

Determine if the benzene concentration data of Table 3.2 are adequately described by the lognormal distribution. Use the Shapiro-Wilk test (see Sec. 3.3.6 for a discussion of the Shapiro-Wilk test). This example represents an extension of Ex. 3.11.

Step 1: Order the data from smallest to largest and take the natural logarithms as listed in the following table.

Step 2: Compute the differences.

i	X_i	$X_{r(i)}$	X_{n-i+1}	$X_{n-i+1} - X_i$	a_{n-i+1}	b_i
1	11.74	2.463	3.314	.851	.4734	.403
2	13.32	2.589	3.229	.640	.3211	.206
3	14.09	2.645	3.119	.474	.2565	.122
4	14.90	2.701	3.103	.402	.2085	.084
5	15.64	2.750	3.065	.315	.1686	.053
6	15.78	2.759	3.019	.260	.1334	.035
7	16.09	2.778	3.014	.236	.1013	.024
8	16.16	2.783	2.929	.146	.0711	.010
9	16.39	2.797	2.897	.100	.0422	.004
10	16.66	2.813	2.912	.099	.0140	.001
11	18.11	2.912	2.813			$b = 16.59$
12	18.12	2.897	2.797			
13	18.71	2.929	2.783			
14	20.36	3.014	2.778			
15	20.47	3.019	2.759			
16	21.43	3.065	2.750			
17	22.27	3.103	2.701			
18	22.63	3.119	2.645			
19	25.25	3.229	2.589			
20	27.50	3.314	2.463			

Step 3: for $n = 20, k = 10$

Step 4: Coefficients a_{n-i+1} from Table A.6; $b = .942$

Step 5: Compute the standard deviation of the log-transformed data as $S = .218$

$$W = \left[\frac{.942}{.218\sqrt{19}}\right]^2 = 0.983$$

Step 6: $W_{.05,20} = 0.905$ from Table A.6. Since $W > 0.905$, the benzene data do not show significant evidence of nonnormality in the log-transformed sense.

4.4 GENERATION OF LOGNORMAL CONCENTRATION DATA

Numerous computer programs provide a simple procedure to generate a sequence of uniformly distributed random numbers and normally distributed random numbers. However, there may be circumstances in which a set of lognormally distributed random numbers is of interest. This can be accomplished as follows:

If $z = \ln y$, where x is the actual concentration of a chemical, then the probability distribution of y is normal with a mean μ_y and variance σ_y^2. The mean and variance of z are readily computed from the mean and variance of y (Aitchison and Brown, 1957).

$$\mu_z = \ln (\mu_y) = 0.5 [\ln \{\sigma_y / \mu_y\}^2 + 1] \quad [4.11]$$

$$\sigma_z = [\ln (\sigma_y / \mu_y)^2 + 1]^{0.5} \quad [4.12]$$

For generating data sets for Monte Carlo experiments, the population mean, μ_y, and the coefficient of variation, σ_y/μ_y, can be specified and used to calculate μ_y and σ_y. Then a data set of true concentrations, y_i, of any number of values n can be generated by

$$y_i = \exp [\mu_z + s_z \, \varepsilon_i] \quad [4.13]$$

where ε_i is a randomly chosen value from a normal distribution with a mean of zero and variance of one.

A linear trend may be added by causing the specified population μ_y, to increase at a constant rate m with time t, as

$$\mu_z = \ln (\mu_y + m\,t) - 0.5 [\ln [(\sigma_y / \mu_y)^2 + 1] \quad [4.14]$$

4.5 REFERENCES

AITCHISON, J., and J.A. BROWN, *The Lognormal Distribution.* Cambridge, MA: University Press, 1957.

CHATFIELD, C., *Statistics for Technology.* New York, NY: John Wiley and Sons, 1978.

CHOW, V.T., ed., *Handbook of Applied Hydrology.* New York, NY: McGraw-Hill, 1964.

DAGAN, G., "Models of Groundwater Flow in Statistically Homogeneous Porous Formations," *Water Resources Research,* 15, no. 1 (1979), 47–63.

DONALD, S.B., and E.A. McBEAN, "Statistical Analyses of Compacted Clay Landfill Liners. Part 1: Model Development," *Canadian Journal of Civil Engineering,* 21 (1994), 872–882.

McBEAN, E.A., and F.A. ROVERS, "Estimation of the Probability of Exceedance of Contamination Concentration Risk Problems," *Ground Water Monitoring Review* (Winter, 1992), 115–119.

SMITH, L., and R.A. FREEZE, "Stochastic Analysis of Steady State Groundwater Flow in a Bounded Domain. Part 2: Two-dimensional Simulations," *Water Resources Research,* 15, no. 6 (1979), 1543–1559.

WARRICK, A.W., and D.R. NIELSEN, "Spatial Variability of Soil Physical Properties in the Field," in *Applications of Soil Physics,* ed. Daniel Hillel, pp. 319–344. New York, NY: Academic Press, 1980.

4.6 PROBLEMS

4.1. Assume that the lognormal distribution is a good descriptor of data on PCB levels. You have collected the following data at a site:

Date/location	MW1	MW2	MW3	MW4	MW5	MW6	MW7	MW8	MW9	MW10
2-Apr-95	5.74	2.36	1.83	1.84	8.65	2.50	3.33	2.61	1.84	1.94
9-Apr-95	2.06	2.21	3.39	3.53	5.99	1.39	2.08	3.15	4.50	6.21
16-Apr-95	1.67	2.89	4.14	1.48	2.23	6.81	2.14	2.84	3.94	2.47
23-Apr-95	4.92	3.26	3.20	3.29	6.22	3.40	1.29	3.74	1.75	2.55
30-Apr-95	1.94	1.30	1.21	2.17	4.39	1.66	4.40	2.81	6.55	3.78
7-May-95	1.81	4.47	2.29	2.65	3.76	2.32	2.53	3.00	2.96	2.44
14-May-95	1.57	1.85	6.35	4.66	2.61	1.33	1.43	3.13	2.02	3.56
21-May-95	4.30	3.49	1.17	5.39	4.26	2.05	2.12	1.32	2.19	2.21
28-May-95	3.28	1.77	4.84	2.99	2.17	6.25	3.95	1.11	1.85	1.18
4-Jun-95	2.53	2.43	1.83	3.30	6.01	1.99	2.12	1.73	2.98	2.92

Calculate the 95 percent exceedance concentration.

4.2. Using the data from Prob. 4.1, calculate the arithmetic mean and the geometric mean. Which one is a better indicator of the central tendency of the data?

4.3. The following data indicate concentration levels of TCE in a monitoring well downstream from an industrial sewage lagoon. Assuming that the data are lognormally distributed, calculate the arithmetic mean of the data transformation.

6.323, 4.71, 5.616, 4.191, 4.836, 6.322, 3.516, 6.042, 3.013, 5.354, 4.106, 2.056, 5.973, 5.157, 3.609.

After log transformation, which of the two measures of central tendency is a better representation of the data, and why?

4.4. The data listed below represent measurements of hydraulic conductivity in soil samples taken from a landfill liner.

Sample	Hydraulic conductivity (cm/s)
1	1.08×10^{-8}
2	8.76×10^{-9}
3	1.01×10^{-8}
4	8.07×10^{-9}
5	5.63×10^{-9}
6	4.60×10^{-9}
7	7.42×10^{-9}
8	1.49×10^{-8}
9	3.48×10^{-9}
10	6.40×10^{-8}

a. Calculate the arithmetic mean and the geometric mean. Which of the two measures of central tendency, the arithmetic mean or the geometric mean, is a better characterization of the data?

b. Plot the data as a histogram to indicate which of the two measurements of central tendency is more appropriate for this circumstance.

c. Is the logarithm of the arithmetic mean the same as the mean of the logarithm?

CHAPTER 5

ADDITIONAL USEFUL DISTRIBUTIONS FOR CHARACTERIZING ENVIRONMENTAL QUALITY DATA

5.1 INTRODUCTION

Numerous useful probability distributions exist (beyond the normal and lognormal distributions) that may be appropriate for characterizing environmental quality data. These distributions include the binomial, Poisson, Gumbel, and log Pearson III, plus many more. By selecting a particular distribution, we don't know of course what the next value might be, but by the distribution that we select, we can assign the probability associated with the various values that might occur. As will be seen, circumstances sometimes require distributional characteristics different from those of the normal and lognormal distributions considered in Chaps. 3 and 4. These other distributions have been presented here on the basis of their ability to "fit" the data due to their greater flexibility. The intent of this chapter is to examine some of the important features of other probability distributions as pertain to their useful application to environmental quality problems.

Statistical distributions in the technical literature are frequently demonstrated in applications in which data lengths number in the thousands. Large data sets such as these seldom exist in environmental quality situations. For example, some of these additional distributions require characterization of the skewness coefficient, which as indicated in Chap. 2 may be considerably in error when the data set is small. The intent of the various sections in this chapter is to identify both positive and negative attributes of these other distributions as they pertain to environmental quality data.

5.2 THE BINOMIAL DISTRIBUTION

The binomial distribution is used to define the probabilities of discrete events. This distribution is applicable to random variables that satisfy the following four assumptions:

1. There are n occurrences or trials of the random variable.
2. The results of the n trials are independent from one another.
3. There are only two possible outcomes for each trial.
4. The probability of each outcome is constant from trial to trial.

To examine the characteristics of the distribution in more detail, consider p as the probability that an event will happen in any single trial (called the probability of success) and $q = 1 - p$ is the probability that it will fail to happen in any single trial (called the probability of failure). The probability $b(\,)$ is the probability that the event will happen exactly x times in n trials (i.e., x successes and n–x failures) for the binomial distribution is given by:

$$b(x;n,p) = \binom{n}{x} p^x (1 - p)^{n-x} = \frac{n!\, p^x}{x!(n - x)!}(1 - p)^{n-x} \qquad [5.1]$$

where $x = 0, 1, 2, \ldots n$ and $n! = n(n\text{–}1)(n\text{–}2)(n\text{–}3)\ldots$ and $0! = 1$.

Note that Eq. [5.2] characterizes the probability of x successes in n trials where the events occur in a specific order:

$$p^x (1 - p)^{n-x} \qquad [5.2]$$

whereas Eq. [5.1] accumulates all possible orders of x successes in n trials.

EXAMPLE 5.1

a. Determine the probability of obtaining exactly two occurrences in four trials, where the probability of a success is 0.3.
From Eq. [5.1] the probability of exactly two occurrences in four trials is:

$$b(2;4,0.3) = \binom{4}{2} 0.3^2 (1 - 0.3)^{4-2} = 0.2646$$

b. Alternatively, what is the probability of obtaining the sequence pass, fail, pass, fail? The calculation proceeds as:

$$(.3)(.7)(.3)(.7) = .3^2\,.7^2 = 0.044$$

The probability of the specific sequence indicated in this part (b) is less than that for part (a) since part (a) accounts for additional sequences.

If there are, on the average, x successes per trial, the average number of successes in n trials must be xn.

It is noteworthy that there is a relationship between the binomial distribution and the normal distribution. If n is large and if neither p nor q is too close to zero, the binomial distribution can be closely approximated by a normal distribution with standardized variables given by

$$z = \frac{x - np}{\sqrt{npq}} \tag{5.3}$$

The approximation becomes better as n becomes larger. As n increases, the skewness and kurtosis for the binomial distribution approach that of the normal distribution. In practice this approximation is very good if both np and nq are greater than five.

5.3 THE POISSON DISTRIBUTION

In 1837, Poisson published the derivation of a distribution that bears his name. Poisson's approach was to derive a distribution for a series of independent events in which the number of trials were large, the probability of occurrence of the outcome was small, and the probability remained constant over the trials. To utilize the Poisson distribution, the random variable takes on only discrete, usually integer values; if this is not the case, the analyses should be carried out using a continuously distributed random variable. Thus, the Poisson distribution describes the number of events or occurrences.

The Poisson distribution is governed by the average rate of occurrence μ, which can be estimated by summing the Poisson events of all samples in the background pool of data and dividing by the number of samples in the pool. Once the average rate of occurrences has been estimated, the formula for the Poisson distribution is given by:

$$P(x;\mu) = \frac{e^{-\mu}\mu^{x}}{x!} \text{ for } x = 0, 1, 2, 3 \ldots \tag{5.4}$$

where x represents the Poisson count and μ represents the average rate of occurrence.

The cumulative distribution is

$$F(x;\mu) = \sum_{i=0}^{x} P\,(i;\mu) \tag{5.5}$$

In other words, the cumulative probability—$F(1;3)$ means the probability of obtaining $x = 0$ or 1, where the mean of the distribution is 3—is

$$F(1;3) = P(0;3) + P(1;3) \tag{5.6}$$

Using Eq. [5.4], then, Eq. [5.6] is calculated as

$$F(1;3) = \frac{e^{-3}3^0}{0!} + \frac{e^{-3}3^1}{1!} = 0.0497 + 0.1494$$
$$= 0.1991$$
$$\cong 0.20$$

In words, if the mean of the Poisson distribution is three, there is a 20 percent probability of having a zero or one integer response.

Examples of environmental engineering applications of the Poisson distribution include the counts of the number of microorganisms of a certain type, as quantified by a lab test, or the number of α-particles emitted per unit of time during radioactive decay.

The probabilities $P(x)$ are nonnegative and sum to one, as

$$\sum_{x=0}^{\infty} P(x) = \sum_{x=0}^{\infty} \frac{e^{-m}m^x}{x!}$$
$$= e^{-\mu}\frac{\mu^x}{x!}$$
$$= e^{-\mu}e^{+\mu}$$
$$= 1 \qquad [5.7]$$

The function of the factor $e^{-\mu}$ which does not vary with x is to ensure that the probabilities add up to unity.

The parameter μ equals the mean of the probability function; the variance is also equal to μ, or

$$\text{Mean} = \mu$$

and

$$\text{Variance} = \sigma^2 = \mu$$

Both the mean and the variance are equal to μ while the skewness is $1/\sqrt{\mu}$ so that the distribution is always skewed to the right. However, the degree of skewness decreases as μ increases.

The Poisson distribution can thus be used to determine probabilities for discrete random variables that satisfy the following three assumptions:

1. The random variable is the number of times that an event occurs in a single trial, which is defined as a basic unit of time, space, or quantity. A trial may be defined as a unit of time (e.g., a millisecond, hour, or year), a unit of space (i.e., length, area, volume) or some other quantity. The number of events in a

trial is usually an integer value from 0 to 1, 2, 3. . . . The random variable is the number of events that occur per trial.

2. The average number of occurrences of the random variable in a single trial remains constant from trial to trial.
3. The discrete values are nonoverlapping, and are independent of one another.

Given these three assumptions, it can be seen that the Poisson distribution is the limiting form of the binomial distribution where there is a large number of trials but only a small probability of success at each of them.

In the context of environmental quality monitoring, further applications of the Poisson distribution include the number of "hits" out of a large number of volatile organic compound measurements. This problem has been modeled by the Poisson distribution (e.g., Silver and Dunn, 1986). Alternatively, Gibbons (1987) considered the molecule as a unit of observation and postulated that the number of molecules of a particular compound out of a much larger number of molecules is the result of a Poisson process. Poisson's approach is justified in that the number of units (i.e., molecules) examined is large, and the probability of the occurrence (e.g., a molecule being classified as benzene) is small and, thus considered, the occurrence of low-level hits of volatile organic priority pollutant compounds can be statistically modeled as a Poisson process.

Other examples of application of the Poisson distribution will be developed in later chapters as specific areas of application.

EXAMPLE 5.2

As a simple example of the Poisson distribution to assist with the explanation, the parts per billion in the sequence of nine monitoring periods (of air mass through a screen), is

3, 1, 10, 2, 4, 6, 8, 2 (units)

At low levels, the values are only reported to specific magnitudes. What is the probability that in the next monitoring period, there will be one or less ppb?

SOLUTION

The mean of the samples is 4.33.

The probability of 0 is $P(0;4.33) = \dfrac{e^{-4.33}\,4.33^0}{0!} = .013$

The probability of 1 is $P(1;4.33) = \dfrac{e^{-4.33}\,4.33^1}{1!} = .057$

The probability that in the next monitoring period there will be one or less ppb measured is .013 + .057 = .070.

5.4 EXTREME VALUE DISTRIBUTIONS

There is a family of extreme value distributions that, as the name implies, have been utilized to characterize extreme events (e.g., what is the flood flow that is exceeded only once on average in 100 years?). The utilization of an extreme value distribution involves the fitting of a continuous distribution such as Pearson or log Pearson and/or an asymptotic distribution such as Gumbel's extreme value. Whereas the normal and lognormal distributions are referred to as central-fitting distributions, the tails of distribution (i.e., the extreme values) are better described by distributions other than the normal or lognormal distributions.

Chemical concentration data are constrained on the low side by the detection limit of the analytical technique but are essentially unconstrained on the high side. Because of this, the probability distribution is often considered as lognormally distributed (as per Chap. 4). On the negative side, the flexibility of the lognormal distribution to fit the data is limited because the skewness is fixed; only two parameters define the entire lognormal distribution. Therefore, for example, in risk assessment studies, in which much of the concern may be focused on a relatively extreme event, the reliance on only two statistics does not allow for much flexibility for the fitting of the distribution.

5.4.1 The Gumbel Distribution

Gumbel (1954, 1958) showed that the exponential function is useful in characterizing the distribution of annual flood peak discharges. Gumbel proposed a double exponential distribution for the largest value. Specifically, in its cumulative form the Gumbel model is defined as

$$F(x) = \exp\left[-e^{-\alpha(x-\beta)}\right] \quad [5.8]$$

in which $F(x)$ is the cumulative distribution of the large value x, and α and β are fitting parameters. The probability distribution is:

$$f(x) = \left[-\alpha(x - \beta)\, e^{-\alpha(x-\beta)}\right] \quad [5.9]$$

The parameters α and β are related to the mean and standard deviation of the data by the relationship

$$\mu = \beta + \frac{0.5772}{\alpha} \quad [5.10]$$

and

$$\sigma = \frac{\pi}{\alpha\sqrt{6}} = \frac{1.282}{\alpha} \quad [5.11]$$

Alternatively, these same two equations may be rewritten as

$$\beta = \mu - 0.450\,\sigma \tag{5.12}$$

and

$$\alpha = \frac{1.282}{\sigma} \tag{5.13}$$

Several features about the Gumbel distribution are noteworthy: The distribution is a two-parameter distribution. The skewness coefficient is fixed at 1.1396, a positive and constant value. The kurtosis coefficient of the Gumbel distribution is also constant as 4.5. Since the kurtosis coefficient is greater than 3.0, the peakedness of the distribution is different from the normal distribution.

The data utilized in defining the fitting parameters of the distribution must be independent values and extremes. Thus, the Gumbel model is commonly adopted to describe hydrologic phenomena, such as the maximum daily flow in a year or the annual peak hourly discharge during a flood (e.g., see Chow, 1951). The rationale for this model is that a single value utilized in the case of maximum daily flow in a year represents the single highest value (i.e., one out of 365) of flow in the year. See Ex. 5.3 on p. 91.

Probability paper for the Gumbel distribution has been developed, as illustrated in Fig. 5.1. All of the features described in Sec. 3.3.3 that have to do with plotting on probability paper, apply to using Gumbel probability paper. The Gumbel probability paper contains an extreme-value probability scale versus either an arithmetic or logarithmic scale so as to test whether the data approximate a Gumbel or log-Gumbel cumulative distribution function.

By a symmetrical argument to that in Eq. [5.8], the distribution of the smallest of many independent variables with a common unlimited distribution with an exponential-like lower tail can be addressed. The cumulative distribution for this case is

$$F(z) = 1 - \exp\left(-e^{-\alpha(z-\beta)}\right) \tag{5.14}$$

with the mean of

$$\mu_z = \beta - \frac{0.5772}{\alpha} \tag{5.15}$$

and

$$\sigma_z = \frac{\pi}{\alpha\sqrt{6}} \tag{5.16}$$

and coefficient of skewness of –1.1396.

Two log-Gumbel distributions may be defined and are known as the extreme-value type II and III distributions; the former may be applied to floods

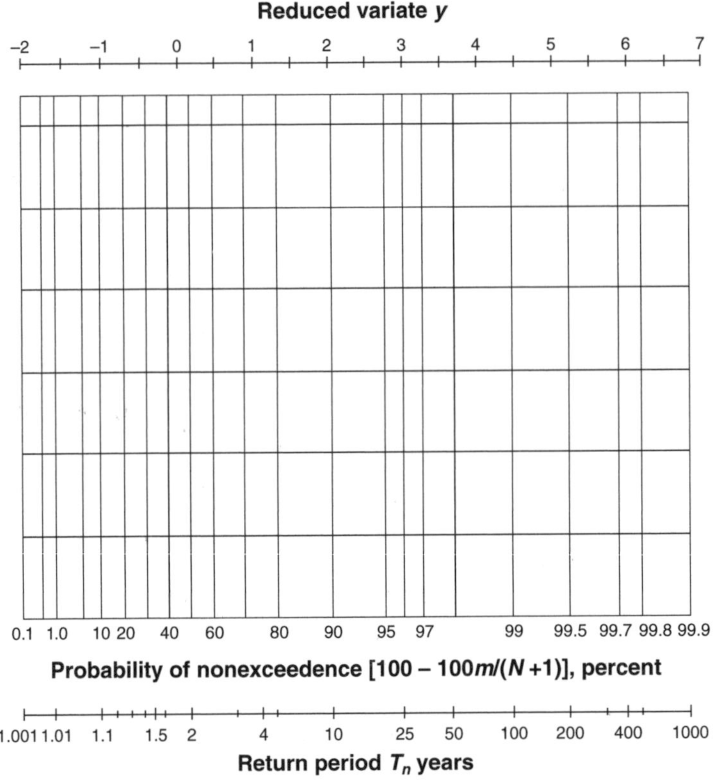

FIGURE 5.1 Gumbel probability paper

(maxima) while the latter distribution (also known as the Weibull distribution) is applied to minima events. However, as noted in Benjamin and Cornell (1970), the extreme-small-value model (minima) is not always a good choice if the variable being characterized is necessarily positive as, for example, stream flows and/or concentrations must be. This is because the distribution is unbounded, extending to infinity in both a positive and a negative direction.

5.4.2 Log Pearson Type III Distribution

Chow (1951) demonstrated that for many types of frequency analyses the extreme value Z_T (Where Z_T is the value having the potential for being equaled or exceeded once on average in T values), can be written as

$$Z_T = \bar{x} + k_T S \quad [5.17]$$

where:

$\bar{x}$ = the mean of a data set

k_T = is a frequency factor

S = standard deviation.

EXAMPLE 5.3

A history of records of the peak annual flows in a stream is analyzed to determine the estimates of the mean and variance as

$$\mu = 18 \text{ m}^3/\text{s}$$
$$\sigma^2 = 16 \text{ or } \sigma = 4 \text{ m}^3/\text{s}$$

Estimate the probability that the flow will exceed 30 m^3/s.

SOLUTION

Solving for the parameters α and β give

$$\alpha = \frac{1.282}{\sigma} = \frac{1.282}{4} = .321$$

and

$$\beta = \mu - 0.45\,\sigma$$
$$= 18 - 0.45\,(4) = 16.2$$

The probability that the peak flow in a particular year will exceed 30 m^3/s is

$$P(x \geq 30) = 1 - F(30)$$
$$= 1 - \exp\left[- 3^{-.321(30-16.2)}\right]$$
$$= 1 - .988 = .012$$

There is only a 1.2 percent chance that the flow will exceed 30 m^3/s in a particular year.

Values for the frequency factor, k_T are a function of the distribution being utilized in the analysis. Equation [5.17] is of a general form. For the case of log-transformed data

$$Y_T = \bar{y} + k_T'\,S \qquad [5.18]$$

where:

$\bar{y}$ = the mean of the log-transformed data
k_T' = the frequency factor for the normal distribution
S = the standard deviation of the log-transformed data.

If the magnitude of the frequency factor is selected in accordance with the log Pearson type III distribution, then Eq. [5.18] becomes the equation for estimating the return frequency distribution assuming the log Pearson type III distribution. Thus, the log Pearson distribution represents one particular form of Eq. [5.17] and others exist.

An important merit of the log Pearson distribution is that k_T' is a function of the frequency factor and the skewness. In other words, the log Pearson III distribution is a three-parameter distribution that provides the flexibility to fit to the tail of the data distribution. The flexibility to provide this extra fitting arises because of the existence of the third parameter.

Values of k_T for the Pearson Type III distribution are included as Table 5.1 and Table A.14.

If the skew coefficient falls between –1.0 and 1.0, approximate values of the frequency factor for the Pearson type III can be obtained from

$$k_T = \frac{2}{C_s}\left[\left(z - \frac{C_s}{6}\right)\frac{C_s}{6} + 1\right]^3 - 1 \qquad [5.19]$$

where z is the standard normal deviate for the selected recurrence interval T, and C_s is

$$C_s = \frac{a}{S^3} = \frac{n}{(n-1)(n-2)}\sum_{i=1}^{n}(x_i - \bar{x})^3 \qquad [5.20]$$

Further discussion on the log Pearson III distribution is presented in Viessman, Lewis, and Knapp (1989).

EXAMPLE 5.4

As a demonstration of the log Pearson distribution (after McBean and Rovers, 1992), consider the monitoring data of .049, .07, .07, .07, .07, .14, .14, .158, .29, .32, 12 μg/L. These concentrations represent monthly groundwater sample concentration data from a monitoring well in close proximity to a stream. The data have been plotted on lognormal probability paper in Fig. 5.2 on p. 94, using the Weibull plotting position. The failure of the data points to plot as a reasonably straight line indicates that the lognormal distribution is not a good characterization of the data (after assurance is obtained that the high value is not an error in sampling).

Figure 5.3 on p. 95 illustrates a plot of the data on log Pearson type III paper. Note how the lower ordinate of the log Pearson paper differs from the lognormal probability paper.

TABLE 5.1 *k* values for Pearson type III distribution

Skew Coefficient C_s	Recurrence Interval In Years						
	1.0101	1.1111	1.2500	5	25	50	100
	Percent Chance						
	99	90	80	20	4	2	1
		Positive Skew					
3.0	−0.667	−0.660	−0.636	0.420	2.278	3.152	4.051
2.8	−0.714	−0.702	−0.666	0.460	2.275	3.114	3.973
2.6	−0.769	−0.747	−0.696	0.499	2.267	3.071	3.889
2.4	−0.832	−0.795	−0.725	0.537	2.256	3.023	3.800
2.2	−0.905	−0.844	−0.752	0.574	2.240	2.970	3.705
2.0	−0.990	−0.895	−0.777	0.609	2.219	2.912	3.605
1.8	−1.087	−0.945	−0.799	0.643	2.193	2.848	3.499
1.6	−1.197	−0.994	−0.817	0.675	2.163	2.780	3.388
1.4	−1.318	−1.041	−0.832	0.705	2.128	2.706	3.271
1.2	−1.449	−1.086	−0.844	0.732	2.087	2.626	3.149
1.0	−1.588	−1.128	−0.852	0.758	2.043	2.542	3.022
0.8	−1.733	−1.166	−0.856	0.780	1.993	2.453	2.891
0.6	−1.880	−1.200	−0.857	0.800	1.939	2.359	2.755
0.4	−2.029	−1.231	−0.855	0.816	1.880	2.261	2.615
0.2	−2.178	−1.258	−0.850	0.830	1.818	2.159	2.472
0.0	−2.326	−1.282	−0.842	0.842	1.751	2.054	2.326
		Negative Skew					
−0.2	−2.472	−1.301	−0.830	0.850	1.680	1.945	2.178
−0.4	−2.615	−1.317	−0.816	0.855	1.606	1.834	2.029
−0.6	−2.755	−1.328	−0.800	0.857	1.528	1.720	1.880
−0.8	−2.891	−1.336	−0.780	0.856	1.448	1.606	1.733
−1.0	−3.022	−1.340	−0.758	0.852	1.366	1.492	1.588
−1.2	−3.149	−1.340	−0.732	0.844	1.282	1.379	1.449
−1.4	−3.271	−1.337	−0.705	0.832	1.198	1.270	1.318
−1.6	−3.388	−1.329	−0.675	0.817	1.116	1.166	1.197
−1.8	−3.499	−1.318	−0.643	0.799	1.035	1.069	1.087
−2.0	−3.605	−1.302	−0.609	0.777	0.959	0.980	0.990
−2.2	−3.705	−1.284	−0.574	0.752	0.888	0.900	0.905
−2.4	−3.800	−1.262	−0.537	0.725	0.823	0.830	0.832
−2.6	−3.889	−1.238	−0.499	0.696	0.764	0.768	0.769
−2.8	−3.973	−1.210	−0.460	0.666	0.712	0.714	0.714
−3.0	−4.051	−1.180	−0.420	0.636	0.666	0.666	0.667

Source: Water Resources Council, *Bulletin No. 15*, Washington, D.C., December 1967.

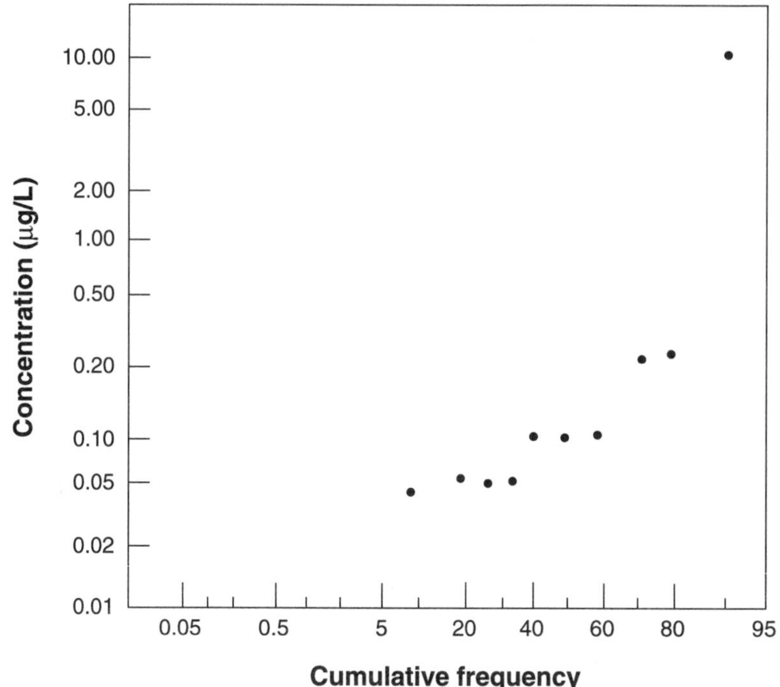

FIGURE 5.2 Lognormal distribution plot of monitoring data

5.5 REFERENCES

BENJAMIN, J.R., and C.A. CORNELL, *Probability, Statistics, and Decisions for Civil Engineers.* New York, NY: McGraw-Hill, 1970.

CHOW, V.T., 1951, "A General Formula for Hydrologic Frequency Analysis," *Transactions American Geophysical Union*, 32, no. 2 (1951), 231–237.

CHOW, V.T., ed., *Handbook of Applied Hydrology.* New York, NY: McGraw-Hill, 1964.

GIBBONS, R.D., "Statistical Models for the Analysis of Volatile Organic Compounds in Waste Disposal Sites," *Ground Water*, 25, no. 5 (1987), 572–580.

GUMBEL, E.J., "Statistical Theory of Droughts," *Proceedings of ASCE*, 80, no. 439 (1954).

GUMBEL, E.J., *Statistics of Extremes.* New York, NY: Columbia University Press, 1958.

McBEAN, E., and F. ROVERS, "Estimation of the Probability of Exceedance of Contaminant Concentration Risk Problems," *Ground Water Monitoring Review* (Winter, 1992) 115–119.

SILVER, C.A. and D. DUNN, *Statistical Analysis of Rare Events in Groundwater*, report prepared for the Environmental Institute for Waste Management Studies. Tuscaloosa, AL: University of Alabama, 1986.

VIESSMAN, W., G.L. LEWIS, and J.W. KNAPP, *Introduction to Hydrology*, 3rd ed., p. 175. New York, NY: Harper and Row, 1989.

WATER RESOURCES COUNCIL, *Bulletin No. 15* (Washington, D.C., December 1967).

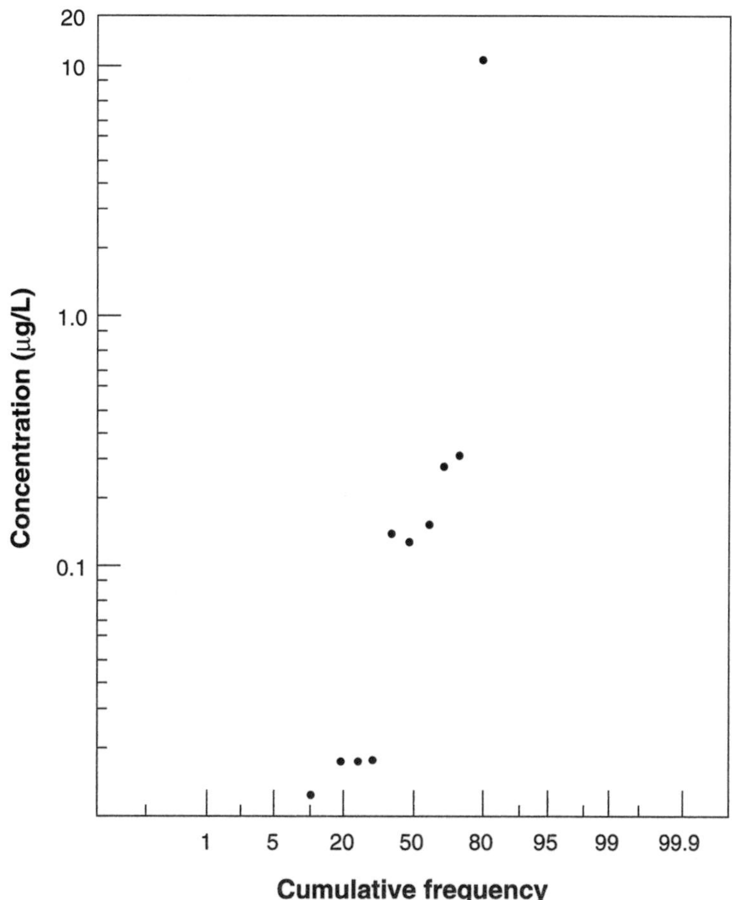

FIGURE 5.3 Log Pearson type III distribution plot of monitoring data

5.6 PROBLEMS

5.1. Given the following concentration data, estimate the once-in-50-samples exceedance value using the normal, lognormal, log Pearson and Gumbel distributions. Comment on why the estimates are different, and indicate which of the estimates you have the most confidence in.

Monitoring data for chloride concentrations are:

Sample number	Chloride concentration (mg/L)	Sample number	Chloride concentration (mg/L)
1	82	11	27
2	64	12	38
3	126	13	91
4	416	14	87
5	125	15	73

Sample number	Chloride concentration (mg/L)	Sample number	Chloride concentration (mg/L)
6	89	16	79
7	97	17	98
8	36	18	47
9	132	19	69
10	146	20	77

5.2. A Poisson distribution is given by

$$P(x) = \frac{(0.72)^x e^{-0.72}}{x!}$$

a. Find $P(0)$
b. Find $P(1)$
c. Find $P(2)$

5.3. If three percent of the concentration measurements are greater than the MCL, find the probability that in a sample of 100 measurements: (a) zero, (b) one, (c) two, (d) three, (e) five, and (f) six measurements will be greater than the MCL.

5.4. In Prob. 5.2, find the probability that (a) more than five (b) between one and three and (c) less than or equal to two samples will be greater than the MCL.

5.5. The following data have been collected: .049, .09, .10, .10, .14, .19, .198, .29, .32, and 12 mg/L. Do not consider the last value as an outlier.

a. Using the lognormal distribution, predict the 95 percent exceedance limit.
b. Estimate the 95 percent exceedance limit using the log Pearson Type III distribution.
c. Comment on how the two values differ and why they differ.

5.6. Use log Pearson probability paper and plot the individual data points to consider the appropriateness of the log Pearson distribution paper in application to the data listed in Prob. 5.

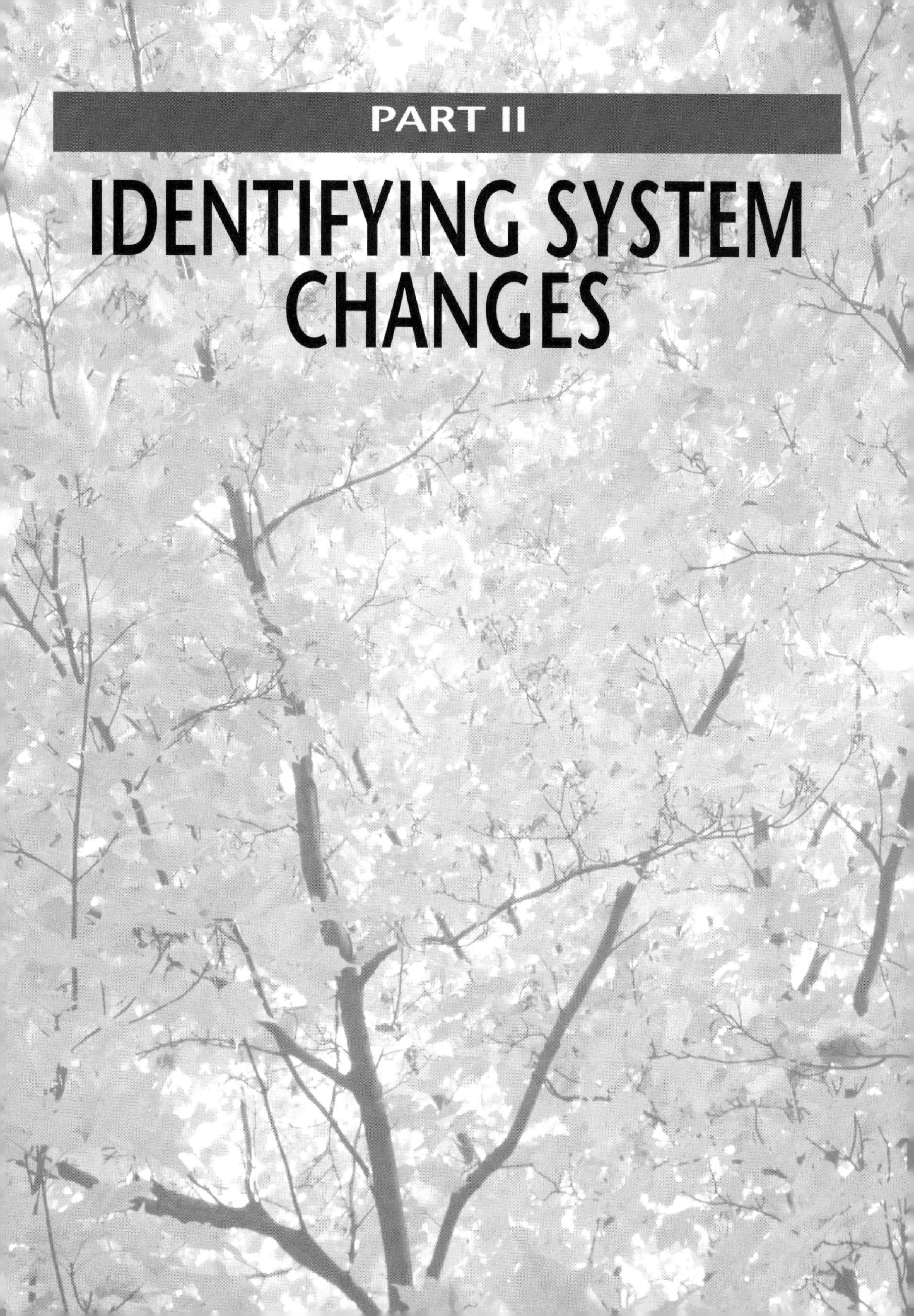

PART II

IDENTIFYING SYSTEM CHANGES

CHAPTER 6

Identification of System Changes and Outliers Using Control Charts and Simple Procedures

6.1 INTRODUCTION

When examining a particular environmental monitoring record, of interest may be whether there has been a change in concentration(s). For example, has the wastewater treatment system encountered upset conditions such that the quality of the treated effluent has deteriorated? Alternatively, there may be one or two sampling results contained within the record that are suspiciously different from the others: Is this due to misreading an instrument; has a transcription error been made in recording the data; or is there a significant reason to believe that changes to the environmental system have occurred? This is when we need statistical procedures to determine whether data outliers are likely present in the data record.

This chapter describes several analytical procedures appropriate to addressing these types of questions. It may well be that more detailed assessments than these are needed; they will be examined in subsequent chapters. At this point we describe only the capabilities of a series of relatively simple procedures.

Of interest initially is the development of three statistical intervals for detecting system changes. The intervals are:

1. Tolerance intervals (see Sec. 6.2): Tolerance intervals are designed to contain a designated proportion of the population (e.g., 95 percent of all possible sample measurements).

2. Confidence intervals (see Sec. 6.3): Confidence intervals are designed to contain a specified population parameter with a designated level of confidence or probability, denoted as $1 - \alpha$. The interval will fail to include the true parameter in approximately α percent of the cases where such intervals are constructed;
3. Prediction intervals (see Sec. 6.4): Prediction intervals are constructed to contain the next sample value(s) from a population or distribution with a specified probability. For example, after sampling a monitoring location for some time, the data may be used to construct an interval that, with a specified probability, will contain the next monitoring result(s), assuming the distribution has not changed. (Prediction intervals can also detect the significance of changes that have already occurred.)

In summary, a tolerance interval usually contains a specified proportion of the population, a confidence interval is created to contain a statistical parameter, and a prediction interval contains one or more future observations. These intervals are all based on sample data (as opposed to populations) and thus have uncertainty associated with them.

The final portion of this chapter (Sec. 6.5) describes a series of different procedures to identify statistical outliers. Data outliers are extreme high (or low) values that diverge widely from the majority of the data. Are these values legitimate members of the data set that therefore contain important information, or do they represent, for example, errors in measurement?

6.2 TOLERANCE INTERVALS

A tolerance interval establishes a concentration range to contain a specified proportion (P percent) of the population with a specified confidence, so that a specified large proportion of monitoring observations should fall within this interval. Tolerance intervals are usually constructed with the assumption that the data (or the transformed data) are symmetrically distributed (specifically, the t-distribution and, as the number of samples becomes large, the normal distribution), but they can also be constructed assuming other distributions.

The steps in developing a tolerance interval are as follows:

1. compute the mean, $\bar{x}$, and the standard deviation, S_x, of the data
2. construct the one-sided upper and lower tolerance limits TL as

$$TL = \bar{x} \pm k\, S_x \qquad [6.1]$$

where k is the one-sided normal tolerance factor assuming, the data are adequately described by the normal distribution. The magnitude of k is selected from tables in accordance with the desired tolerance limit. For example, a tolerance limit with 95 percent coverage gives an upper and lower bound between which 95 percent of the individual observations of the distributions should fall. The magnitude of k is determined from Table A.3 for different tolerance limit probabilities and degrees of freedom. It is useful here to emphasize that the t-distribution approaches the normal distribution as the number of sample points becomes large and becomes identical to it as the number of samples become infinite. For further discussion of this feature, see Sec. 3.4.

EXAMPLE 6.1—DEVELOPMENT OF TOLERANCE LIMITS

Determine the 95 percent tolerance limits for the following data: 13.55, 6.39, 13.81, 11.20, 13.88. Thus, $n = 5$, $\overline{x} = 11.77$ and $S_x = 3.20$. The value of k is determined from Table A.4 for (5–1) = 4 degrees of freedom and for a two-sided tolerance limit (i.e., $k = 2.776$).

Then the 95 percent tolerance limit on the individual values is

$$TL = 11.77 \pm 2.776\ (3.20) \text{ or a tolerance interval from 2.89 to 20.65}$$

In words, 95 percent of measurements should be within the range of 2.89 to 20.65.

Figure 6.1(a) indicates the tolerance limits and plotted monitoring values.

Figure 6.1(a) 95 percent tolerance limits derived from five sampling results

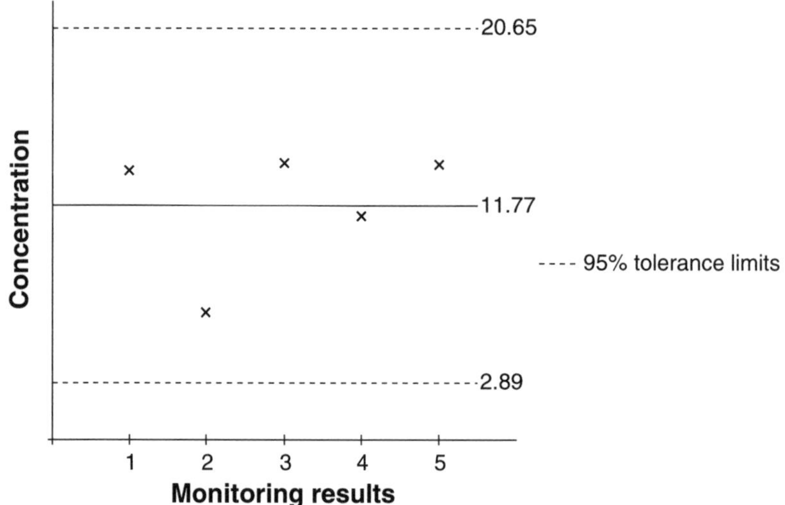

Assume that an additional five samples are now collected. Construct the resulting 95 percent tolerance limits, where the additional data are 5.88, 10.22, 6.86, 12.43, 7.94.

Now $n = 10$, $\overline{x} = 10.22$, and $S_x = 3.22$.

Incorporating this additional information, the 95 percent tolerance limit on the individual values becomes

$$TL = 10.22 \pm 2.262\ (3.22) \text{ or a tolerance interval from 2.94 to 17.50}$$

The width of the 95 percent tolerance limits has decreased as a result of the additional sampling results. Figure 6.1(b) indicates the new tolerance limits and plotted monitoring values.

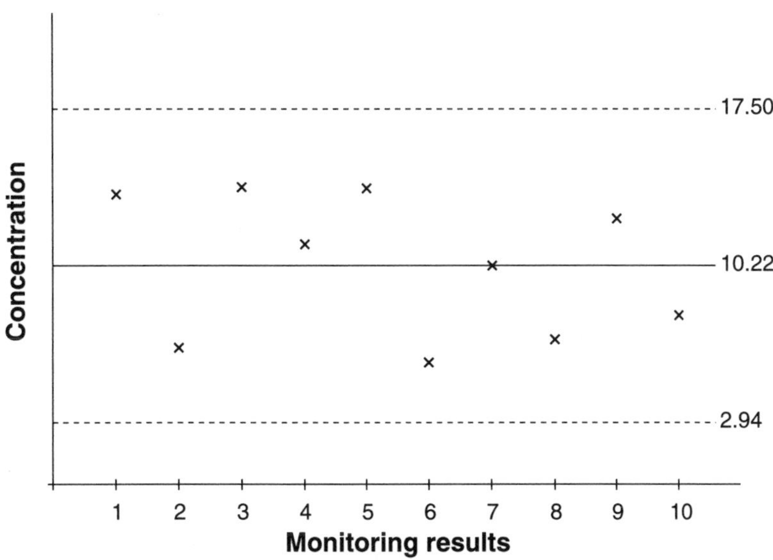

Figure 6.1(b) 95 percent tolerance limits derived from 10 sampling results

If the data are not normally distributed, a possible approach involves transforming the data using, for example, a logarithmic transformation prior to the determination of the tolerance interval. In this situation, the resulting tolerance intervals will not be equidistant from the mean when transformed back to the original domain as demonstrated in Ex. 6.2.

EXAMPLE 6.2—DEVELOPMENT OF TOLERANCE LIMIT FOR NON-GAUSSIAN DATA

The data as follows were obtained for air quality for sulfur oxides, namely 10.2, 147.6, 8.1, 12.2, 8.6, 54.3, 12.4, 15.6, 34.6.

a. Determine the 95 percent tolerance interval without using any data transformation. The mean of the data is 33.73, the standard deviation is 45, $n = 9$, $k = 2.31$ from Table A.4, and the 95 percent tolerance interval is

$$TL = 33.73 \pm 2.31\ (45) = -70.2 \text{ to } 137.6$$

The results are plotted in Fig. 6.2(a). The inappropriateness of the lower tolerance limit is apparent since air quality concentrations cannot assume negative values.

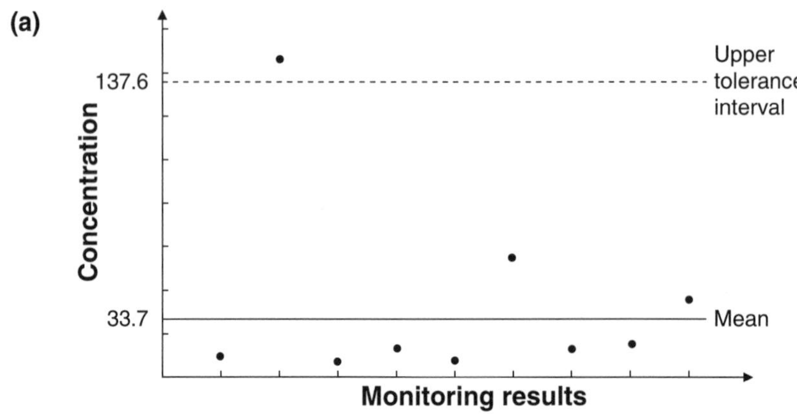

Figure 6.2(a) Data and tolerance interval with no data transformation

b. Determine the 95 percent tolerance interval following a logarithmic transformation (i.e., using the logarithms of the data, $\overline{y} = 2.98$ and $S_y = .988$).

$$TL = 2.98 \pm 2.31\ (.988) = .70 \text{ to } 5.26$$

Figure 6.2(b) Data and tolerance interval with logarithmic transformation

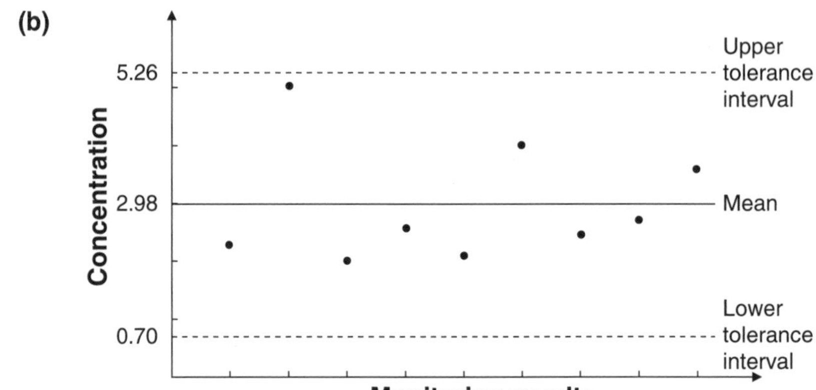

The results are shown in Fig. 6.2(b) where the ordinate is plotted as a logarithmic scale.

$$\text{Mean} = e^{2.98} = 19.7$$

The tolerance limit extends from $e^{.70}$ to $e^{5.26}$ or 2.0 to 192.

The results are shown in Fig. 6.2(c) where the ordinate is plotted on an arithmetic scale. This figure demonstrates how the tolerance limits are asymmetrical when transformed back to the arithmetic scale.

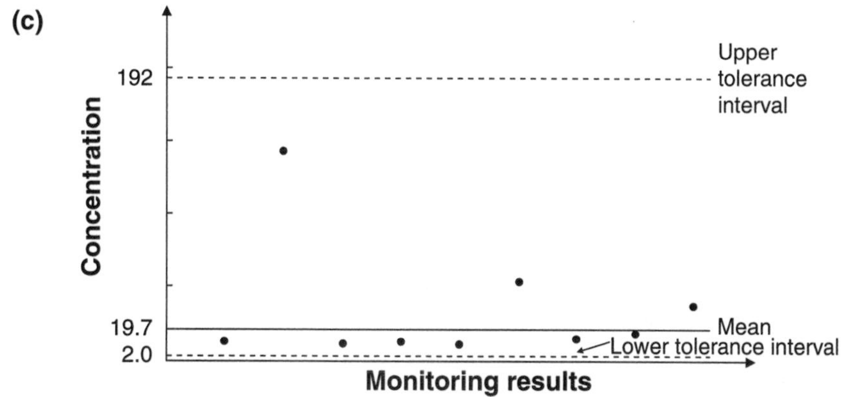

Figure 6.2(c) Data and tolerance interval calculated after logarithmic transformation and back-transformed onto arithmetic scale

EXAMPLE 6.3—DEVIATION OF AN UPPER TOLERANCE LIMIT

The concentrations of trichloroethylene in soil at a contaminated site are as listed below. Define the 95 percent upper tolerance limit on the data.

Number of sample	Concentration (µg/kg)	Weibull plotting position	ln of concentration
1	730	.031	5.677
2	800	.063	5.769
3	810	.094	5.781
4	1,000	.125	5.992
5	1,700	.156	6.523
6	1,900	.188	6.634
7	2,000	.219	6.685
8	2,400	.250	6.868
9	2,800	.281	7.022
10	5,600	.313	7.715
11	6,100	.344	7.801
12	6,700	.375	7.894
13	8,700	.406	8.156
14	10,000	.438	8.294
15	12,000	.469	8.477
16	12,000	.500	8.477
17	16,000	.531	8.765
18	17,000	.563	8.826

Number of sample	Concentration (μg/kg)	Weibull plotting position	ln of concentration
19	25,000	.594	9.211
20	31,000	.625	9.426
21	41,000	.656	9.706
22	46,000	.688	9.821
23	63,000	.719	10.316
24	100,000	.750	10.598
25	240,000	.781	10.780
26	290,000	.813	10.969
27	290,000	.844	10.969
28	360,000	.875	11.186
29	960,000	.906	12.167
30	1,000,000	.938	12.207
31	6,100,000	.969	14.016
Mean	309,920		9.804
Standard deviation	1,103,425		2.37

a. Is the normal distribution a reasonable means of characterizing the data? A check of the coefficient of variation, namely,

$$\frac{1{,}103{,}425}{309{,}920} = 3.6$$

indicates that the value is much greater than unity and that the normal distribution is not a good description of the data. Similarly, plotting the data on normal probability paper, as depicted in Fig. 6.3(a), clearly shows that the data do not plot as a straight line. Conclusion: The normal distribution is not a good descriptor of the data.

b. Is the lognormal distribution a reasonable means of characterizing the data? Plotting the data as indicated in Fig. 6.3(b) as a lognormal distribution, the data plot as a straight line and consequently the lognormal distribution is an appropriate description of the data. As a result, the 95 percent UTL can determined as,

$$UTL_{95} = \bar{x} + kS$$

where $\bar{x}$ = the mean of the log-transformed data = 9.804
= from Table A.4 $k = 1.697$ corresponding to the t-value for 5 percent (one-sided) and the number of samples $n = 31$ and thus, degrees of freedom = 30
S = standard deviation of the log-transformed data
= 2.37

or

$$UTL_{95} = 9.804 + 1.697(2.37) = 13.826$$

or

$$UTL_{95} = e^{13.826} = 1{,}010{,}433\ \mu g/kg$$

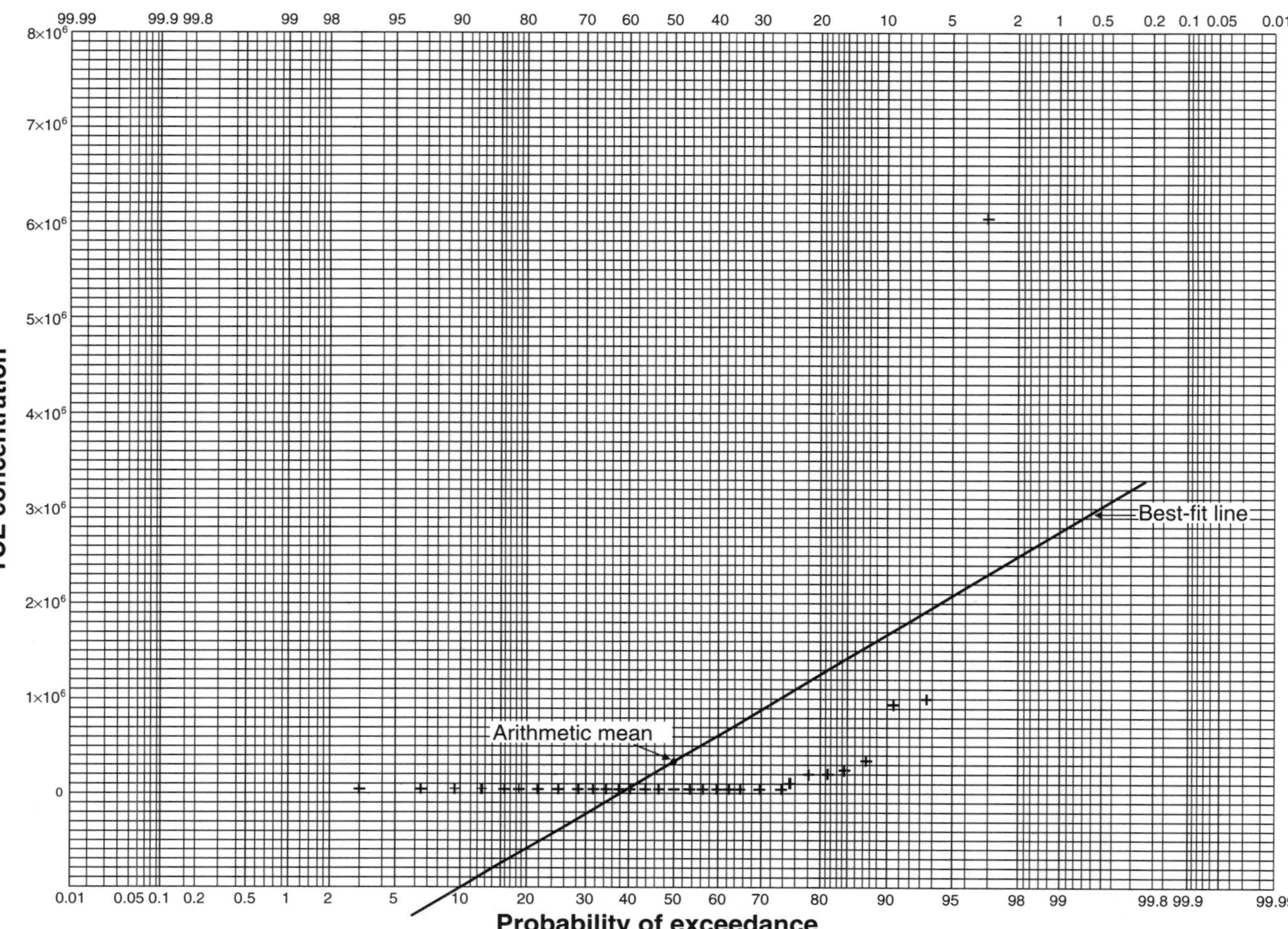

Figure 6.3(a) Normal probability plot of TCE data from Ex. 6.3

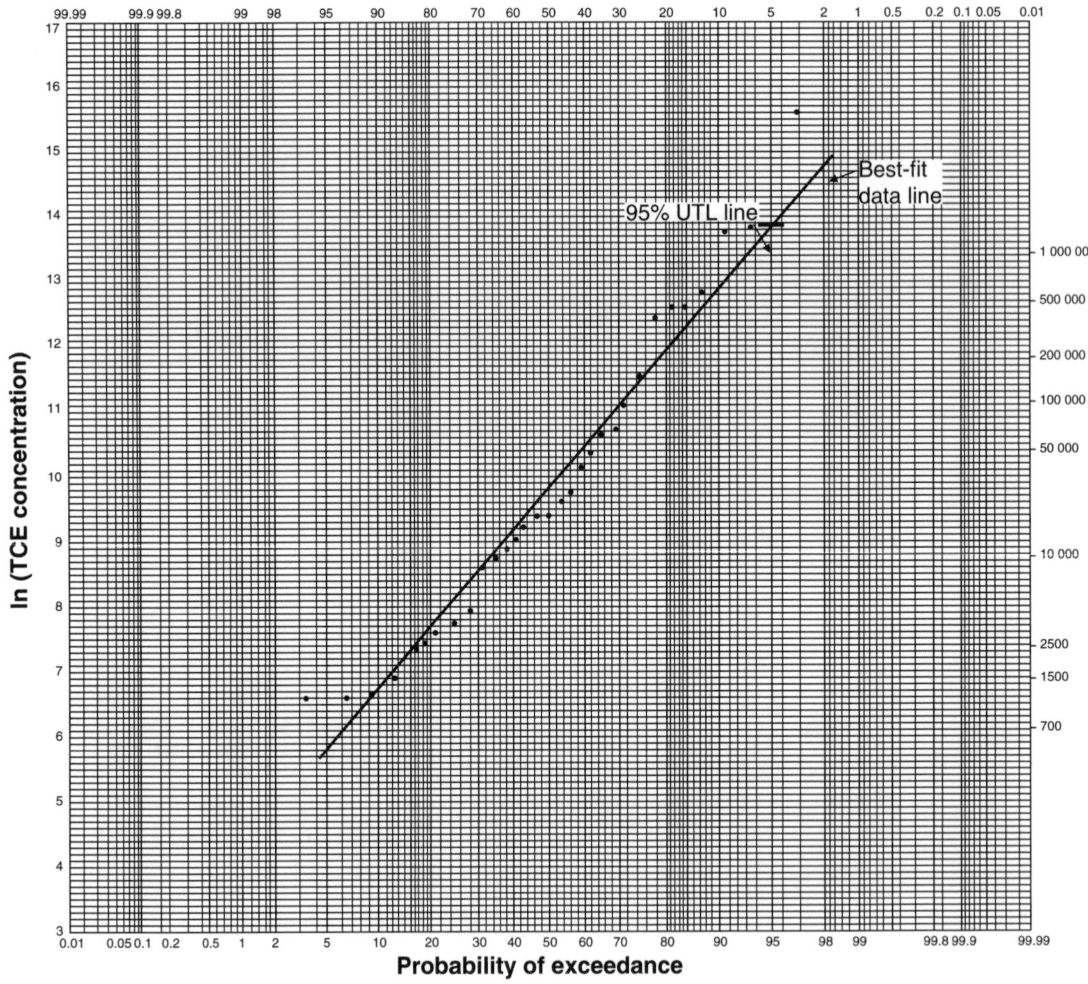

Figure 6.3(b) Lognormal probability plot of TCE data from Ex. 6.3

In words, we would expect to see only approximately five percent of the individual values exceeding 1,010,433 µg/kg. These results appear reasonable, since only one of the values out of 31 total are in excess of the 95 percent UTL.

An extension of the tolerance interval procedure involves construction of an interval that contains a specified fraction of the concentration measurement with a specified degree of confidence. For example, we may want to have the tolerance interval contain 95 percent of the measurements of concentration with a confidence of at least 95 percent.

EXAMPLE 6.4

Calculate the upper limit of the tolerance interval that contains 95 percent of the measurements of concentration with confidence at least 95 percent for the data from Ex. 6.3.

SOLUTION

From Ex. 6.3, $\bar{x} = 9.804$, $S = 2.37$ and from Table A.3,

$$k = 2.19$$

The upper tolerance limit with 95 percent confidence is calculated as

$$= 9.804 + 2.19(2.37)$$

$$= 15.0$$

or, in arithmetic terms, this value is $e^{15.0} = 3{,}269{,}000$ µg/kg. This value is larger than the 95 percent UTL indicated in Ex. 6.3. This occurs because in this case, we now have the tolerance interval that contains 95 percent of the measurements of concentration with confidence of at least 95 percent.

The procedure to quantify a tolerance interval (or tolerance limit) is similar to that indicated in Eq. [6.1] except that the value of k is selected to reflect the degree of confidence. To construct a one-sided upper tolerance interval with an average coverage of (1–β) percent, the k multiplier can be computed directly from the t-distribution table, as

$$k = t_{n-1,1-\beta}\sqrt{1 + \frac{1}{n}}$$

where the t-value represents the (1–β)th upper percentile of the t-distribution with $(n - 1)$ degrees of freedom.

Tabulated values of k for different lengths of record of data are listed in Table A.3.

6.3 CONFIDENCE INTERVALS

Confidence intervals use historical information for the construction of one or more lines or limits to characterize the uncertainty associated with a particular parameter, such as the mean. Confidence intervals are constructed from sample data and thus are random quantities. In other words, each set of sample data will generate a different confidence interval. The limits on the confidence intervals may be updated periodically as additional information becomes available.

Construction of confidence intervals for the population mean in general requires the assumption that the observations are drawn from a statistically convenient distribution, usually the normal distribution.

6.3.1 Confidence Limits Using the Normal Distribution (and the *t*-Distribution)

Statistical confidence limits define an interval such that there is a specified probability $(1 - \alpha)$, that a parameter will be contained in the interval (e.g., 95 percent). The probability that the confidence interval contains a specified population parameter is written as

$$Pr\{x_{(l)} \le x \le x_{(u)}\} = 1 - \alpha \quad [6.2]$$

where $100\,(1 - \alpha)$ is the level of confidence, and for integers l and u, a confidence interval $\{x_{(l)}, x_{(u)}\}$ specifies the confidence interval.

If x is normally distributed, then the precision with which the sample mean $\overline{x}$ approximates μ_x can be assessed with the t-statistic as

$$t = \frac{\overline{x} - \mu_x}{\sqrt{S_x^2/n}} \quad [6.3]$$

If the individual members of the set of x are independent and normally distributed with the same mean and variance, then t from Eq. [6.3] follows the t-distribution with $n - 1$ degrees of freedom. Knowing the distribution of t, one can construct a confidence interval for μ_x. Rearranging Eq. [6.3] yields

$$\overline{x} - \frac{tS_x}{\sqrt{n}} \le \mu_x \le \overline{x} + \frac{t\,S_x}{\sqrt{n}} \quad [6.4]$$

If the set of x is not normally distributed, the exact distribution of t in Eq. [6.4] is generally not known. However, the central limit theorem can be used to prove that if x has a distribution with finite mean and variance, then for large n, the distribution of t approaches that of the normal distribution with zero mean and a variance of unity.

As a result, in mathematical terms when the degrees of freedom are infinite (the t-distribution becomes the normal distribution),

$$\text{Probability}\left(-1.96 < \frac{\overline{x} - \mu}{\sigma/\sqrt{n}} < 1.96\right) = 0.95 \quad [6.5]$$

or, rewriting

$$\text{Probability}\left(\overline{x} - \frac{1.96\,\sigma}{\sqrt{n}} < \mu \le \overline{x} + \frac{1.96\,\sigma}{\sqrt{n}}\right) = 0.95 \quad [6.6]$$

TABLE 6.1 Indication of changes in the *t*-statistic with the number of samples and the confidence limits

No. of samples[a] n	Degrees of freedom[a] $df = (n-1)$	95% confidence limit t_{95}	99% confidence limit t_{99}
4	3	3.18	5.84
5	4	2.78	4.60
6	5	2.57	4.03
7	6	2.45	3.71
8	7	2.36	3.50
9	8	2.31	3.36
10	9	2.26	3.25
12	11	2.20	3.11
20	19	2.09	2.86
		1.96	2.58

[a]Values assume a two-sided characterization.

In practice, the true standard deviation σ is unknown. Then, if a random sample of size n is drawn, an estimate of the standard error of the mean $\bar{x}$ is given by $S_x / \sqrt{n}$. For small samples (e.g., $n<30$), the confidence limits are increased (since S_x is only an estimate of σ), by use of the t-distribution.

For environmental quality phenomena, the population parameters are rarely known; thus Eq. [6.4] is utilized to specify the confidence interval.

The values of t thus reflect both the level of significance α and the number of samples n. Values for 95 percent confidence interval for the value of t are as listed in Table 6.1 for a two-sided characterization (e.g., for a 95 percent confidence limit, 2 1/2 percent in the one tail of the distribution and 2 1/2 percent in the other tail of the distribution). As is apparent from the values listed in Table 6.1, as n increases, the t-value decreases. A comprehensive tabulation for a one-sided characterization is contained in Table A.2.

EXAMPLE 6.5—Confidence Limits on the Mean

Using the data from Ex. 6.1 calculate the 95 percent confidence limits on the mean.

Solution

For the first five samples,

$$\text{Confidence limit} = 11.77 \pm \frac{2.76\,(3.20)}{\sqrt{5}} = 11.77 \pm 3.97$$

or 7.80 to 15.74.

For the ten samples,

$$\text{Confidence limit} = 10.22 \pm \frac{2.26\ (3.22)}{\sqrt{10}} = 10.22 \pm 2.31$$

or 7.91 to 12.53.

The width of the 95 percent confidence limits on the mean have decreased as a result of the additional sampling. As would be expected, the confidence limits on the mean are much narrower than are the 95 percent tolerance intervals on the individual values, as summarized below.

Sample size	Tolerance interval	Confidence limit
5	2.89 to 20.65	7.80 to 15.74
10	2.94 to 17.50	7.91 to 12.53

For small sample sizes and/or for data with large spread (large standard deviations), the confidence intervals will be wide. As a result, when the data have a large spread, a substantial shift in, for example, water quality levels will be required for the subsequently calculated mean to fall outside the confidence limits.

As more data become available, the confidence interval will likely become narrower, as noted in Ex. 6.5, and the test will become more sensitive to smaller shifts in water quality.

It is noteworthy that there is no particular reason that a 95 percent confidence interval is used. The level of confidence required is a function of the degree of certainty required to make a detection on the result of the statistical test. For example, a 99 percent confidence limit is considerably larger than the 95 percent confidence limit (there is only one percent likelihood that the mean would be expected to lie outside the 99 percent confidence limits). To be more "certain" of an environmental change, reduce the chance of a calculated mean falling outside the confidence interval when no change had actually occurred. In other words, reduce the false positive or Type I error to $1-\alpha = 0.99$.

EXAMPLE 6.6

Using the data in Example 6.1, develop the 95 percent and 99 percent confidence limits for the mean.

For n = five samples from Ex. 6.1,

- the 95 percent confidence limit is $11.77 \pm \frac{2.776\ (3.20)}{\sqrt{5}}$ or 7.80 to 15.74
- the 99 percent confidence limit is $11.77 \pm \frac{4.60\ (3.20)}{\sqrt{5}}$ or 5.15 to 18.35

For n = ten samples from Ex. 6.1,

- the 95 percent confidence limit is $10.22 \pm \frac{2.262\,(3.20)}{\sqrt{10}}$ or 7.91 to 12.53
- the 99 percent confidence limit is $10.22 \pm \frac{3.25\,(3.22)}{\sqrt{10}}$ or 6.91 to 13.53

The false positive error, the risk of the mean being outside the confidence interval, is less for the 99 percent lines, but obtained at the expense of a larger range.

The 99 percent confidence limit will be less sensitive in detecting changes than the 95 percent confidence limit. This loss of sensitivity and power is the tradeoff associated with increased confidence in conclusions that changes have occurred.

6.3.2 Confidence Limits for Lognormally Distributed Data

If data are lognormally distributed as determined by the procedures described in Chap. 4, confidence limits for the mean can be completed using the log-transformed data. In other words, following a logarithmic transform, the same calculation procedure described in Sec. 6.3.1 is followed.

It is noteworthy that the distribution of the estimate of the mean can usually be described by the normal distribution, almost irrespective of the distribution of the data.

6.3.3 Distribution-Free or Nonparametric Confidence Limits

Distribution-free confidence limits may be developed that do not depend on the assumption of normality or lognormality. Distribution-free tolerance intervals are, however, wider and therefore less sensitive to changes in environmental quality than their normal counterparts. The interested reader is referred to Conover (1980) for further consideration of distribution-free confidence limits.

6.4 PREDICTION INTERVAL CHARACTERIZATIONS

Prediction intervals are intended to describe an interval within which future samples should lie, assuming the distribution or system has not changed.

6.4.1 The *t*-Distribution Prediction Intervals

The *t*-distribution may be utilized as the basis for developing prediction intervals. The prediction interval can be across time or space. The spatial prediction interval might involve checking whether results at one monitoring location are representative of those at another monitoring location.

EXAMPLE 6.7

Evidence of contamination is indicated if subsequent data indicate deterioration in quality by being above the upper prediction interval.

The data listed below are nickel concentrations measured at a groundwater monitoring well. Calculate the prediction interval and determine whether there is evidence of contamination as demonstrated by the monitoring results determined in months 4 and 5.

SOLUTION

The initial data (for the first three months) from which the upper prediction interval is determined:

Sample date	Nickel concentration (μg/L)	
July	10.7	mean = 15.3
	11.4	
	20.6	
	18.4	
August	13.1	mean = 16.3
	22.6	
	11.4	
	18.1	
September	17.1	mean = 18.6
	26.4	
	12.3	

$n = 11$

$\bar{x} = 16.55$

$S = 5.23$

Subsequent monitoring results for the next two months are

October	22.1
	18.6
	27.1

$n = 3$

$\bar{x} = 22.6$

$S = 4.27$

November	21.2
	18.6
	29.2

$n = 3$

$\bar{x} = 23.0$

$S = 29.2$

The individual data points are plotted in Fig. 6.4.

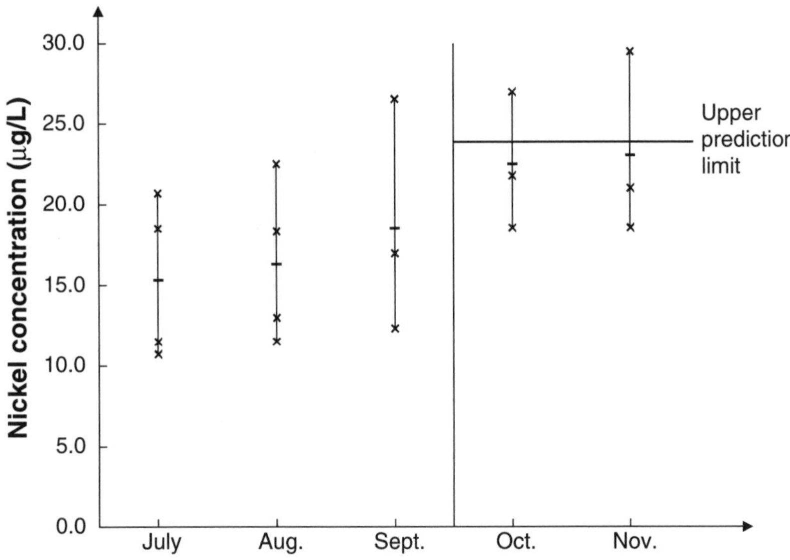

Figure 6.4 Plot of nickel concentrations over time, for data from Ex. 6.6

Step 1. Test the first three months of "background" data for approximate normality. Only the data up to the end of the first three months are included, since these are used to construct the upper prediction interval.

The Shapiro-Wilk test is utilized to test for normality of the data as described below.

Shapiro-Wilk test

i	$x_{(i)}$	$X_{(n-i+1)}$	a_{n-i+1}	b_i
1	10.7	26.4	.5601	14.79
2	11.4	22.6	.3315	7.49
3	11.4	20.6	.2260	4.66
4	12.3	18.4	.1429	2.63
5	13.1	18.1	.0695	1.26
6	17.1	17.1		$b = 30.83$
7	18.1	13.1		
8	18.4	12.3		
9	20.6	11.4		
10	22.6	11.4		
11	26.4	10.7		

$$W = \left[\frac{b}{S\sqrt{n-1}}\right]^2 = \left[\frac{30.83}{5.23\sqrt{10}}\right]^2 = 1.86$$

The critical value at the five percent level for the Shapiro-Wilk test on 11 observations is .850. Since the calculated $W = 1.86$ is well above the critical value, there is no evidence to reject the assumption of normality.

Step 2. Compute the prediction interval using the background data. Since there are two future months of data to be compared to the upper prediction limit, the number of future sampling periods is $k = 2$. At each sampling period, a mean of three independent samples will be computed so $m = 3$ in the upper prediction interval formula. The Bonferroni t-statistic is $t_{10,2,.95}$ with $k = 2$ and 11 degrees of freedom (*df*) and is equivalent to the t-statistic at .975 level with 10 *df*. In other words, $t_{10,.975} = 2.23$.

Compute the upper one-sided upper prediction limit using the formula

$$UL = \bar{x} + t_{n-1,k,.95}\, S\sqrt{\frac{1}{m} + \frac{1}{n}}$$

Then the upper prediction limit is given by

$$UL = 16.55 + (2.23)\,5.23\sqrt{\frac{1}{3} + \frac{1}{11}} = 24.1$$

Now compare the upper prediction limit to the compliance data; the means of the observations for months 4 and 5 are 22.60 and 23.0 µg/L, respectively.

Since the mean concentrations for months 4 and 5 are below the upper prediction limit, there is no evidence of recent contamination at the monitoring facility to indicate that a significant change has occurred.

6.4.2 Poisson Prediction Intervals

When a sizable percentage of the data are reported as "less than" (censored), using the t-distribution to develop prediction intervals is inappropriate—there is considerable difficulty in estimating the mean and standard deviation required in Eq. [6.1]. An alternative approach, appropriate for the situation where many of the monitoring results are censored, involves the Poisson distribution. To estimate a prediction limit using the Poisson model, the approach described by Gibbons (1987) and based on the work of Cox and Hinckley (1974) will be described.

Specifically, an upper limit is estimated for an interval that will contain all k future measurements for a particular constituent with confidence level $(1 - \alpha)$, given n previous background measurements. To do this, let T_n represent the sum of the Poisson counts of n historically obtained samples. The goal is to predict T_k^* representing the Poisson count of the next k sample measurements. As Cox and Hinckley show, if T_n has a Poisson distribution with mean μ and if no contamination has occurred, it is reasonable to assume that T_k^* will have a Poisson distribution but with mean $c\mu$ where c depends on the number of future measurements being predicted.

In particular, Cox and Hinckley demonstrate that the quantity,

$$\frac{\left[T_k^* - \dfrac{c(T_n + T_k^*)}{1 + c}\right]^2}{\dfrac{c\,(T_n + T_k^*)}{(1 + c)^2}} \qquad [6.7]$$

has an approximate standard normal distribution. For this reason, an upper prediction limit for T_k^* is calculated by Gibbons (1987) to be approximately

$$T_k^* = c\,T_n + c\,t^2 + t\,c\sqrt{T\left(1 + \frac{1}{c}\right) + \frac{t^2}{4}} \qquad [6.8]$$

where $t = t_{n-1,\alpha}$ is the upper $(1 - \alpha)$ percentile of the t-distribution with $(n - 1)$ degrees of freedom, and $c = k/n$ where k = number of future samples. The upper limit corresponds to the sum of the next k samples; if the sum exceeds the upper limit the test is triggered.

EXAMPLE 6.8

Use the following monitoring data for cadmium from four background wells to estimate an upper 99 percent Poisson prediction limit for the next three measurements from a single downgradient well as configured on Fig. 6.5.

Cadmium concentration (μg/L)

Sample number	Well 1	Well 2	Well 3	Well 4
1	<5	8	<5	<5
2	<5	<5	<5	<5
3	<5	<5	12	<5
4	<5	<5	<5	10
5	<5	<5	<5	<5
6	<5	<5	<5	<5
7	<5	<5	<5	<5

SOLUTION

Step 1. Pooling the background data yields $n = 4 \times 7 = 28$ samples of which 25 (89 percent) are less than the detection level. Because the rate of detection is so infrequent (i.e., $< 10 - 15$

Figure 6.5 Configuration of monitoring wells for Ex. 6.8

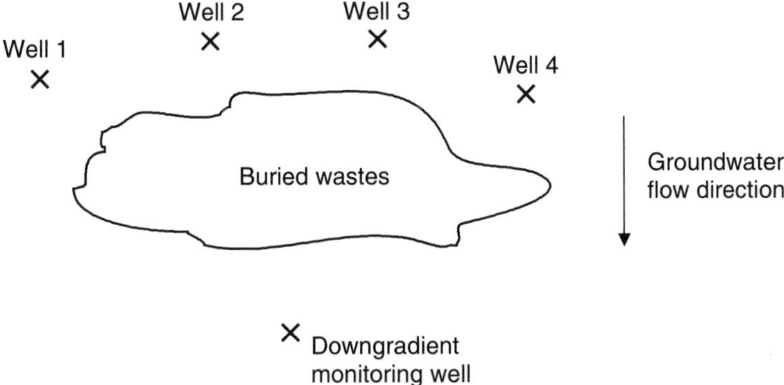

percent), a Poisson-based prediction limit may be appropriate. Since three future measurements are to be predicted at the downgradient well, $k = 3$ and hence $c = k/n = 3/28$.

Step 2. Set each nondetect to one-half the detection limit or 2.5 μg/L. (Alternative procedures for treating nondetect data are described in Chap. 10.) Then compute the Poisson count of the sum of all the upgradient samples. In this case, it corresponds to 25 samples at 2.5 and the three samples reported as greater than 5, or,

$$T = 25\,(2.5) + (8 + 12 + 10) = 92.5$$

Step 3. To calculate an upper 99 percent prediction limit, the upper 99th percentile of the t-distribution with $(n–1) = 27$ df must be taken from a reference table. From Table A.2 $t_{27,.01} =$ 2.473.

$$T_k^* = \frac{3}{28}(92.5) + \frac{(2.473)^2}{2}\frac{3}{28} + 2.473\frac{(3)}{28}\sqrt{92.5\left(1 + \frac{28}{3}\right) + \frac{(2.473)^2}{4}}$$

$$= 9.91 + .33 + .265\,(\,955.8 + 1.53)^{0.5}$$

$$= 18.44$$

Step 4. To test the upper prediction limit, the Poisson count of the sum of the monitoring results for the next three samples is calculated. If this sum is greater than 18.44 μg/L there is significant evidence of contamination at the downgradient well. If not, the well may be regarded as not yet impacted.

It is potentially of interest to construct the Poisson-based intervals both for single pollutants (as above) and the sum of several similar pollutants (e.g., volatile organic carbons). As a general guideline, however, the detection rate for each compound should be less than (approximately) 20 percent to invoke the Poisson modeling assumption.

6.5 DETECTION OF DATA OUTLIERS

Data outliers are extreme (high or low) values that diverge widely from the main body of a data set. The presence of one or more outliers within a data set may greatly influence any calculated statistics and yield biased results. By increasing the variability within a sample, outliers will decrease the sensitivity of subsequent statistical tests. However, there is also the possibility that the outlier is a legitimate member of the data set; that is, it could actually represent the environmental impact that the data collection effort of an environmental monitoring program is supposed to identify. Outlier detection tests are to determine whether there is sufficient statistical evidence to conclude that an observation appears extreme and does not fit the distribution of the rest of the data or that the observation should remain in the data set.

Before concluding that data outliers kept in the data set necessarily indicate environmental change, bear in mind that they may arise from a faulty instrument—poor calibration of a water quality probe that measures dissolved oxygen, for example. Sometimes, too, the values in a data set include errors in the transcription of data. Other sources of error include unobserved technician or laboratory error, contaminated sampling equipment, misreading of instruments, blunders in experimental procedures, intentional cover-up because no value was taken, inconsistent sampling of analytical chemistry methodology, and so on. Data outliers may also be due to the vagaries of nature (the tendency for some features to frustrate the analyst). How can we decide that the monitoring result could be a consequence of a catastrophic unnatural occurrence such as a spill? The answer hinges on an estimate, to the extent the data set will allow, whether the observation should be included or excluded in a particular circumstance.

If one of the error-producing processes indicated above can be pegged to an outlier, the value may be safely deleted from consideration. Therefore, the first task is to trace the conditions and experimental procedure of the aberrant observation through lab notebooks. Obviously the suspect value will be excluded if a rational basis for an exclusion can be identified. The subsequent data analysis can then be performed without the value.

A number of procedures have been developed as alternative methods for detecting outliers. As will be seen, many of the procedures premise the exclusion of a suspect data point on the basis of an assumed underlying normal distribution. As well, when the data set is small, the outlier must be substantially different to be identified as an outlier. With these caveats in mind the various procedures for identifying an outlier include:

1. *Probability plotting approach for outlier detection.* This outlier identification procedure involves the preparation of a probability plot of the data followed by visual inspection. However, it is difficult to assign specific rules that ensure replicability of approach. As a result, this procedure is more commonly utilized to identify the need for more detailed analyses by one of the more formal procedures outlined below. Regardless, the data illustrated in Fig. 6.6 certainly imply that the single very high value is a data outlier worthy of closer examination.
2. *Scattergram analysis test for outlier detection.* The use of a scattergram or figure indicating interparameter correlation between two constituents involves development of a diagram such as depicted in Fig. 6.7. For example, if there tends to be correlative behavior between two constituents, a single monitoring result with inconsistent behavior, such as identified by the point lying a significant distance from the general trend of the data, is indicative that the individual value may be an outlier.
3. *Standard deviation test for outlier detection.* Another procedure for detecting outliers involves examining the changes in magnitude of a parameter estimating the dispersion of the data. A typical parameter utilized for this test is the standard deviation, or variance. Inflation or increase in the variance can result from including outliers in the data set. It is noteworthy that this test is

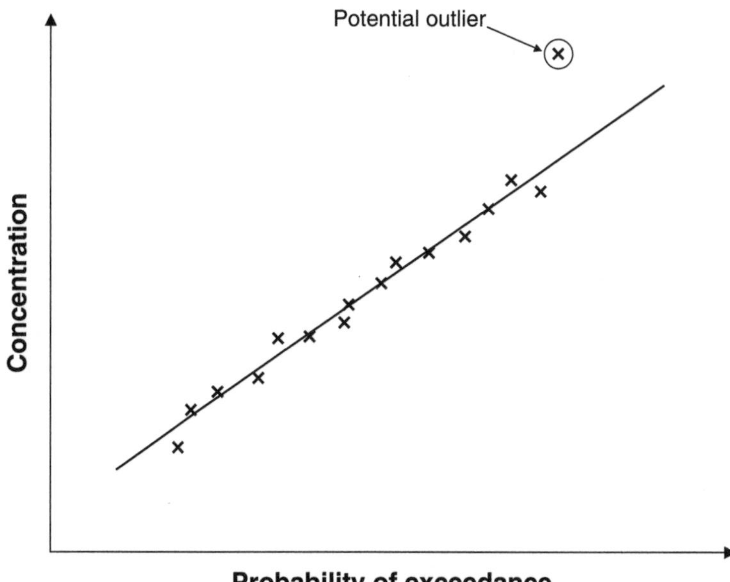

Figure 6.6 Use of probability plot as a means of identifying possible data outliers

Figure 6.7 Scattergram plot for purposes of identifying possible data outliers

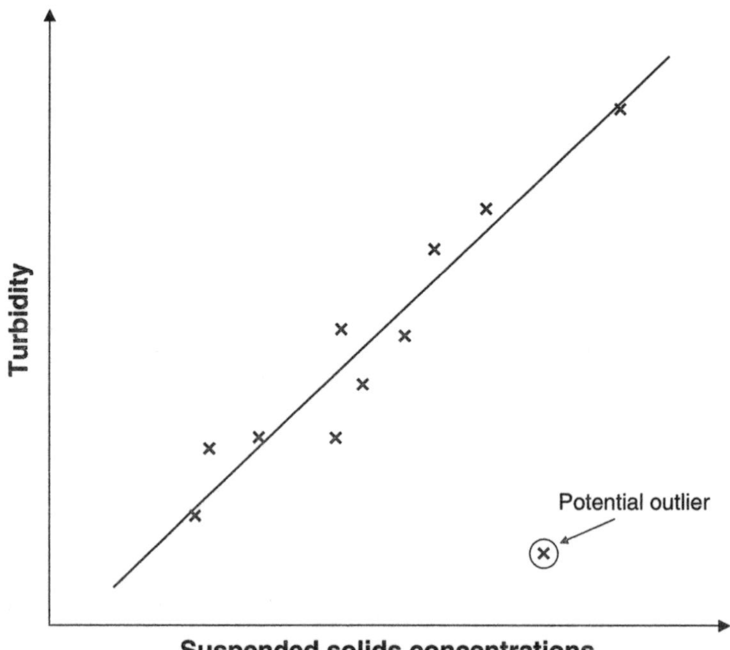

useful only when the data set is not large, because a single value will not likely create a significant change in the standard deviation when the data set is greater than 20 to 30. The steps for this procedure:

a. rank-order the data; and
b. calculate the cumulative mean and standard deviation with and without the suspect observations.

EXAMPLE 6.9

For the aluminum concentration data listed below, utilize the standard deviation test for outlier detection to determine if there is likely an outlier amongst the data. The aluminum concentrations (μg/L) are 56, 27, 85, 87, 362.

Rank-ordered concentration	Cumulative means	Cumulative standard deviations
27	27	—
56	41.5	20.5
81	54.7	27.0
86	62.3	26.8
362	122.2	136.0

Conclusion: the single large concentration data point has a considerable influence on the calculated standard deviation. The results suggest an outlier.

4. *Standard normal test of outlier detection.* A more formalized procedure for detecting outliers than Procedure 3 uses the following steps:
   a. rank-order the data;
   b. calculate the mean and the standard deviation of all the data;
   c. calculate T_n = (largest value – mean)/standard deviation; and
   d. determine if T_n is really large, utilizing the standard normal distribution tables. "Really large" is typically interpreted as being three standard deviations (i.e., if the distribution is normal), so a single value that is three standard deviations larger than the mean is indeed an unusual value. See Ex. 6.10.
5. *Dixon's test for outlier detection (after Dixon and deMassey, 1969).* In Dixon's test for outlier detection, the set of observations is first ordered according to the individual magnitudes. The ratio of the difference of an extreme value from one of its nearest neighbor values in the range sample values is then calculated, using a formula that varies with sample size. This ratio is then compared to a tabulated value and, if found equal or greater, the extreme value is considered an outlier at the $p \leq 0.05$ level or $p \leq 0.01$ level. The formula for the ratio varies with sample size and according to whether the suspected outlier is the smallest or largest value.

EXAMPLE 6.10

For the aluminum concentration data listed in Ex. 6.9, use the standard normal test to determine if the value 362 is likely an outlier.

From the data, the following statistics result

$$\text{Mean} = 122.2$$

$$\text{Standard deviation} = 136.0$$

$$t_c = \frac{362 - 122.2}{136.0} = 1.76$$

From the normal distribution table (see Table A.1) the probability of a value exceeding 1.76 standard deviation (the area under the curve) is 1 – (0.50 + 0.46) = 0.04. The single value does not lie more than three standard deviations from the mean and is thus not considered an outlier by this procedure.

The tables associated with Dixon's test are included as Table A.16. The test may be utilized on sample sizes less than 25.

EXAMPLE 6.11

Assume the data listed in the table below, where the second column of the table indicates the reported concentration of the ordered observations.

Utilize Dixon's test to determine whether the highest value, 57.0, is an outlier.

Ordered data point	Concentration (mg/L)	Ordered data point	Concentration (mg/L)
1	0.5	12	12.1
2	0.6	13	12.4
3	0.8	14	12.5
4	1.0	15	13.1
5	1.1	16	14.5
6	1.9	17	20.2
7	2.9	18	22.1
8	4.6	19	24.7
9	8.8	20	24.9
10	9.2	21	44.0
11	11.1	22	46.9
		23	57.0

For the highest value recorded, x_{23}, the Dixon's coefficient is calculated from

$$r = \frac{x_{23} - x_{21}}{x_{23} - x_3} = \frac{57 - 44}{57 - 0.8} = 0.23$$

The critical value for 95 percent is, from Table A.16, a magnitude of 0.42. The Dixon coefficient is less than the critical value, thus x_{23} is not considered an outlier at the 95 percent level of significance. Thus x_{23} is not considered an outlier.

Dixon's test also allows testing for two outliers (e.g., testing whether both x_{23} and x_{22} are outliers).

It is apparent from the preceding discussion on Dixon's test that the procedure uses the extremes of the data (i.e., both the highs and the lows). Thus, when portions of the data set are censored (to be discussed in Chap. 10), Dixon's procedure cannot be utilized.

Since Dixon's test uses individual values (the lowest and the highest), Dixon's test differs from the tests described previously which consider deviations from the mean.

EXAMPLE 6.12

Concentration measurements for benzo(a)pyrene are 2.77, 2.80, 2.90, 2.92, 3.45, 3.95, 4.44, 4.61, 5.21, 7.46.

Utilize Dixon's test to examine whether the highest value is an outlier.

$$r = \frac{x_{10} - x_9}{x_{10} - x_2} = \frac{7.46 - 5.22}{7.46 - 2.80} = 0.48$$

Since $r = 0.48$ is larger than the critical value of 0.477 at five percent (see Table A.16), then 7.46 is considered as an outlier (but not for one percent).

6. *Maximum normal residual and the extreme studentized deviate tests for outlier detection.* The maximum normal residual procedure (after Stefansky, 1972) examines the data set for the maximum deviation for sample:

$$\text{Maximum normal residual } (MNR) = \frac{\max |x_i - \bar{x}|}{\sqrt{\sum (x_i - \bar{x})^2}} \qquad [6.9]$$

Apparent from Eq. [6.9] is that the MNR procedure, like the first procedures described in this section, requires estimation of the mean. If the data set includes numerous censored values, such as nondetect measurements, the ability to quantify the mean is influenced.

EXAMPLE 6.13

Use the extreme studentized deviate test to determine if there is an outlier in the arsenic concentrations listed below.

Observation	Concentration	Deviation	Studentized deviate
1	12.0	–19.1	–1.11
2	22.1	–9.0	–0.52
3	18.6	–29.5	–1.72
4	75.0	43.9	2.55*
5	36.1	5.0	0.29
6	19.2	–11.9	–0.69
7	20.4	–10.7	–0.62
8	31.7	0.6	0.03
9	52.4	21.3	1.24
10	41.0	9.9	0.58
11	17.9	–13.2	–0.77
12	27.0	–4.1	–0.24

The asterisk (*) identifies the maximum deviate. The mean and standard deviations of the concentration data are as follows:

$$\bar{x} = 31.1$$

$$S_x = 17.97$$

$$n = \text{number of observations} = 12$$

The deviation is calculated as:

$$\text{Deviation} = \text{concentration} - \text{mean concentration}$$

$$\text{Studentized deviation} = \text{studentized residual}$$

$$= \frac{\text{Deviation}}{S_x\left(1 - \dfrac{1}{n}\right)^{0.5}} \quad [6.10]$$

where Eq. [6.10] has been derived by manipulating Eq. [6.9], into a simpler form.

Consider the outlier ($Z_{97.5} = 1.96$) where Z is the standard normal deviate. Now, test whether the highest value namely, 75.0, is an outlier. Since $Z_{97.5} = 1.96$ (from Table A.1), and $2.55 > 1.96$ then, the value of 75.0 is considered as an outlier. It is noted that this procedure becomes identical to the standard normal test in Procedure 4 above, if the same condition for rejection is selected.

7. *Barnett and Lewis' outlier detection test.* An objective test for outliers which turns out to be the same as the maximum normal residual test described above, is described by Barnett and Lewis (1984). If a set of data is ordered from low to high, $x_{low}, x_2, \ldots x_{high}$ and the mean and standard deviation are calculated, then suspected high or low outliers can be tested by the following procedure

$$t^* = \frac{x_H - \overline{x}}{S_x} \text{ for a high value} \qquad [6.12]$$

or

$$t^* = \frac{\overline{x} - x_L}{S_x} \text{ for a low value} \qquad [6.13]$$

Compare the value of t^* for the five percent or one percent level of significance. If the calculated t is larger than the table value for the number of measurements n the x_H or x_L is an outlier at that level of significance.

Number of measurements	Critical value		Number of measurements	Critical value	
	5 percent	1 percent		5 percent	1 percent
3	1.15	1.15	16	2.44	2.75
4	1.46	1.49	18	2.50	2.82
5	1.67	1.75	20	2.56	2.88
6	1.82	1.94	30	2.74	3.10
7	1.94	2.10	40	2.87	3.24
8	2.03	2.22	50	2.96	3.34
9	2.11	2.32	60	3.03	3.41
10	2.18	2.41	100	3.21	3.60
12	2.29	2.55	120	3.27	3.66
14	2.37	2.66			
15	2.41	2.71			

EXAMPLE 6.14

Utilize the Barnett and Lewis outlier detection test on the following data to determine if the highest value is an outlier:

0.5, 0.6, 0.8, 1.0, 1.1, 1.9

From the data,

mean $\overline{x} = 0.98$
standard deviation $S_x = 0.50$

From Eq. [6.12],

$$t^* = \frac{1.9 - 0.98}{0.50} = 1.84$$

Since $n = 6$, $t_c = 1.82$ for the five percent level of significance, therefore, $t^* > t_c$, $1.84 > 1.82$ and the high value is determined to be an outlier. It is to be noted that if we adopted a one-percent level of significance,

$$t_c = 1.94 \text{ and thus}$$
$$t^* < t_c$$
$$1.84 < 1.94$$

which does not identify the high value as an outlier.

8. *Chauvenet's criterion for outlier detection.* For data that are distributed in accord with the normal distribution, with a single extreme value a simple method such as Chauvenet's criterion may be employed (Meyer, 1975). Chauvenet's criterion states that if the probability of a value deviating from the mean is greater than 0.5 n, expressed as a percentage, there are adequate grounds for rejection.

EXAMPLE 6.15

For the monitoring values

1,6,7,8,8,9,9,9,10,10,10,10,10,11,11,11,12,12,13,14

consider whether the 1 is an outlier.

The mean of the data is 9.55 and the standard deviation is 2.80, and n, the number of samples is 20. One approach involves rejecting the value of one if its probability of occurrence were less than 0.5 (20) = 10 percent. From the table of Z scores in Table A.1, we see that ten percent of the values in a normal distribution are beyond ± 1.645 standard deviations of the mean or, in other words, (1.645) (2.80) = 4.61

This means we would reject values beyond this range from the mean; for example, 9.55 – 4.61 = 4.94 or greater than 9.55 + 4.61 = 14.16. We therefore would reject the value of one but accept the remainder of the data.

It is to be noted that as the sample size gets larger, the rejection zone for Chauvenet's criterion will also increase. A value of n of approximately 20 is essentially a maximum for use of this procedure.

9. *Winsorization of data as an approach to deal with outliers.* If an outlier is rejected, it can be completely removed from the data set. Alternatively, in some situations, it is reasonable to replace the outlier with the next largest (or smallest) observation. This latter technique, called Winsorization (see Dixon, 1960, for further details), is often used to calculate the sample mean when one observation is significantly higher or lower than the rest.

   Another variation on this approach is to replace the highest and lowest values in a set of data.

EXAMPLE 6.16

Assume a grouping of data consists of the values 75, 56, 47, 42, 39, 29, 11. We would replace the high value 75 with a second, 56, and the lowest value, 11, with a replicate 29. This would give a modified data set as 56, 56, 47, 42, 39, 29, 29, which is then treated as if it were the original data set.

Winsorizing should not be performed if the extreme values constitute more than a small minority of the entire data set.

10. *Regression modeling to identify outliers.* This approach involves first, the fitting of a particular regression model to the data (this is a subject developed in Chap. 7). The degree to which the individual data points deviate from the assumed regression model is then considered. For example, one approach is to reject individual data points whose residual is more than four times the residual standard deviation. However, further discussion of this regression-based procedure is delayed to Chap. 7.

6.6 SUMMARY OF APPROACHES FOR IDENTIFYING DATA OUTLIERS

By increasing the variability within a sample, outliers decrease the sensitivity of any subsequent statistical tests. Outliers are observations that appear inconsistent with the rest of the data. The magnitude of the reported value may be in error, or it may be genuine and indicate a valuable data point—one of the reasons we are collecting the monitoring data.

The principal safeguard against errors in data are vigilance in carrying out the operations, the measurements, and the recording of the data. Different pros and cons exist for the various outlier tests, so a test selected for one circumstance may not be appropriate for alternative circumstances. While many of the tests require that data be characterized by the normal distribution, a data transformation may be feasible to broaden their applicability. The sample size and presence/absence of censored data are probably the most important features determining which outlier detection test is the most appropriate.

Additional techniques to identify outliers are described in Grubbs (1969) and Beckman and Cook (1983).

6.7 REFERENCES

BARNETT, V., and T. LEWIS. *Outliers in Statistical Data*, 2nd ed. New York, NY: John Wiley and Sons, 1984.

BECKMAN, R. J., and R. D. COOK. "Outliers," *Technometrics*, 25 (1983), 119–163.

CONOVER, W. J. *Practical Nonparametric Statistics*, 2nd ed. New York, NY: John Wiley and Sons Inc., 1980.

COX, D. R., and D. V. HINCKLEY. *Theoretical Statistics*. London: Chapman and Hall, 1974.

DIXON, W. J., and F. J. DEMASSEY. *Introduction to Statistical Analyses*, 3rd ed. New York, NY: McGraw-Hill, 1969.

DIXON, W. J. "Processing Data for Outliers," *Biometrics*, 9 (1953), 74–89.

DIXON, W. J. "Simplified Estimation from Censored Normal Samples," *Annals of Mathematical Statistics*, 31 (1960), 385–391.

GIBBONS, R. D. "Statistical Models for the Analysis of Volatile Organic Compounds at Waste Disposal Sites," *Ground Water*, 25 no. 5 (1987), 572–580.

GRUBBS, F. E. "Procedures for Detecting Outlying Observations in Samples," *Technometrics*, 11 (1969), 1–21.

HAHN, G. J. "Statistical Intervals for a Normal Population Part I: Tables, Examples and Applications," *Journal of Quality Technology* 2 no. 3 (1970), 115–125.

MEYER, S. L. *Data Analysis for Scientists and Engineers*, pp. 17–18. New York, NY: John Wiley, 1975.

STEFANSKY, W. "Rejecting Outliers in Factorial Designs," *Technometrics*, 14 (1972), 469–479.

STARKS, T. H., and G. T. FLATMAN. "RCRA Ground-Water Monitoring Decision Procedures Viewed as Quality Control Schemes," *Environmental Monitoring and Assessment*, 16 (1991), 19–37.

6.8 PROBLEMS

6.1. The results of a program of measuring depths in three monitoring wells are listed below:

Monitoring Well		
MW1	**MW2**	**MW3**
164.08	177.58	209.84
163.32	168.28	223.46
162.98	172.78	214.46
167.80	180.94	213.14
165.16	110.44	204.46

a. Use a scattergram analysis test to identify any potential outliers in the three monitoring well results;
b. Use the standard deviation test to identify any outliers in the depths in the individual monitoring wells.
c. Use the standard normal test for outlier detection to identify any outliers in the depths in the individual monitoring wells.
d. Utilize Dixon's test for outlier detection to identify any outliers in the depths in the individual monitoring wells.
e. Utilize the maximum normal residual test to identify any outliers in the depths in the individual monitoring wells.
f. Utilize Chauvenet's criterion to identify any outliers in the depths in the individual monitoring wells.
g. Utilize Barnett and Lewis' outlier detection test to identify any outliers in the depths in the individual monitoring wells.

6.2. On the diskette are 20 groupings of 25 monitoring results contained in a file called 'NORM'.

a. Use the standard normal test for outlier detection to identify any outliers for individual sampling locations.
b. Use Dixon's test for outlier detection to identify any outliers for individual sampling locations.
c. Use Barnett and Lewis' outlier detection test to identify any outliers for individual sampling locations.
d. Redo (a) through (c) using only the first ten values from each data set. Comment on any trends you see in which the data sets have outliers, by the different procedures. In this respect, consider the fact that the data for sampling locations A through J are data selected randomly from a normal distribution and that data sets K through T are from a lognormal distribution.

6.3. Given the following arsenic concentrations, .45, 1.4, 1.9, 10, determine the 90 percent upper confidence limit. Use the Shapiro-Wilk test to check any data transformation you may employ.

6.4. The data listed below represent simultaneous measures of biochemical oxygen demand (BOD) and total organic carbon (TOC) obtained from grab samples obtained from a river.

Sampling date	BOD (mg/L)	TOC (mg/L)
January	16.4	12.3
February	12.3	10.0
March	16.7	14.9
April	8.2	8.3
May	7.1	8.9
June	7.6	9.3
July	10.1	9.8
August	11.2	10.8
September	14.6	12.6
October	12.6	17.6
November	8.8	9.4
December	12.4	13.6

a. Calculate the upper 95 percent tolerance limits for the BOD and TOC concentrations.

b. Calculate the upper 95 percent confidence limits on the mean for the BOD and TOC concentrations.

c. Use the probability plot procedure to examine whether the TOC sample of October appears to be an outlier.

d. Use the scattergram procedure to assess whether the October TOC value is likely an outlier.

e. Use the standard normal test of outlier detection to indicate whether the October TOC value is likely an outlier (assuming 3S as the criterion).

f. Use Dixon's test for outlier detection to indicate whether the October TOC value is likely an outlier.

g. Use Chauvenet's criterion to determine if the October TOC value is likely an outlier.

h. Use the Barnett and Lewis outlier detection test to determine if the October TOC value is likely an outlier.

6.5. Estimate the 95 percent upper prediction limit for the next two measurements, given the following monitoring data obtained to date:

<15	17	<15
<15	26	<15
<15	<15	16
<15	<15	<15

CHAPTER 7

Characterizing Coincident Behavior—Regression and Correlation

7.1 INTRODUCTION

In assessing environmental quality, we are often interested in examining coincident behavior, that is, quantifying the relationship between two or more variables. This may allow us, for example, to fill in missing data for one constituent, using correlated information from the other variable(s). It also may help us predict future levels of a constituent (although this must be done very carefully), given the current monitoring results for the constituent. Both of these applications utilize regression and correlation techniques. Regression and correlation techniques are the simplest, most commonly used approaches for generating statistical models to assess mathematical relationships (e.g., similarity in magnitude between different constituents and/or for one constituent over time). Thus, the focus in this chapter is how to use regression and correlation procedures to quantify the relationships among characteristics of environmental constituents.

The relationship between two (or more) variables can be characterized in one of two ways:

1. *Regression analysis (or dependency).* Regression analysis is focused on the degree to which one variable (the dependent variable) is dependent upon one or more other variables, the independent variable(s). Thus, regression is a means of calibrating the coefficients of a predictive equation.

EXAMPLE 7.1

The surface cover on land influences the extent to which surface runoff occurs. Extensive vegetation causes an initial abstraction that, for minimal rainfall, may result in zero surface runoff. However, as rainfall quantities became larger, the influence of the initial abstraction becomes minimal so that additional rainfall increments result in equivalent increases in surface runoff. The surface runoff versus rainfall quantities, both measured in depth of water, are depicted in Fig. 7.1. A best-fit curve to the data would involve a regression model, since flow (the dependent variable) is functionally dependent on rainfall (the independent variable).

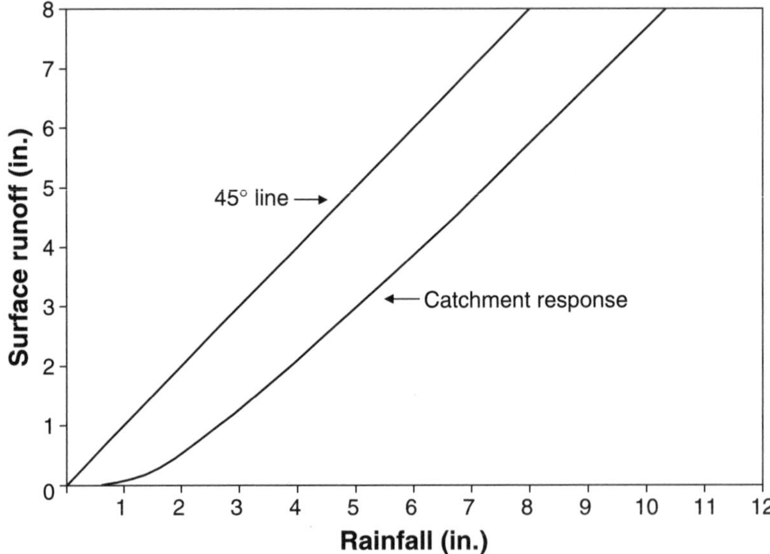

FIGURE 7.1 Rainfall-runoff relationship

2. *Correlation analysis.* In correlation analysis neither of the variables is identified as more important than the other, but the investigator is interested in their interdependence or joint behavior. Correlation or association is not causation (e.g., the rooster who thinks his crowing makes the sun rise is deluded). Correlation provides a measure of the goodness of fit.

EXAMPLE 7.2

The flows in the two branches of a river, at locations A and B as depicted in Fig. 7.2, may be caused by the same rainfall event. Therefore, if the flow at A is high because of a particular rainstorm, then there is a strong likelihood that the flow at B is also high. Note that high flow at A does not cause high flow at B, so there is not a direct dependency; instead, high flows at A and B tend to be caused by the same high rainfall events.

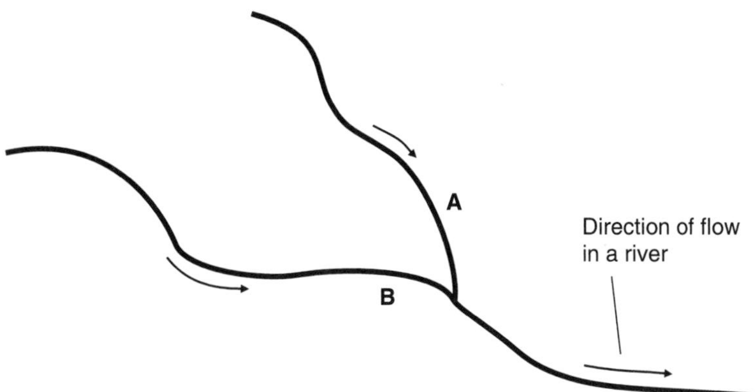

FIGURE 7.2 Plan view of a drainage pattern that will show a strong correlation between flow locations at A and B

In the sections to follow, it will become apparent that the theory of regression has many points in common with the theory of correlation, although they answer rather different questions. Many books have been written on the subjects of regression and correlation, and the interested reader is referred to these for information that goes beyond what is presented here (e.g., Draper and Smith, 1981). This chapter emphasizes how these two procedures may be applied to environmental engineering concerns.

7.2 THE CORRELATION COEFFICIENT

When n measurements are made simultaneously on two random variables, x_i and y_i, for $i = 1, \ldots n$, regardless whether the objective is to demonstrate either a dependent behavior or similarity in behavior, a useful statistical measure is the correlation coefficient. The first step in investigating correlation is to plot the data as a scattergram to illustrate whether a relationship exists between x and y. An example of a scattergram is illustrated in Fig. 7.3, which depicts the individual monitored results of the biochemical oxygen demand at five days (BOD_5) versus the total organic carbon (TOC). The depicted BOD_5 versus TOC monitored results shown in Fig. 7.3 are water quality measures associated with influent wastewater to a wastewater treatment facility (after Constable and McBean, 1979).

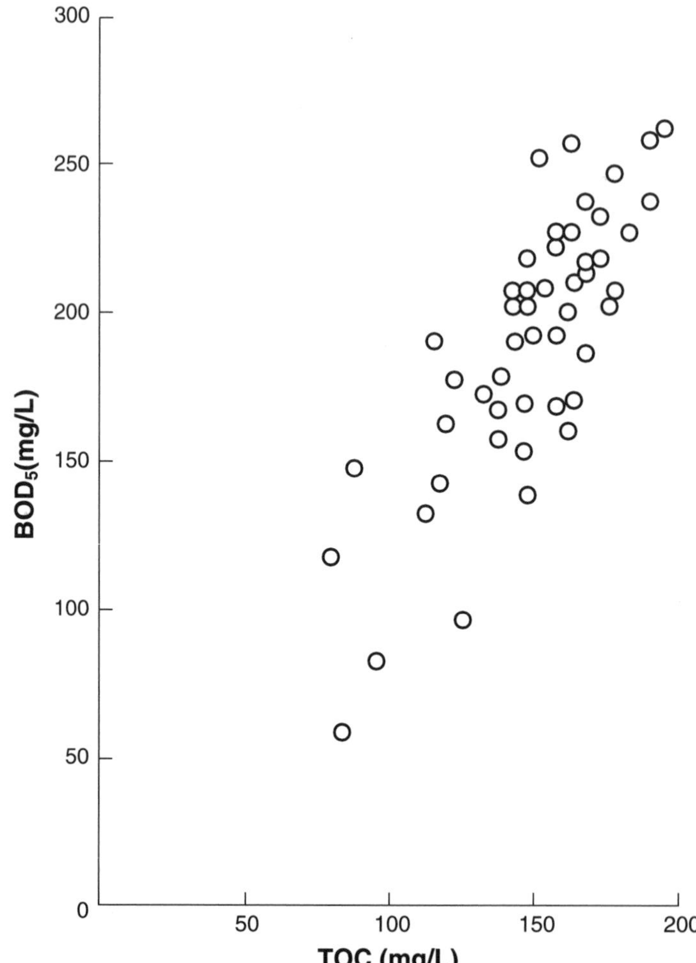

FIGURE 7.3 Scattergram plot of BOD_5 versus TOC

An alternative example of a scattergram is illustrated in Fig. 7.4 which depicts hydraulic conductivity levels of soil in two different regions of the soil strata, namely the alluvium and the bedrock. A third example of correlation, and one that is also a dependency relationship, is demonstrated in Figs. 7.5(a) through (c) (after McBean and Al-Nassri, 1987), in which sediment concentrations in rivers are demonstrated to be at least in part a function of the river flow levels.

Frequently, a simple examination of the scattergram depicting the nature and distribution of data can suggest unanticipated patterns and results. The scattergram might indicate, for example, whether a transformation of one of the variables would be appropriate before analyzing the data in terms of an assumed associative model and also whether there are isolated, very discrepant

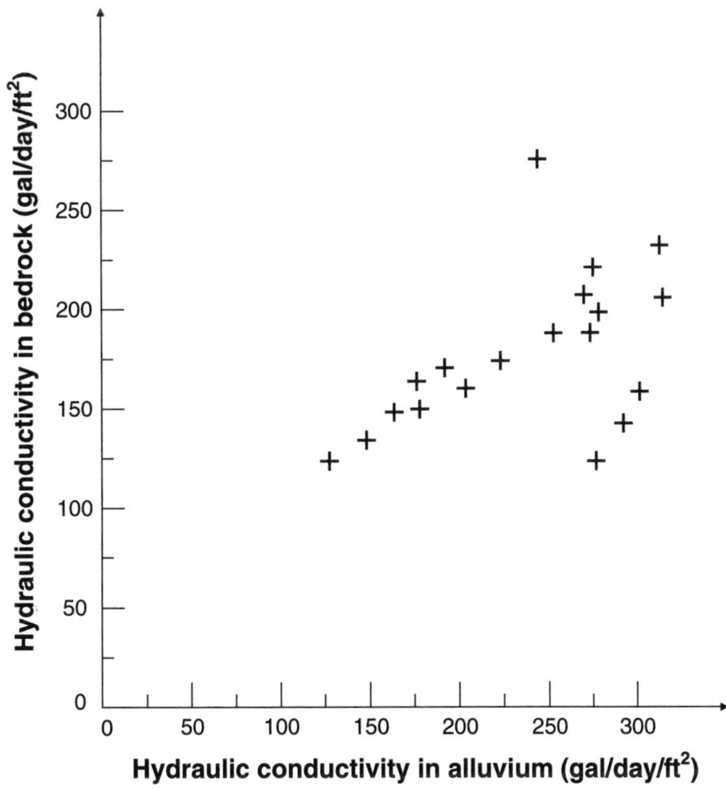

FIGURE 7.4 Example of correlated data for hydraulic conductivity levels in alluvium and bedrock strata

FIGURE 7.5a Astore River, January 1–December 1, 1975: suspended sediment concentration versus discharge

Source: Adapted from McBean, E., and Al-Nassri, S., "Uncertainty in Suspended Sediment Transport Curves," *ASCE Journal of Hydraulics Division*, 114, no. 1 (January, 1987), pp. 63–74.

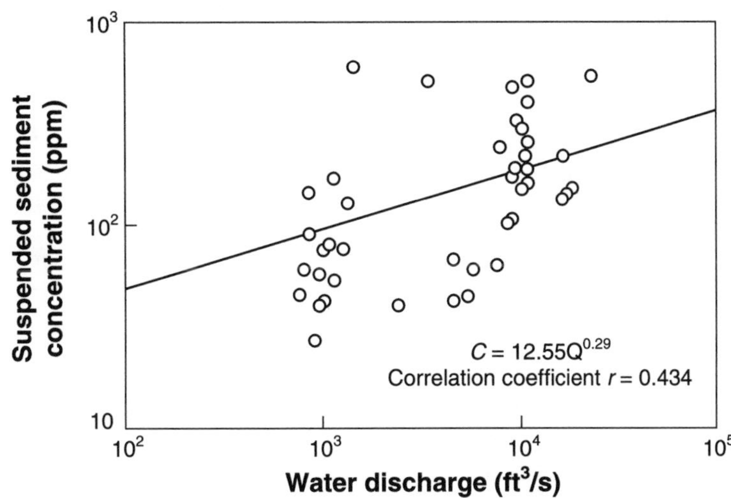

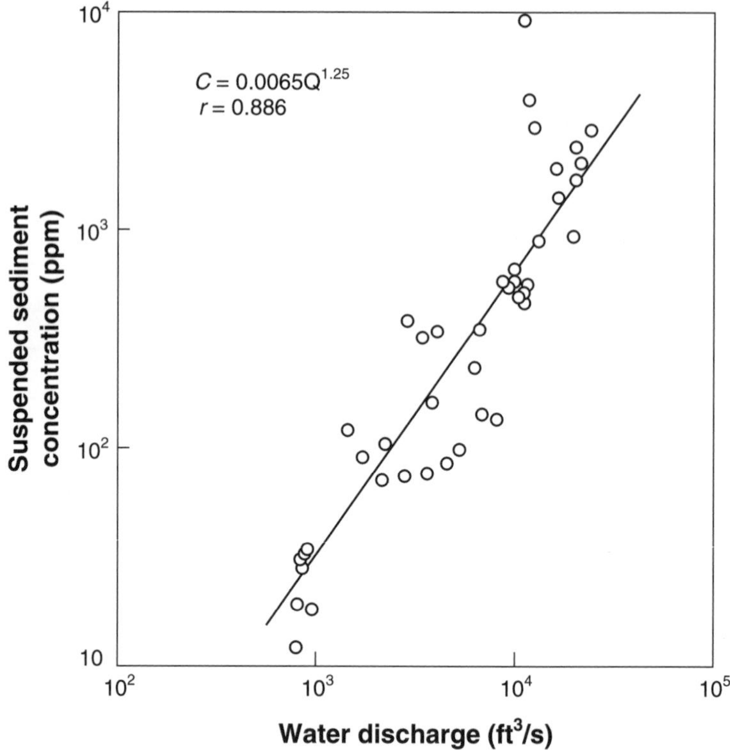

FIGURE 7.5b Astore River, January 27–September 27, 1978: suspended sediment concentration versus discharge

Source: Adapted from McBean, E., and Al-Nassri, S., "Uncertainty in Suspended Sediment Transport Curves," *ASCE Journal of Hydraulics Division*, 114, no. 1 (January, 1987), pp. 63–74.

observations whose inclusion or exclusion needs special consideration as outliers. Perhaps a particular data point seems to be an outlier because of a laboratory mistake, in which case the data point should not be considered in subsequent analyses.

To quantify the correlation, we are interested in identifying the degree of association or correlation (e.g., can a smooth curve—linear or nonlinear—be fitted to the data?). The population correlation coefficient ρ_{xy} between the two random variables x and y is defined in terms of the covariance of x and y, σ_{xy}, and the standard deviations of x and y, σ_x and σ_y, respectively, as

$$\rho_{xy} = \frac{\sigma_{xy}}{\sigma_x \sigma_y} \qquad [7.1]$$

The sample estimate of the correlation coefficient, r_{xy}, is written as

$$r_{xy} = \frac{S_{xy}}{S_x S_y} \qquad [7.2]$$

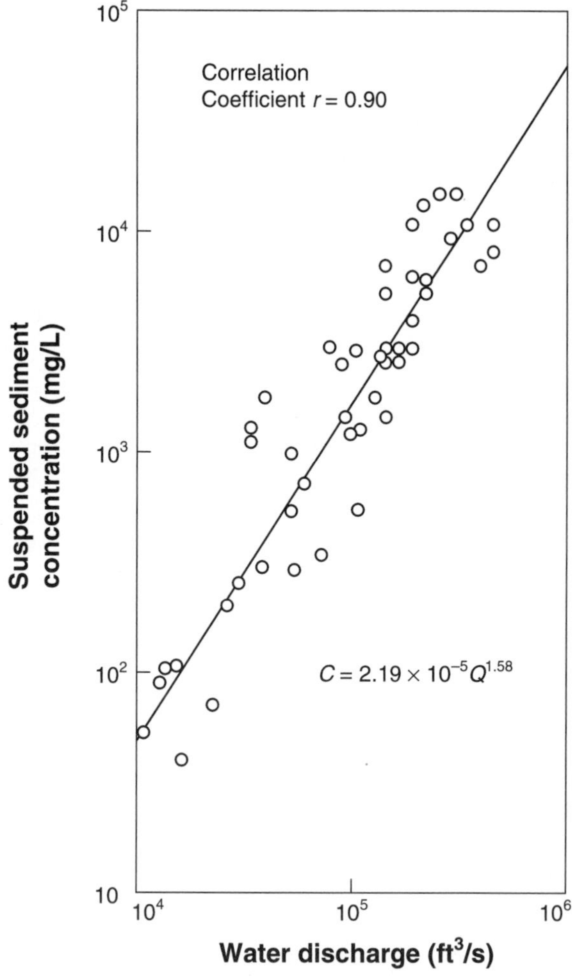

FIGURE 7.5c Indus River, April 16–December 18, 1976: suspended sediment concentration versus discharge

Source: Adapted from McBean, E., and Al-Nassri, S., "Uncertainty in Suspended Sediment Transport Curves," *ASCE Journal of Hydraulics Division,* 114, no. 1 (January, 1987), pp. 63–74.

Mathematically, the calculation for Eq. [7.2] is written as

$$r_{xy} = \frac{\sum_{i=1}^{n}(x_i - \bar{x})(y_i - \bar{y})}{\left[\sum(x_i - \bar{x})^2\right]^{1/2}\left[\sum(y_i - \bar{y})^2\right]^{1/2}} \qquad [7.3]$$

and quantifies the strength of the linear relationship between the two variables. The correlation coefficient r is referred to as the Pearson product moment (parametric) correlation coefficient. It (r) is unitless and varies between minus one and plus one, where –1.0 represents a perfect negative correlation in which all measured points fall on a line having a negative slope, through 0.0 (i.e., absolutely no linear relationship between the variables exists), to +1.0 (i.e., a perfect correlation of points on a line having a positive slope). The correlation is positive if large values of both variables tend to occur together, and is negative if large values of one variable tend to occur with small values of the other variable. The larger the *absolute* magnitude of r, the stronger the degree of linear relationship. Figure 7.6 demonstrates three situations: Figure 7.6(a) demonstrates a high negative correlation between the two noted variables, indicating that when one value is large the other is small and vice versa. Figure 7.6(b) demonstrates a very low correlation, indicating that there is no relation between the two variables. Figure 7.6(c) demonstrates a high positive correlation, indicating that when one value is low the other is also low, and when one value is high the other is also high. If the contribution to the scatter in y, for example, comes predominantly from factors unrelated to x, then the correlation coefficient is small so that the value of r approaches zero.

FIGURE 7.6a Example of negative correlation

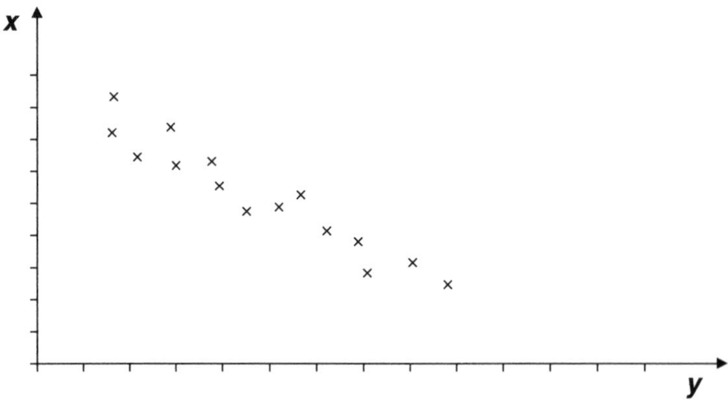

FIGURE 7.6b Example of zero correlation

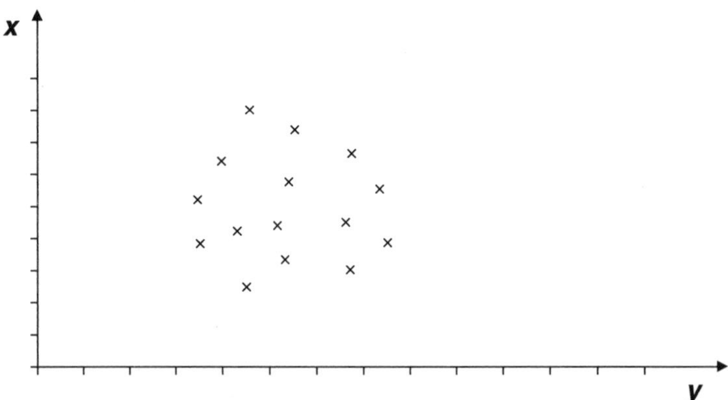

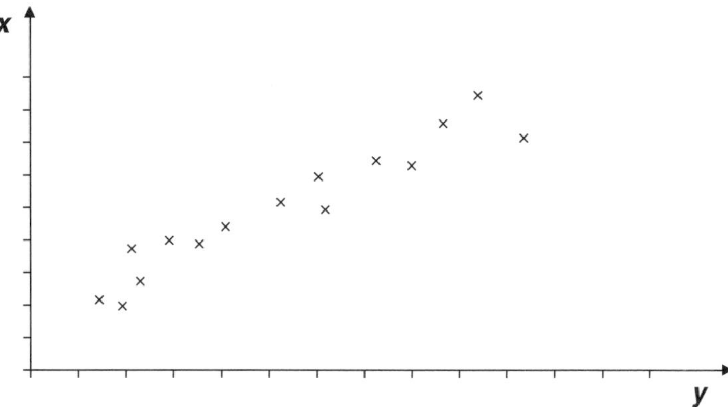

FIGURE 7.6c Example of positive correlation

Regardless of the magnitude of r, the value of r may or may not be significant. A significance test must be used to determine if the observed correlation coefficient is significantly different from zero. If there is really no correlation between the two variables, it is still possible that a spuriously high (positive or negative) sample correlation value may occur by chance. When the true correlation is zero, it can be shown that the statistic

$$t^* = r\frac{\sqrt{n-2}}{\sqrt{1-r^2}} \qquad [7.4]$$

has a t-distribution with $n-2$ degrees of freedom, provided that both variables are normally distributed.

EXAMPLE 7.3

A correlation coefficient based on a sample of size 18 was computed to be 0.32. Can we conclude at significance levels of (a) 0.05 and (b) 0.01 that the corresponding population correlation coefficient differs from zero?

Solution

To answer this question, we must assess a particular hypothesis. The first hypothesis will be the null hypothesis, denoted by H_0, in which we assume the correlation is zero. The second hypothesis, denoted by H_1, is that the correlation is greater than zero. To determine which situation applies for this problem, the characteristics of the problem are inserted into Eq. [7.4] as

$$t^* = 0.32\frac{\sqrt{18-2}}{\sqrt{1-(0.32)^2}} = 1.35$$

a. Using a one-tailed test of the t-distribution at the 0.05 level of significance, we would reject H_0 if $t^* > t_{.95} = 1.75$ for (18-2) = 16 degrees of freedom (see Table A.4 for the t_c, the critical value). Thus, we cannot reject the null hypothesis H_0 at the 0.05 level since $t^* < t_{crit}$ or $1.35 < 1.75$ and thus we cannot conclude the population correlation coefficient differs from zero.

b. Since we cannot reject H_0 at the 0.05 level, we certainly cannot reject it at the 0.01 level. $t_{.99} = 2.58$ for $(18 - 2) = 16$ degrees of freedom (see Table A.4 for the t_c critical value). Thus, we cannot conclude the population correlation coefficient differs from zero.

EXAMPLE 7.4

What is the minimum sample size necessary to conclude that a correlation coefficient of 0.32 differs significantly from zero at the 0.05 level?

Using the one-tailed test of the t-distribution at the 0.05 level, the minimum value of n must be such that

$$0.32 \frac{\sqrt{n - 2}}{\sqrt{1 - (0.32)^2}} = t_{.95}$$

for $n - 2$ degrees of freedom. Since the calculations must be accomplished in an iterative manner, we start knowing that for an infinite number of degrees of freedom, $t_{.95} = 1.645$ (Table A.4) and hence $n = 25.6$, which provides a useful starting point. We then make incremental adjustments.

Iteration number 1
For $n = 26$, $df = 24$, $t_{.95} = 1.711$

$$t^* = 0.32 \frac{\sqrt{24}}{\sqrt{1 - (0.32)^2}} = 1.65$$

or $t^* < t_c$

Iteration number 2
For $n = 27$, $df = 25$, $t_{.95} = 1.708$

$$t^* = 0.32 \frac{\sqrt{25}}{\sqrt{1 - (0.32)^2}} = 1.69$$

or $t^* < t_c$

Iteration number 3
For $n = 28$, $df = 26$, $t_{.95} = 1.706$

$$t^* = 0.32 \frac{\sqrt{26}}{\sqrt{1 - (0.32)^2}} = 1.72$$

$t^* > t_c$

Thus the minimum sample size is $n = 28$ to conclude that a correlation coefficient of 0.32 differs significantly from zero at the 0.05 level.

Note that the above discussion has been focused on a one-sided test. If we are interested in positive or negative correlation, then analysis using a two-tailed test is appropriate.

Table 7.1 gives the 95 percent critical points for the absolute value of the correlation coefficient for different sample sizes. When the sample size is small, a fairly large absolute value of r is required to show statistically significant correlation coefficients.

Note that the correlation coefficient should only be calculated when the relationship between two random variables is thought to be linear. In the event of nonlinear behavior we can often transform the data to obtain linear behavior (e.g., use a log transform to transform the data in Fig. 7.7(a) to the data in Fig. 7.7(b). The magnitude of r is very sensitive to the presence of nonlinear trends such as depicted on Fig. 7.7(a) (which will cause the magnitude of the relationship to be underestimated) and outliers (which may cause the relationship to be underestimated or overestimated). Nonlinear trends and outliers are usually detectable in scattergram plots (i.e., x versus y).

Note that a high correlation coefficient between two variables does not necessarily indicate a causal relationship. There may be a third variable that is causing the simultaneous change in the first two variables.

The square of the correlation coefficient, also called the *coefficient of determination*, is an estimate of the proportion of variance in the dependent variable that is accounted for by the independent variable(s). Thus, r^2 represents the proportion of the variance of the dependent variable, say y, that can be attributed to its linear regression on the independent variable x while $(1 - r^2)$ is the proportion free from x, or

$$\text{Coefficient of determination} = r^2 = \frac{\text{explained variation}}{\text{total variation}} \qquad [7.5]$$

TABLE 7.1 Critical value of the correlation coefficient for alternative sample sizes for 95% confidence interval

Sample size	Critical value
5	0.75
10	0.58
15	0.48
20	0.42
25	0.38
30	0.35
50	0.27
100	0.20

Source: Adapted from D. Ebdon, *Statistics in Geography*, 2nd ed. (Cambridge, MA: Blackwell Publishers, 1985), p. 218. Reprinted with permission.

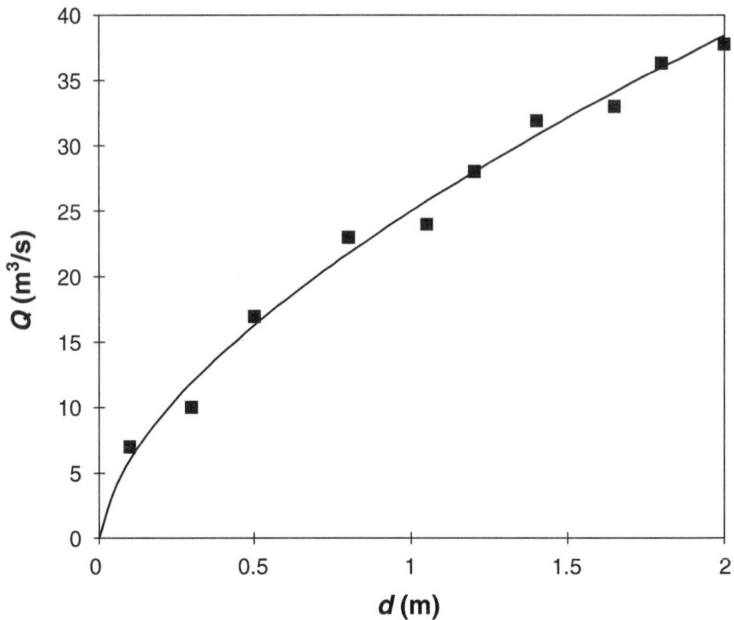

FIGURE 7.7a Relation between depth of flow and flow

FIGURE 7.7b Relation between logarithms of depth of flow and flow

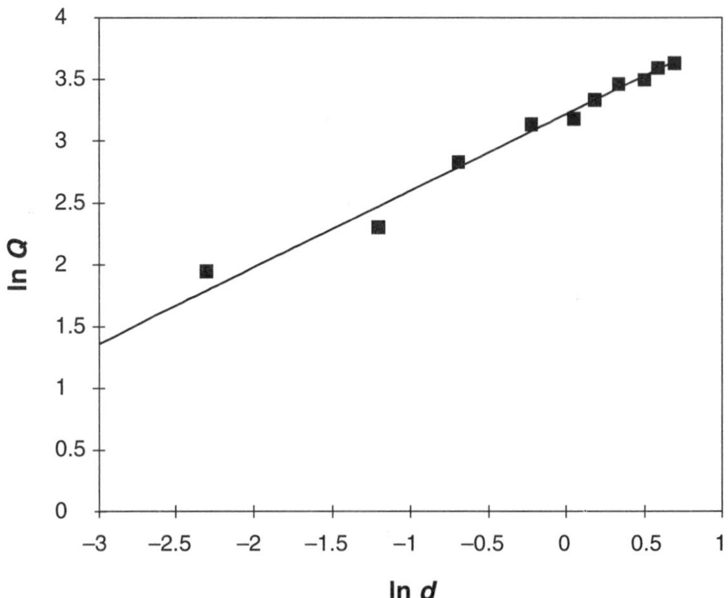

Alternatively, one can write

Total variance = variance due to regression + variance not explained by regression

$$S_y^2 = S_y^2 r^2 + S_y^2(1 - r^2) \quad [7.6]$$

Note in Eq. [7.6] that it is the variance, not the standard deviation, that is the additive parameter.

EXAMPLE 7.5

Water quality monitoring records for turbidity and suspended solids concentrations are as tabulated below:

Sample number	Turbidity	Suspended solids
1	30	60
2	25	61
3	30	75
4	45	65
5	15	70
6	41	90
7	31	77
8	33	100
9	40	58
10	21	89
11	130	300

Using the first ten of these sample results, plot the scattergram and compute the correlation coefficient.

The resulting scattergram is depicted in Fig. 7.8.

Excluding the eleventh or last value, the correlation coefficient value

$$r = -.04$$

yields the conclusion that there is not a strong relationship between the turbidity and concentrations of suspended solids.

A large value for the correlation coefficient does not necessarily guarantee that two variables are related. The value calculated for r will tend to be inflated if there are only a few data pairs. A statistical test is usually required to verify that the correlation coefficient is significantly different from zero.

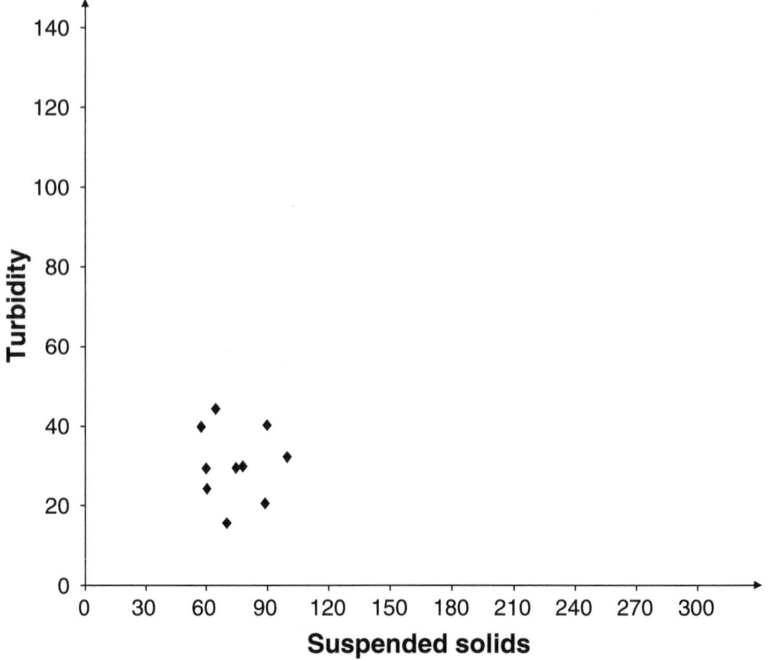

FIGURE 7.8 Turbidity versus suspended solids, using first ten data points in Ex. 7.5

EXAMPLE 7.6

Figure 7.9 illustrates a scattergram that shows one data point distant to the northeast. Much of the variation in the data is explained by variance associated with the single, divergent, data point. When the single point is excluded (e.g., perhaps because it is an outlier) the r drops precipitously.

For the complete data set

$$n = 11, \quad r = .94, \quad r^2 = .88$$

$$t^* = 8.07$$

With $n - 2 = 9$ degrees of freedom, $t_c = 1.83$, $t^* > t_c$, which indicates that the correlation is significant.

For the data set with the single, divergent, data point not included,

$$n = 10, \quad r = -.04, \quad r^2 = .002$$

$$t = .124$$

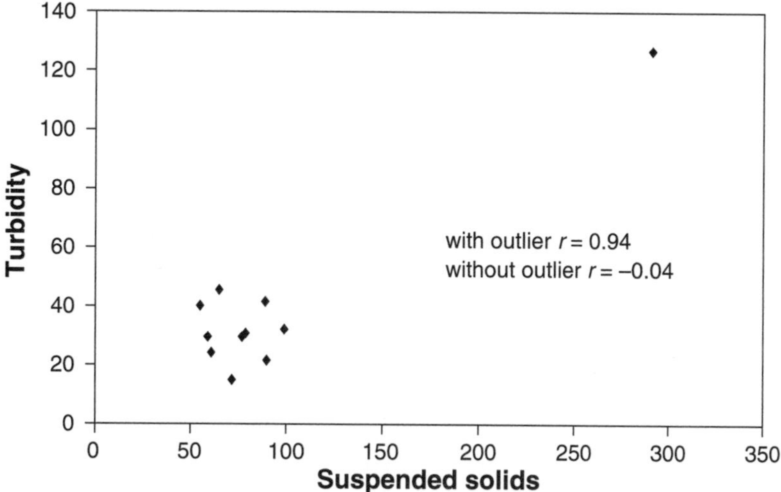

FIGURE 7.9 Turbidity versus suspended solids, using complete data set from Ex. 7.5

With $n - 2 = 8$ degrees of freedom, $t_c = 1.86$, $t^* < t_c$, which indicates insufficient evidence to regard the correlation as significant.

Clearly, the sensitivity of the results to a single, divergent monitoring point is substantial.

The variability of the correlation coefficient with environmental monitoring results is examined in Tyagi et al. (1996). The correlation coefficient is used within both correlation and regression studies.

7.3 REGRESSION ANALYSES

7.3.1 CURVE FITTING

A major goal in many engineering investigations is to make predictions using mathematical equations. Usually such predictions employ a formula that relates the dependent variable (whose value we want to predict) to one or more independent variables, establishing a causal (or dependent) relationship.

The objective of regression is to determine the equation of a line that best represents the linear trend in a set of raw or mathematically transformed data points. The notation utilized here designates y as the dependent, random variable whose magnitude depends on x, the controlled, independent or regressor variable. The resulting curve is called the regression curve of y on x.

The approximating curves frequently utilized as regression lines are given here with their respective equations:

straight line $$y = a_0 + a_0 x \quad [7.7]$$

parabola or quadratic curve $$y = a_0 + a_1 x + a_2 x^2 \quad [7.8]$$

cubic curve $$y = a_0 + a_1 x + a_2 x^2 + a_3 x^3 \quad [7.9]$$

quartic curve $$y = a_0 + a_1 x + a_2 x^2 + a_3 x^3 + a_4 x^4 \quad [7.10]$$

where x is the independent variable and y is the dependent variable.

Additional equations include:

hyperbola $$y = \frac{1}{a_0 + a_1 x} \quad [7.11]$$

exponential curve $$y = ab^x \quad [7.12a]$$

or equivalently $$\ln(y) = \ln(a) + x\ln(b) = a_0 + a_1 x \quad [7.12b]$$

logistic curve $$y = \frac{1}{ab^x + g} \quad [7.13]$$

gompertz curve $$y = pq^{bx} \quad [7.14a]$$

or, equivalently $$\ln(y) = \ln(p) + bx\ln(q) = a_0 + a_1 x \quad [7.14b]$$

To decide which curve should be used in a particular application, it is helpful to utilize a scattergram of the transformed variables.

Typically, the equation selected is developed by minimizing the squares of the residuals, where the residuals are the differences between the observed values of the dependent variable and the values of the dependent variable predicted by the model. In many problems of this kind, the independent variable is assumed to be observed without error, or with an error that is negligible when compared with the error (chance variation) in the dependent variable. Thus, although the independent variable may be fixed at x, repeated measurements of the dependent variable may lead to y-values that differ considerably. Such differences among y-values can be attributed to several causes, but are attributed chiefly to errors of measurement and to the existence of other uncontrolled variables that may influence the value of y when x is fixed.

Consider the fluctuations on random variables

$$y_i = \alpha + \beta x_i + \epsilon_i \qquad [7.15]$$

where α and β are constants whose values we wish to determine, and ϵ_i is a random "error" term. In general, an observed y_i will differ from an assumed regression line, with the difference denoted by ϵ_i. The ϵ_i are assumed to be independent, normally distributed random variables. The "error" term resulting in the observed value ϵ_i is conventionally taken to have expectation zero (any constant nonzero expectation could in any case be incorporated into α) and variance σ^2.

The parameters α and β in Eq. [7.15] completely define the regression line, and the estimation of α and β is equivalent to finding the equation of the straight line that best fits the data points. However, we do not know the magnitudes of α and β (since these are population parameters) and we look to estimate a and b from the sample data we have available. We are therefore looking to quantify a and b for the regression equation as

$$y' = a + bx \qquad [7.16]$$

The estimates of a and b are equivalent to finding the equation of the straight line that best fits the available data points. One way of doing this is by the method of least squares.

The common variance σ^2 is usually estimated by the vertical deviation of the sample points from the least-squares line. The ith such deviation is

$$y_i - y_i' = y_i - (a + bx_i) \qquad [7.17]$$

where y_i' is the value estimated using the regression equation.

The estimate of the variance σ^2 is

$$S_e^2 = \frac{1}{n-2}\sum_{i=1}^{n}[y_i - (a + bx_i)]^2 \qquad [7.18]$$

where S_e is termed the *standard error of estimate*. The divisor n-2 is used to make the resulting estimator for σ^2 unbiased; the two regression coefficients α and β had to be replaced by their least-squares estimates a and b.

To find the line that minimizes S_e^2 we must solve the pair of simultaneous equations

$$\frac{\partial S_e^2}{\partial a} = -2\sum(y_i - a - bx_i) = -2\sum y_i + 2na + 2b\sum x_i = 0 \qquad [7.19]$$

and

$$\frac{\partial S_e^2}{\partial b} = 2x_i\sum(y_i - a - bx_i) = -2\sum x_i y_i + 2a\sum x_i + 2b\sum x_i^2 \qquad [7.20]$$

The solution of Eq. [7.19] is

$$a = \overline{y} - b\overline{x} \qquad [7.21]$$

where the ($^-$) indicates the mean of the respective parameters. This indicates that the line passes through the point ($\overline{x}$, $\overline{y}$). Substituting this expression into Eq. [7.20] we find

$$b = \frac{\sum x_i y_i - n\overline{x}\,\overline{y}}{\sum x_i^2 - n\overline{x}^2} = \frac{\sum(x_i - \overline{x})(y_i - \overline{y})}{\sum(x_i - \overline{x})^2} \qquad [7.22]$$

If we define

$$S_{xx} = n\sum x_i^2 - \left(\sum x_i\right)^2 \qquad [7.23]$$

$$S_{yy} = n\sum y_i^2 - \left(\sum y_i\right)^2 \qquad [7.24]$$

$$S_{xy} = n\sum x_i y_i - \left(\sum x_i\right)\left(\sum y_i\right) \qquad [7.25]$$

we can then find

$$b = \frac{S_{xy}}{S_{xx}} \qquad [7.26]$$

Using the variables defined in Eqs. [7.24] through [7.26], an equivalent formula for the estimate of σ^2 is

$$S_e^2 = \frac{S_{xx}S_{yy} - (S_{xy})^2}{n(n-2)S_{xx}} \qquad [7.27]$$

Under the assumptions stated, the sampling distributions of the statistics are

$$t = \frac{a - \alpha}{S_e}\sqrt{\frac{nS_{xx}}{S_{xx} + (n\overline{x})^2}} \qquad [7.28]$$

and

$$t = \frac{b - \beta}{S_e}\sqrt{\frac{S_{xx}}{n}} \qquad [7.29]$$

which are t-distributions with $n - 2$ degrees of freedom.

To test the hypothesis that the regression coefficient a is equal to some specified value α, we can construct a confidence interval for α, solving the double inequality

$$-t_{\alpha'/2} < t < t_{\alpha'/2} \qquad [7.30]$$

Note that α' is utilized in this equation (and elsewhere in this chapter) instead of the term α (used in the other chapters to denote the level of significance) to differentiate it from the coefficient α in the regression equation.

The limits of the resulting confidence interval for α are

$$\alpha = a \pm t_{\alpha'/2}S_e\sqrt{\frac{S_{xx} + (n\overline{x})^2}{nS_{xx}}} \qquad [7.31]$$

Similar arguments lead to the following confidence limits for β as

$$\beta = b \pm t_{\alpha'/2}S_e\sqrt{\frac{n}{S_{xx}}} \qquad [7.32]$$

EXAMPLE 7.7

Chloride concentrations as a function of time are as listed below. Determine the best-fit regression equation for the chloride concentrations versus time.

x (time)	10	20	30	40	50	60	70	80	90	100
y (concentration)	17	26	21	22	28	39	38	46	52	56

The number of data points, $n = 10$

$$\sum x_i = 550 \qquad \sum x_i^2 = 38500 \qquad \sum x_i y_i = 22560$$

$$\sum y_i = 345 \qquad \sum y_i^2 = 13595$$

$$S_{xx} = n\sum x_i^2 - \left(\sum x_i\right)^2 = 10(38500) - (550)^2 = 82500$$

$$S_{yy} = n\sum y_i^2 - \left(\sum y_i\right)^2 = 10(13595) - (345)^2 = 16925$$

$$S_{xy} = n\sum x_i y_i - \left(\sum x_i\right)\left(\sum y_i\right) = 10(22560) - 550(315) = 35850$$

$$S_e^2 = \frac{82500(16925) - (35850)^2}{(10)(8)(82500)} = 168$$

or $S_e = 13.0$

Then

$$b = \frac{S_{xy}}{S_{xx}} = \frac{35850}{82500} = .434$$

$$a = y - bx = \frac{345}{10} - \frac{b(55)}{10} = 34.5 - .434(55) = 10.6$$

The value $t_{.025}$ equals 2.306 for 10 – 2 = 8 degrees of freedom, that is, 95 percent confidence interval for α and β.

The confidence limits are

$$a \pm t_{\alpha'/2} S_e \sqrt{\frac{S_{xx} + (n\overline{x})^2}{nS_{xx}}} = 10.6 \pm 2.306(13)\sqrt{\frac{82500 + (10(55))^2}{10(82500)}} = 10.6 \pm 20.48$$

or

$$-9.88 < \alpha < 31.08$$

$$b \pm t_{\alpha'/2} S_e \sqrt{\frac{n}{S_{xx}}} = .434 \pm 2.306(13.0)\sqrt{\frac{10}{82500}} = .434 \pm .330$$

$$0.104 < \beta < 0.764$$

β is the slope of the regression line. If $\beta = 0$ the regression line is horizontal and the mean does not depend linearly on x; however, for the data of this example, the 95 percent confidence interval for β does not include zero. The resulting interpretation is that there is an increase in chloride concentration with time.

The foregoing equations can be used to establish criteria for testing hypotheses concerning α and β. For example, the value of β gives the change in the mean of y corresponding to a unit increase in x. The results for the example problem indicate that the regression line of y on x is not a horizontal line (the slope of zero is not included). On the other hand, if the confidence limit on β includes zero, knowledge of x does not help in the prediction of y (for the specified 95 percent).

To test the null hypothesis H_0 ($\beta = \beta_0$), Eq. [7.32] can be rewritten as

$$t = \frac{b - \beta_0}{S_e}\sqrt{\frac{S_{xx}}{n}} \quad [7.33]$$

Of interest also is the prediction of y', a future value of y when $x = x_0$. An interval can be constructed in which y can be expected to lie with a given probability when $x = x_0$. If α and β were known, we could use the fact that y is a value of a random variable having a normal distribution with the mean $\alpha + \beta x_0$ and variance σ^2. However, we don't know the values of α and β, so we must consider the quantity $y - a - b x_0$ where y, a, and b are all values of random variables. The resulting theory leads to the following limits of prediction for y when $x = x_0$ as

$$(a + bx_0) \pm t_{\alpha'/2} S_e \sqrt{1 + \frac{1}{2} + n\frac{(x_0 - \overline{x})^2}{S_{xx}}} \quad [7.34]$$

where the number of degrees of freedom for $t_{\alpha'/2}$ is again $n - 2$.

Note that although the mean of the distribution of all y can be estimated fairly closely, the value of a single future observation cannot be predicted with good precision. Even as n approaches infinity, the width of the interval doesn't go to zero; the limiting width depends on S_e, which expresses the inherent variability of the data. Example applications of regression analysis are numerous and include Snedecor and Cochran (1980), Bowker and Lieberman (1972), and McBean and Rovers (1984).

EXAMPLE 7.8

An industrial outfall is discharging to a stream. Groundwater recharge appears to be discharging to the stream at a relatively constant rate with distance, causing dilution of the concentrations of the instream constituents. Are these findings statistically significant, indicating that concentrations are decreasing with increasing distance along the stream?

Distance downstream from outfall (m)	Concentration (mg/L)	
	Calcium	Iron
0	86	50
80	79	50
160	94	59
240	85	51
360	90	60
480	84	29
620	87	31
760	87	29
900	74	29
1060	75	31

The following regression equations expressing concentration as a function of distance were determined from the data.

$$\text{Calcium } y = 88.753 - 0.10x \qquad r^2 = 0.30$$

where y is the concentration and x is distance in meters. From Eq. [7.4],

$$t^* = 1.85$$

From Table A.4, for 8 degrees of freedom and a two-sided test, this result is significant at a ten percent level of significance.

$$\text{Iron } y = 55.33 - 0.029x \qquad r^2 = 0.62$$

From Equation [7.4]

$$t^* = 3.61$$

From Table A.4 for eight degrees of freedom and a two-sided test, this result is significant at a one percent level of significance.

The results of the linear regression indicate a statistically significant decrease in concentration with distance

---

EXAMPLE 7.9

The biochemical oxygen demand (BOD) test measures the oxygen required by aerobic microorganisms in the stabilization of the decomposable organic matter. Since the BOD test is dependent on microbial action, it is subject to wide biological variability caused by many factors. As a result of these factors, the reproducibility of BOD results is often only of the order of ±10 to 20 percent (Ballinger and Lishka, 1962; Constable and McBean, 1977). In addition to the biological interferences that can significantly alter the magnitude of the measured BOD, the time involved between sample collection and final analysis is often too long to provide meaningful information for use in process control and effluent monitoring decisions at wastewater treatment plants.

The development of carbon analyzers provided an alternative technique for characterizing water quality. The analyzers utilize the concept of complete combustion of all organic matter to carbon dioxide and water. Measurements are thus made for total carbon and inorganic carbon and the difference between these is the total organic carbon (TOC) concentration. It is important to note that the TOC tests do not measure the same constituent. The BOD test measures organic pollution concentrations indirectly in terms of an equivalent oxygen demand, that is, the amount of oxygen consumed by biological oxidation of the organic matter. However, not all organic materials are biologically degradable. Conversely, the TOC evaluation measures the total concentration of organic carbon present.

While the reproducibility of BOD results is of the order of ±10 to 20 percent, the precision of the TOC analyzer is often ±2 to 6 percent, depending on the amount of carbon and the size of the particulate matter present in the sample. Second, while the use of replicate analyses in the BOD test is expensive and time consuming, the ease and speed with which TOC analyses are conducted makes TOC replicate sampling an inexpensive and rapid procedure. The result is that correlations between measured BOD and TOC values can be established for particular situations, such as an individual wastewater.

For these reasons, a number of investigators have developed regression equations for different types of wastes, including:

Regression equation	Nature of wastewater	Reference
$BOD_5 = 11.6 + 1.875$ TOC	Domestic wastewater	Schaefer et al. (1965)
$BOD_5 = 86.15 + 0.84$ TOC	Domestic wastewater	Chandler et al. (1976)
$BOD_5 = 0.247 + 0.0708$ TOC	Reservoir impoundment	Emery et al. (1971)
$BOD_5 = -55.43 + 1.507$ TOC[a]	Raw domestic wastewater	Constable and McBean (1979)
$BOD_5 = 2.54 + 1.336$ TOC	Effluent following primary treatment	Constable and McBean (1979)

[a]Depicted in Fig. 7.3 are the individual plotted results and regression results.

The implications of a small sample size can be evaluated by calculating the power of the *t*-test for various sample sizes and trend magnitudes. The power $(1 - \beta)$ of the test is the probability of detecting a trend, given that one exists. The probability of not detecting a trend given that one exists is β, the type II error (Type I error is the probability of concluding that a trend is present when one does not exist.). These aspects of type I and II errors are described more fully in Chap. 8.

7.3.2 Significance of the Correlation Coefficient

The correlation coefficient r is based on the logic that one would expect a line of good fit to have a large percentage of the sum of the squares about the mean explained by the sum of squares due to regression; that is,

$$r^2 = \frac{\sum(y_i' - \overline{y})^2}{\sum(y_i - \overline{y})^2} = \frac{\sum(y_i - \overline{y})^2 - \sum\epsilon_i^2}{\sum(y_i - \overline{y})^2} = 1 - \frac{\sum\epsilon_i^2}{(y_i - \overline{y})^2} \qquad [7.35]$$

where y_i' is the value estimated from the regression equation, y_i is the measured value, and $\overline{y}$ is the mean of the measured values.

In the case of an exact fit, r^2 will equal unity and r^2 will decrease in magnitude as the quality of the fit of the model to the data diminishes. If r^2 approaches zero, the best estimate of the dependent variable, given the magnitude of the independent variable, is not better than the overall mean of the dependent variable.

The value of r may be inflated, since much of the variance may be associated with a single data point. For example, the variance explained by a regression line for turbidity versus suspended solids, as depicted in Fig. 7.9, is largely attributable to the single high value for each constituent. The absence of that data point dramatically changes the extent of the correlation. As a consequence, a large value for r does not necessarily guarantee that two variables are related.

Thus, it is possible to have a poor model with a high r^2 and, alternatively, to have a low r^2 but still have an acceptable model. The magnitude of r^2 depends somewhat on the steepness in slope of the relationship being studied, thus causing the r^2 quantity to be greater in the case of steeper line, given the same residuals. As a result, the correlation coefficient should only be used as a general indicator of goodness of fit.

The standard error of estimate of y, S_e, is defined as:

$$S_e = S_y\sqrt{1 - r^2} \qquad [7.36]$$

S_e is a measure of the variability in the y values that remains unexplained by the fact that there is a true regression between y and x, as per Eq. [7.18]. In other words, if the correlation coefficient is not significant, then whatever value x has, the best estimate of y would be $\bar{y}$ with a standard error of estimate of S_e. In approximately 95 percent of the cases, the actual values will lie within ±1.96 standard errors of the estimated values given by the regression equation.

EXAMPLE 7.10

In projecting future groundwater quality in the vicinity of a landfill, an equation was fit for data collected over the period 1968 to 1980, as

$$\text{Sulfates} = 18.1 + .80t$$

where t represents time in years, measured from 1968. The equation had an $r = 0.77$ and $S_y =$ 11 mg/L. Thus, a projection to 1995 implies a best estimate of sulfates to be

$$\begin{aligned} S_e &= 11\sqrt{1 - .77^2} \\ &= 7.02 \text{ and for 95\% tolerance interval } 7.02 \times 1.96 = 13.8 \end{aligned}$$

Consequently, sulfate concentrations in the year 2000 would be

$$\text{Sulfates (2000)} = 18.1 + 0.80(2000 - 1968) \pm 13.8 = 43.7 \pm 13.8$$

In words, when the correlation coefficient r is very high, the 95 percent tolerance interval is narrow (in 95 percent of the cases, the actual values lie between ±2 standard errors of estimate given by the regression equation).

7.3.3 Regression Analysis

To check the validity or reliability of an equation's goodness of fit there are alternative goodness-of-fit tests. These alternative tests include:

1. r^2. This is a good preliminary indicator of the degree to which the regression relationship explains the variation.
2. *Scattergrams.*
3. *A study of residuals.* This involves the investigation of the ith dependent observation (y_i) and the predicted variable (y_i'). Large differences ($y_i - y_i'$) indicate that the dependent observation does not follow the trend of the model and deviates from the predicted line.
4. *Cook's d residuals.* This residual measures the influence of the ith observation on the regression beta variables. Large Cook's d values indicate that the observation for that particular event significantly affects the beta coefficients of the model.
5. *Leverage residuals.* The leverage is a measure of the distance between the ith observation of the independent (regressor) variables and the mean of the regressor variables. A large leverage value indicates that the independent variable is significantly larger, or smaller, than the other independent observations. A large leverage can substantially increase or decrease the r^2.

7.3.4 Independence versus Dependence for Statistical Analyses of Sampling Results

When samples are dependent, prediction capability of the model will be a function of the correlation structure of the time series. Dependent samples (i.e., samples with concentration measurements that are correlated) will exhibit less variability than really exists. Most statistical tests depend on having a good estimate of the true variability in order to make accurate decisions between competing hypotheses.

1. *Replicate or duplicate samples.* Data field replicates or laboratory split samples (samples divided into two or more samples to allow testing of laboratory precision) are not statistically independent measurements. Consequently, the samples should not be treated as independent results in statistical procedures.

2. *Step changes with dependency in the data.* Lettenmaier (1976) and Lettenmaier and Burges (1977) show that the classical t-test against step or abrupt changes (which arise, for example, as the result of upgrading in wastewater treatment processes) and linear trends may be modified for dependent time series if the equivalent or effective independent sample size $n_b{}^*$ defined by Bayley and Hammersley (1946) is used in place of the actual sample size n.
3. *Autocorrelation for time series.* Consider a time series such as a time sequence of river flows. For observations measured at daily intervals, there is a likelihood of correlation between them (if the flow is high one day for many rivers, we would expect the flows to be high the next day). The effect of this correlation is called autocorrelation (or self-correlation). This autocorrelation increases the estimate of variance of the mean when the autocorrelation is positive (the usual case in environmental data).

Calculation of autocorrelation involves the analysis of data collected over successive days for each pollutant and then estimating the lag-one autocorrelation (correlation between measurements for two consecutive days), as

$$r_{x_i x_{i-1}} = \frac{\sum \text{Covariance } (x_i x_{i-1})}{S_{x_i} S_{x_{i-1}}} \qquad [7.37]$$

for $i = 2, \ldots, n$.

The variance of the mean of the time series should be adjusted by a factor involving the autocorrelation estimate and sample size. If x is normal with parameters $\overline{x}$, and S_x and lag-one autocorrelation r_1, the variance of the sample mean $\overline{x}$ is given by

$$\text{Var } \overline{x} = \frac{S_x{}^2}{n}\left[1 + \frac{2}{n}\sum_{k=1}^{n-1}(n - k)r^k\right] \qquad [7.38]$$

If x is lognormally distributed, that is $y = \ln x$ is normal, with parameters $\overline{x}$, and S_x and lag-one autocorrelation r, then variance of the same mean is given by

$$\text{Var } \overline{x} = \frac{1}{n}\text{Var}(x)f_n(p) \qquad [7.39]$$

where

$$\text{Var}(x) = \exp(2\overline{x} + S_x{}^2)[\exp(S_x{}^2) - 1] \qquad [7.40]$$

and

$$f_n(p) = 1 + \frac{2}{n}\sum_{k=1}^{n-1}(n - k)\frac{(\exp(r^k S^2) - 1)}{\exp(S^2) - 1} \qquad [7.41]$$

7.4 INTERPARAMETER CORRELATION

A desirable monitoring program might involve monitoring only a specific set of contaminants at a given location, with the selection focusing on those constituents that are easy and inexpensive to monitor and analyze. Of interest is whether other constituents could then be characterized by a smaller number of easily monitored constituents (as surrogate measures).

The procedure for assessing the potential for use of surrogate measures starts out by determining the interparameter correlation. For example, Thomson et al. (1994a and b) found high interparameter correlations between heavy metals in highway stormwater runoff, examples of which are listed in Table 7.2. A significant correlation does not necessarily imply a causal relationship between the two random variables but the constituents may, for example, have the same source and may thus also have related behavior. This common source is certainly relevant in the highway stormwater runoff situation.

7.5 MULTIPLE VARIABLES: MULTIPLE REGRESSION

Problems involving more than two variables can be treated in a manner analogous to problems with two variables. For example, there may be a relationship between the three variables y, x_1, and x_2 that can be described by the equation

$$y = \alpha + \beta x_1 + \gamma x_2 + \varepsilon_i \qquad [7.42]$$

The multiple correlation coefficient indicates the strength of the relationship between a dependent variable and two or more independent variables. The partial correlation coefficient indicates the strength of the relationship between a dependent variable and one or more independent variables with the effects of other independent variables held constant.

TABLE 7.2 Matrix of interparameter correlations between heavy metals in urban runoff

	Chromium	Copper	Iron	Lead	Zinc	Nickel
Iron	.73	.49				
Lead	.69	.51	.85			
Zinc	.74	.53	.89	.86		
Nickel	.71	.48	.73	.67	.82	
Cadmium	.62	.35	.60	.55	.65	.73

Source: Adapted from Thomson, N., McBean, E., Mostrenko, I., and Snodgrass, W., "Characterization of Stormwater Runoff from Highways," in *Current Practices in Modelling the Management of Stormwater Impacts*, ed. W. James (Boca Raton, FL: Lewis Publishers, 1994); and Thomson, N., McBean, E., and Mostrenko, I., "Prediction and Characterization of Highway Stormwater Runoff Quality," report to Ministry of Transportation, Research and Development Branch, Toronto, Ontario (August, 1994).

EXAMPLE 7.11

For the chemical reaction A + B forming products, the equation for a mole balance on a constant-volume batch reactor is

$$\frac{-dC_A}{dt} = rA = kC_A^{\alpha}C_B^{\beta}$$

where α and β are both unknowns. For the initial rates (i.e., at time $t = 0$), then

$$\left(\frac{-dC_A}{dt}\right)_0 = r_{A_0} = kC_{A_0}^{\alpha}\, C_{B_0}^{\beta}$$

Taking logarithms of both sides we have

$$\ln\left(\frac{-dC_A}{dt}\right)_0 = \ln(k) + \alpha \ln(C_{A_0}) + \beta \ln(C_{B_0})$$

Let $y = \ln(-dC_A/dt)_0$, $x_1 = \ln C_{A_0}$, $x_2 = \ln C_{B_0}$ and $a_0 = \ln k$ and $a_2 = \beta$. Then

$$y = a_0 + a_1x_1 + a_2x_2$$

If we now carry out n experimental runs, for the jth run, we get

$$y_j = a_0 + a_1x_{1j} + a_2x_{2j}$$

For n runs $j = 1, 2, \ldots n$

$$\sum y_j = na_0 + a_1\sum x_{1j} + a_2\sum x_{2j}$$

$$\sum x_{1j}y_j = a_0\sum x_{1j} + a_1\sum x_{1j}^2 + a_2\sum x_{1j}x_{2j}$$

$$\sum x_{2j}y_j = a_o\sum x_{2j} + a_1\sum x_{1j}x_{2j} + a_2\sum x_{2j}^2$$

We have three linear equations and three unknowns a_0, a_1, a_2 for which we must solve. If we set $a_2 = 0$ and consider only two variables, y and x, the equations reduce to the familiar least-squares equations.

EXAMPLE 7.12

Highway stormwater runoff contains numerous water quality constituents, some of which are generated by the use and existence of the roadway (e.g., vehicular traffic, materials used in highway construction, or maintenance practices) and some that are the result of the surrounding land use (e.g., fertilizers and pollutants associated with commercial zones) and transported to the highway drainage area by atmospheric deposition. Shaheen (1975) has shown that although they are highly correlated, less than five percent of many of the solids originate directly from the vehicles themselves. Therefore, the vehicular traffic acts as both a minor source and a transport mechanism for many solids.

The solids in highway stormwater runoff are carriers for other sorbing constituents. Thomson et al. (1994b and 1997) used multiple regression analyses to predict heavy metal concentrations in highway runoff using total suspended solids (TSS), total dissolved solids (TDS), total volatile solids (TVS) and total organic carbon (TOC) as independent variables. Thus TSS, TDS, TVS and TOC are effective surrogate parameters for numerous metals, ionic species, and nutrients. In other words, sampling for these four constituents can tell a great deal about the others. Examples of the results of the analyses produced the following models:

Cadmium	$= 0.233 + 0.00768\ (TSS) + 9.08 \times 10^{-5}\ (TDS)$	$[r^2 = 0.726]$
Zinc	$= 1.004\ (TSS) - 0.00216\ (TDS) + 2.066\ (TOC)$	$[r^2 = 0.910]$
Iron	$= 36.82\ (TSS) - 0.06377\ (TDS)$	$[r^2 = 0.846]$
Arsenic	$= 0.0054\ (TDS)$	$[r^2 = 0.808]$
Chloride	$= 0.575\ (TDS)$	$[r^2 = 0.996]$
Sulphate	$= 0.00819\ (TDS) - 0.0555\ (TSS) + 0.995\ (TOC)$	$[r^2 = 0.976]$
Total P	$= 0.002924\ (TVS) + 0.001064\ (TSS)$	$[r^2 = 0.777]$
Total N	$= 0.751 + 0.0664\ (TOC) + 0.000961\ (TVS)$	$[r^2 = 0.463]$
COD	$= 0.0479\ (TDS) - 0.344\ (TSS) + 4.213\ (TOC)$	$[r^2 = 0.816]$

7.6 REFERENCES

BALLINGER, D.G., and G. LISHKA. "Reliability and Precision of BOD and COD Determinations," *Journal of Water Pollution Control Federation*, 34, no. 5 (1962), 470–474.

BAYLEY, G.U., and J.M. HAMMERSLEY. "The Effective Number of Independent Observations in an Autocorrelated Time Series," *Journal of the Royal Statistical Society*, 8, 1-B (1946), 184–197.

BOWKER, H., and G.I. LIEBERMAN. *Engineering Statistics*, 2nd ed. Upper Saddle River, NJ: Prentice-Hall, 1972.

CHANDLER, R.L., J.C. O'SHAUGHNESSY, and F.C. BLANC. "Pollution Monitoring with Total Organic Carbon Analysis," *Journal of Water Pollution Control Federation*, 48 (1976), 2791–2803.

CONSTABLE, T., and E. McBEAN. "Parameter Estimation for the First-Order BOD Equation Using Nonlinear Techniques," *Canadian Journal of Civil Engineering*, 4 (1977), 462–470.

CONSTABLE, T., and E. McBEAN. "BOD/TOC Correlations and Their Application to Water Quality Evaluation," *Water, Air, and Soil Pollution*, 11 (1979), 363–375.

DRAPER, N., and H. SMITH. *Applied Regression Analysis*, 2nd ed. New York, NY: Wiley-Interscience, 1981.

EMERY, R.M., E.B. WELCH, and R.F. CHRISTMAN. "The Total Organic Carbon Analyzer and Its Application to Water Research," *Journal of Water Pollution Control Federation*, 43 (1971), 1834.

LETTENMAIER, D.P. "Detection of Trends in Water Quality from Records with Dependent Observations," *Water Resources Research*, 12, no. 5 (1976), 1037–1046.

LETTENMAIER, D.P., and S.J. BURGES. "Design of Trend Monitoring Networks," *ASCE-Journal of the Environmental Engineering Division*, 103, no. EE5 (October 1977), 785–802.

McBEAN, E., and S. AL-NASSRI. "Uncertainty in Suspended Sediment Transport Curves," *ASCE-Journal of Hydraulics Division*, 114, no. 1 (January 1987), 63–74.

McBEAN, E., and F. ROVERS. "Alternatives For Assessing Significance of Changes in Concentration Levels," *Groundwater Monitoring Review* (Summer 1984), 39–41.

SCHAEFFER, R.B., C.E. VAN HALL, G.N. McDERMOTT, D. BARTH, V.A. STENGER, S.J. SEBESTA, and S.H. GRIGGS. "Application of a Carbon Analyzer in Waste Treatment," *Journal of Water Pollution Control Federation*, 37 (1965), 1545.

SHAHEEN, D.G. "Contributions of Urban Roadway Usage to Water Pollution," *Environmental Protection Series*, report EPA-600/2-75-004. EPA, 1975.

SNEDECOR, G., and W. COCHRAN. *Statistical Methods*, 2nd ed. Ames, IA: The Iowa State University Press, 1980.

THOMSON, N., E. McBEAN, and I. MOSTRENKO. "Prediction and Characterization of Highway Stormwater Runoff Quality," report to Ministry of Transportation, Research and Development Branch, Toronto, Ontario, August 1994(a).

THOMSON, N., E. McBEAN, W. MOSTRENKO, and W. SNODGRASS. "Characterization of Stormwater Runoff from Highways," in *Current Practices in Modelling the Management of Stormwater Impacts*, ed. W. James. Boca Raton, FL: Lewis Publishers, 1994(b).

THOMSON, N., E. McBEAN, W. SNODGRASS, I. MOSTRENKO. "Highway Stormwater Runoff Quality: Development of Surrogate Parameter Relationships," *Water, Air and Soil Pollution*, 94 (1997), 307–347.

TYAGI, A., M. SHARMA, E. McBEAN. "Best Subset Modelling of Phosphorus in the Grand River Using Correlated Variables," *Canadian Journal of Civil Engineering*, 23 (1996), 893–903.

WEISBERG, S., *Applied Linear Regression*. New York, NY: John Wiley and Sons, 1980.

7.7 PROBLEMS

7.1. The following are measurements of chloride, arsenic, and nitrates over time. Calculate the correlation coefficients between the variables and determine whether the correlations are statistically significantly different from zero.

Sample number	Chloride (mg/L)	Arsenic (mg/L)	Nitrate (mg/L)
1	112	.07	15
2	87	.06	13
3	93	.06	14
4	115	.09	16
5	126	.17	23
6	105	.15	20
7	72	.10	17
8	61	.05	12

Sample number	Chloride (mg/L)	Arsenic (mg/L)	Nitrate (mg/L)
9	57	.07	11
10	75	.07	12
11	89	.09	14
12	105	.10	15

7.2. Calculate the linear regression equation for chloride concentration versus time for the data tabulated below. Use the resulting model to predict the concentrations in the year 1999, including the 95 percent confidence bounds.

Year	Chloride concentration (mg/L)
1981	12
1982	11
1985	14
1986	17
1988	16
1989	21
1990	22
1991	19
1993	24
1995	22
1997	27

7.3. The following data have been measured as "depth of flow" and "flow" for utilization in the development of a rating or measurement curve to predict flow from the depth for an instream weir. Using a logarithmic transformation, determine the least squares regression line relating the flow to the depth.

Flow (m^3/s)	Depth of flow (m)
.29	.05
1.38	.14
1.97	.18
3.895	.28
6.08	.38

7.4. The following data are monitoring results characterizing the quality of stormwater runoff from a highway:

Sample number	Total dissolved solids	Total suspended solids	Total organic carbon
1	710	87	23
2	1900	33	7.3
3	1500	190	48
4	730	100	15
5	450	160	21
6	520	180	43
7	360	130	34
8	210	230	26
9	610	30	16
10	510	53	13

a. Develop a simple interparameter correlation matrix.
b. Test the significance of each correlation to determine if they are significantly different from zero.

7.5. Two additional data points beyond those indicated in Prob. 7.4 were obtained for total dissolved solids and total suspended solids. Utilize single and multiple regression formulas to estimate the missing value for Total Organic Carbon.

Sample number	Total dissolved solids	Total suspended solids
11	470	51
12	230	92

7.6. Monitoring records from a river downgradient of a landfill that is leaking chemicals to the river are as indicated in the following:

Sample number	Arsenic	1,1,1-TCE
1	38	6
2	14	2
3	110	12
4	34	4.7
5	62	7.9
6	110	9.5
7	85	8.5
8	292	10
9	12	1.4
10	11	1.1
11	21	2.4
12	41	3.6

Utilize a scattergram to identify if it appears that a sampling or laboratory error was made for one of the analyses.

7.7. The monitoring record for two constituents is as indicated in the tabulated results. Note the monitoring values that are below the detection level: Use the coincident record to generate a regression equation to determine the best estimates for the censored data.

Sample number	TCE (μg/L)	DCE (μg/L)
1	15	17
2	18	14
3	15	26
4	11	<14
5	12	18
6	13	21
7	15	20
8	13	22
9	9	<14
10	13	22
11	14	28
12	15	20

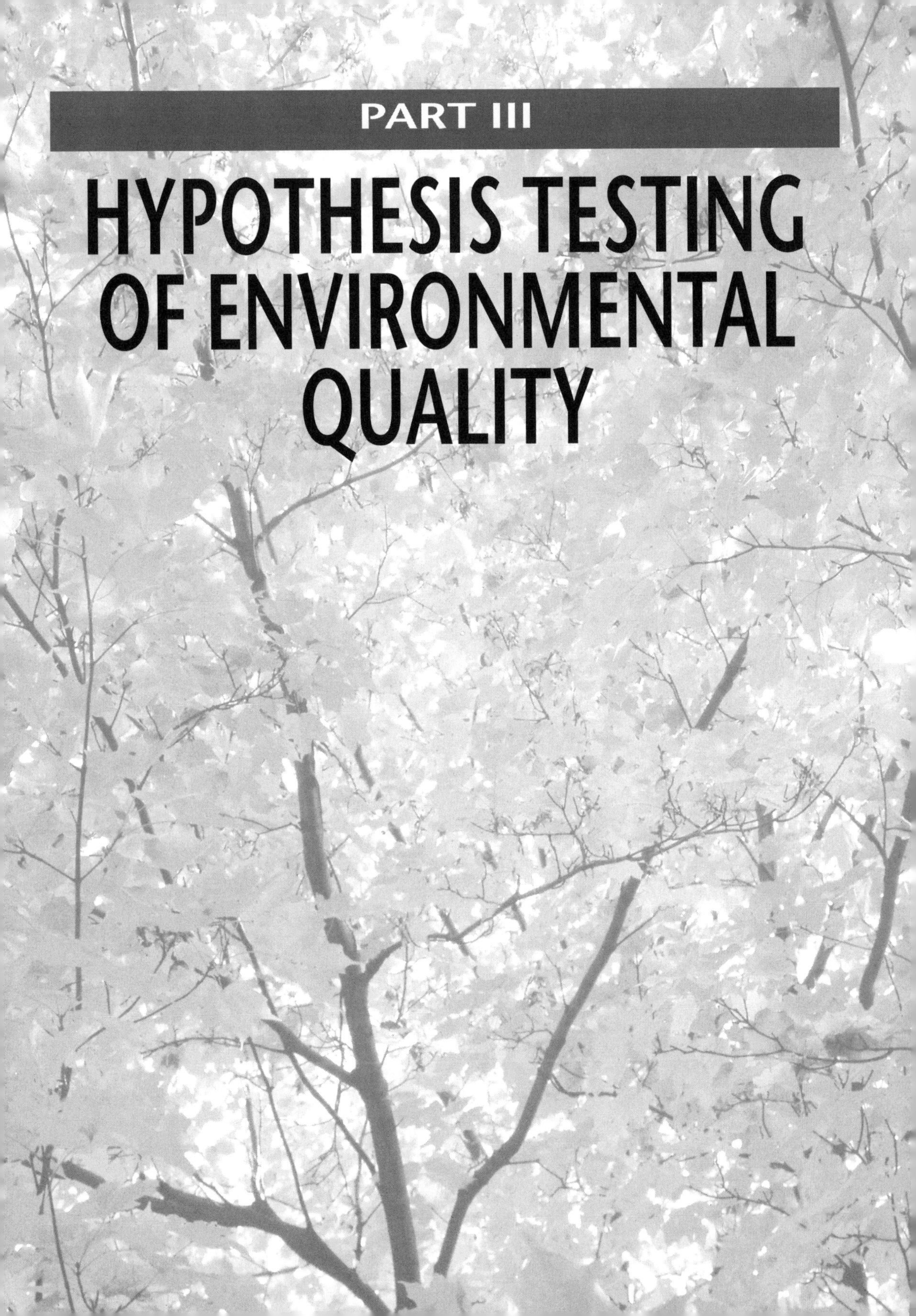

PART III

HYPOTHESIS TESTING OF ENVIRONMENTAL QUALITY

CHAPTER 8

TESTING DIFFERENCES BETWEEN MONITORING RECORDS: DIFFERENCES BETWEEN TWO LOCATIONS FOR SINGLE CONSTITUENTS

8.1 INTRODUCTION

Consider the following situation in which you flip a coin a total of ten times, and get seven heads and three tails. What are your chances of tossing the coin and getting a head on the next throw? If you consider the sample measurements at your disposal, you would estimate a 70 percent chance of getting a head. Alternatively, if you believe the coin to be fair and consider the total population of possible tosses, you would estimate a 50 percent chance of getting a head on the next throw.

The two alternative scenarios illustrate the basic difference between probability and statistical inference. Specifically,

1. with probability, which is based on deductive reasoning, you begin by knowing something about the population (e.g., that the coin is fair), and then reason from the population to the sample (e.g., how likely will the next toss of the coin be a head?);
2. on the other hand, statistical inference is based on inductive reasoning. In this situation, you begin knowing only about the sample results (seven heads out of ten tosses), and not the population. You then reason from the sample to the population (e.g., to determine whether the coin is fair, based on the sample results).

For environmental quality problems, we do not know the population, only the sample monitoring results we collect. We are forced to make statistical inferences in order to estimate the behavior of the population from the sample results. Even if the average behavior of the coin is to land 50 percent heads, specific sample results will vary from experiment to experiment. To estimate the coin's true behavior, we must account for between-sample variability. This is, then, precisely the problem we face in the environment. We have only sample measurements and we utilize statistical inference to gain an understanding of the population.

Carrying these ideas further, a relevant question in environmental quality data assessment is whether there has been a significant change in the concentration of an individual chemical constituent. Questions that may be of interest are included in Examples 8.1 and 8.2.

In the type of assessment addressed in this chapter, a series of alternative procedures are examined to decide whether a significant change in concentration for a particular constituent has occurred. Later chapters address additional issues associated with problems that entail monitoring results with numerous constituents and situations where the data sets are censored (e.g., reported as less than some concentration; information is available about the value, but not the value itself). As the concerns change, the statistical technique(s) appropriate for application may also change, as will become apparent in the later chapters; the focus in this chapter is on the simplest problems, with later chapters dealing with more complex issues.

EXAMPLE 8.1

Are there significant differences between sampling records collected at different locations? For example, the objective of a groundwater monitoring program at a manufacturing facility may be to determine whether a facility is impacting the groundwater. To determine whether contamination is occurring, monitoring wells are placed upgradient and downgradient of the facility, as depicted in Fig. 8.1(a). Samples are analyzed at regular intervals. The data illustrated in Fig. 8.1(b) indicate the time history of water quality data as reported at the two locations. Do the data indicate that the facility has detrimentally impacted the groundwater quality?

EXAMPLE 8.2

Are there significant changes in chemical concentrations collected over time at a particular location? Environmental quality data prior to and following the construction of a manufacturing facility in 1987 are available. Do the monitoring data indicate that the facility has detrimentally impacted the groundwater quality? By segmenting the data into two groups, one preconstruction and the other postconstruction, an assessment can be made to determine if a significant change in groundwater quality has occurred.

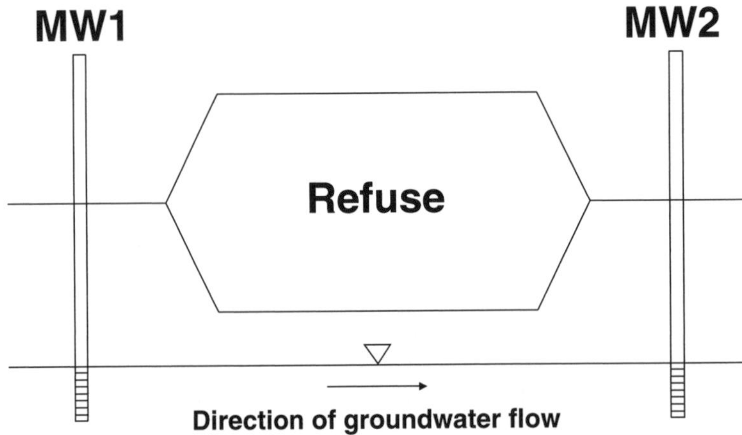

FIGURE 8.1(a) Schematic of upgradient and downgradient monitoring of groundwater

The mathematical analyses to determine whether there has been a change in concentration must reflect a number of considerations, including the following:

1. the available data are samples, not populations;
2. variabilities in concentrations may be due to influences such as seasonality; and
3. uncertainties exist in the sampling and laboratory analysis resulting in noise in the data set.

A statistical inference procedure must account for the types of influences listed above, influences that complicate the problem of detecting differences over time and space.

The most frequent approach to examine whether a difference exists in an environmental monitoring record is to determine if there is a difference in the mean concentration over space or time. In words, the analyst develops an hypothesis and then tests to determine whether the hypothesis is reasonable. An hypo-

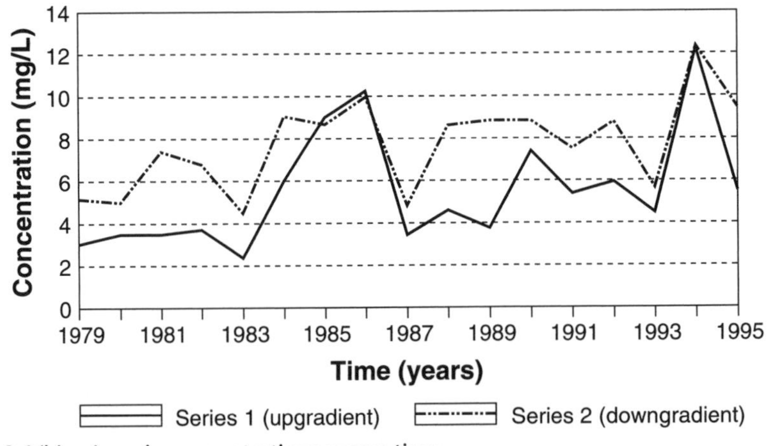

FIGURE 8.1(b) Arsenic concentrations versus time

thesis test may be, therefore, a way of mathematically examining whether a population of measurements of a random variable A (as represented by a sample from the population) overlaps another random variable B to a sufficiently small degree (or not at all), such that A and B must be considered statistically different or, alternatively, that A simply represents another sample beyond B from the same population. An important element of the hypothesis testing is that these procedures represent a class of statistical analysis techniques that are designed to extrapolate information from samples of data in order to make inferences about populations.

8.2 DETAILS OF HYPOTHESIS TESTING

Consider Example 8.3 as an example of an hypothesis.

Statistical tests thus begin with,

H_0, the null hypothesis, which is a statement such as *there is no difference*, or no impact of the landfill on the stream; and,

H_1 or H_a, the **alternative hypothesis**, that *there is a difference.*

If the objective is to compare two or more specific parameters such as the means of two populations, the hypothesis will be statements formulated to indicate the absence or presence of differences in the means.

If there is a significant difference between the mean concentrations, is it due to (a) the inherent randomness associated with estimating a mean from two sets or samples of data taken from the same population (i.e., the same population is sampled twice), or (b) does an actual difference exist between the two sample means because of contaminants discharged by the landfill to the stream?

EXAMPLE 8.3

A landfill located immediately adjacent to a river, as depicted in Fig. 8.2, is suspected of discharging arsenic to the river. The formulation of the hypothesis is, "There is no statistically significant difference between the instream arsenic concentrations as measured upstream and downstream of the site."

The testing of a statistical hypothesis is then the application of a methodology for deciding whether

(a) to accept a particular hypothesis (termed the *null hypothesis*) that is equivalent to concluding that there is **not sufficient evidence** to say there is a difference in the concentrations as measured at the two locations, or

(b) to reject the null hypothesis. This is equivalent to concluding that there **is sufficient evidence** to say there is a difference in the concentrations as measured at the two locations. Specific to this hypothesis is that the results indicate the landfill is increasing the arsenic concentrations in the river.

Accepting the null hypothesis means there is insufficient evidence to show the landfill is impacting the stream; rejecting the hypothesis means that the landfill is impacting the arsenic levels in the stream.

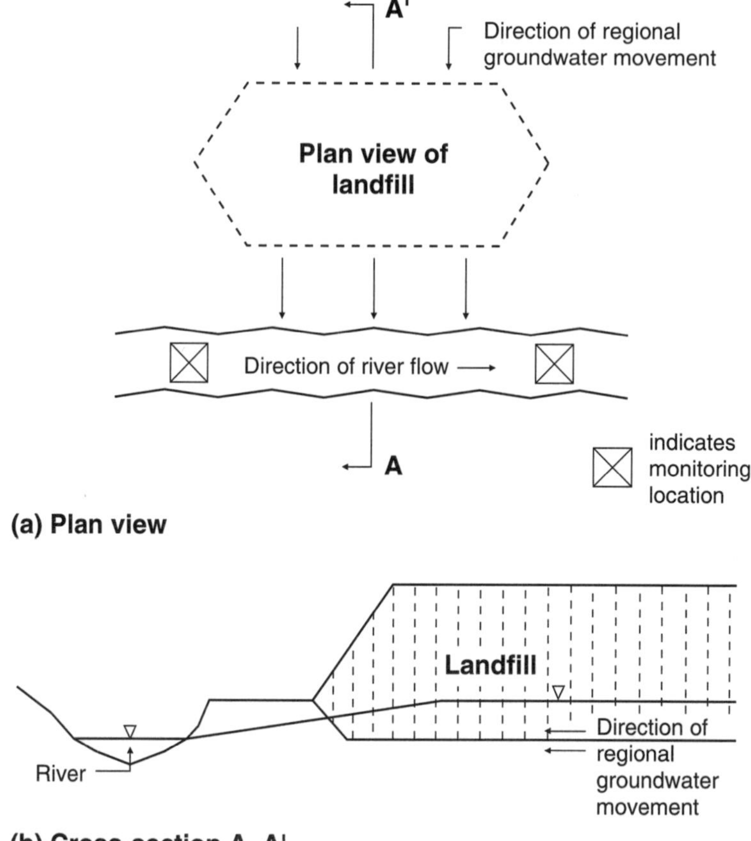

FIGURE 8.2 Plan view schematic of a situation where a landfill may be contaminating a stream

Assume we are interested in determining if a change in the mean of the instream concentration is a consequence of the landfill. This is accomplished by conducting an hypothesis test that compares the differences between the means of the random variables at the upstream and downstream locations to the standard error of the mean (the standard deviation divided by the square root of the number of samples). We accomplish this by calculating a ratio of difference between the means divided by the standard error of the mean; if the ratio is smaller than a tabulated critical value then the null hypothesis cannot be rejected, and the conclusion is that there is no statistically significant difference. Alternatively, if the calculated ratio is larger than the tabulated value, then the null hypothesis is rejected and the conclusion is that there is a statistically significant difference.

Nevertheless, none of these determinations can be stated with 100 percent confidence. Remember, we only have a sample from the population, so our conclusion is only a probability statement. When a result is very unlikely to have arisen by chance, the result is said to be statistically significant; that is, it is difficult to ascribe the findings to chance, and the difference must in all common sense be accepted as a real difference. The observation was probably not just luck, although there is still the possibility that the result happened by chance. Since the judgment is based on probability, it falls short of absolute certainty; but the degree of certainty is α and we have control over the magnitude of α. The investigator decides what probability represents the limit beyond which he/she is unwilling to believe that a random event has occurred (typically these values are chosen as 0.05 and 0.01, or equivalently, five percent and one percent). Any event with a calculated probability of occurrence less than that limit is called statistically significant (although it is not an impossible event, it is considered sufficiently unlikely that the hypothesis must be doubted, so a real difference is presumed). For example, the difference between the mean of a water quality constituent is presumed to be a real difference. Thus, α is referred to as the *level of significance.*

If a significance test results in the acceptance of the null hypothesis, it does not follow that we have grounds for supposing this hypothesis to be true, but merely that we have no grounds for supposing it to be false.

8.3 STEPS FOR SIGNIFICANCE TESTING

Recall that the t-distribution was presented in Chap. 3, Sec. 3.5. It was demonstrated there that the normal distribution is a subset of the t-distribution. The t-distribution is the basis for the significant-difference testing examined in this chapter.

The procedure to carry out significant-difference testing consists of the following steps:

Step	Example of the Step
1. Formulate a suitable (in terms of the problem context) null hypothesis, H_0	Assume there is no difference between the upgradient and downgradient groundwater monitoring results;
2. Calculate an appropriate test statistic, t^*	t-value: $t^* = \dfrac{\text{Difference between the means}}{\text{Standard error of the mean}}$
3. Define an acceptance/rejection region; and	Define a critical statistic, t_c
4. Draw a conclusion based on whether the calculated test statistic is in the acceptance or rejection region.	Reject the null hypothesis if $t^* > t_c$. In words, if $t^* > t_c$ there is evidence to indicate the hypothesis is incorrect or that there is a significant difference between the upgradient and downgradient monitoring results.

Note that if the result proves nonsignificant ($t^* < t_c$), it is equivalent to a verdict of "not proven"; there is still the opportunity to consider further evidence.

In response to the significance testing problem, a number of alternative procedures have been proposed. The selection of the best procedure involves careful scrutiny of the characteristics of the problem at hand and of the assumptions implicit in the particular discrimination test being considered. This involves not just the choice of significance level but also the choice of a statistic test and the requirements of the number of samples. The following sections develop each of these features in ways specific to each of the various tests.

8.4 STUDENT'S *t*-TEST

8.4.1 DEVELOPMENT OF THE EQUATIONS

(I) COMPARING ONE SAMPLE WITH THE POPULATION MEAN

The original derivation of the Student *t*-test was presented by W.S. Gosset in 1908, who published his results in the scientific literature under the pseudonym of "Student." Gosset did this to prevent his findings from being used by his competitors in the brewery industry. His formulation revolutionized the statistics of small samples.

The Student *t*-test (henceforth referred to as the *t*-test) expresses

$$t^* = \frac{|\bar{x} - \mu|}{S/\sqrt{n}} \qquad [8.1]$$

where t^* is the deviation of the estimated mean from the population mean, measured in terms of the unit $S/\sqrt{n}$. The quantity $S/\sqrt{n}$ is the sample estimate of $\sigma/\sqrt{n}$. Note that the current analysis regards a sample value as a deviation from the population mean. Subsequent analyses will compare one sample relative to another, which is of greater interest in the environmental field. However, the derivations in this section are restricted to deviation of sample mean from the population mean.

A low value of t^* from Eq. [8.1] indicates little difference between the means, and a high value of t^* indicates a large difference; the latter provides more justification for rejecting the null hypothesis.

Associated with the determination of the critical statistic t_c, against which t^* is compared, is the number of *degrees of freedom*. Then,

if $t^* > t_c$ we reject the null hypothesis and conclude there is a difference;

if $t^* < t_c$ we cannot reject the null hypothesis. The evidence suggests the sample has not deviated from the population.

Degrees of freedom is a value representing the size of the sample involved in the test. When t_c is calculated from a single random variable of size n, the degrees of freedom are $n - 1$. Note also that the magnitude of t_c is a function of the level of significance α. A table of the values of the critical statistic t_c is in Table 8.1 for the five percent level of significance. The value for t_c is selected from the table associated with the level of significance and the sample size (as quantified by the number of degrees of freedom). When the number of degrees of freedom are infinite, $t_c = 1.645$ when $\alpha = 5$ percent for a one-sided test of significance. That is, the five percent is located in the one tail of the distribution, as noted in Fig. 8.3(a); an alternative consideration may involve a two-sided test of significance, as Figure 8.3(b) shows. Further discussion of the differences between one and two-sided tests will be presented in Subsec. (ii) to follow. With 30 degrees of freedom, t_c has increased to 1.697; as the size decreases, the differences between the means of the sample and the population must be larger before the hypothesis is rejected. See Ex. 8.4.

TABLE 8.1 Values of critical statistic t_c for five percent level of significance

Number of degrees of freedom	Value of t_c One-sided test	Two-sided test
1	6.314	12.706
2	2.920	4.303
3	2.353	3.182
4	2.132	2.776
5	2.015	2.571
6	1.943	2.447
7	1.895	2.365
8	1.860	2.306
9	1.833	2.262
10	1.812	2.228
11	1.796	2.201
12	1.782	2.179
13	1.771	2.160
14	1.761	2.145
15	1.753	2.131
16	1.746	2.120
17	1.740	2.110
18	1.734	2.101
19	1.729	2.093
20	1.725	2.086
21	1.721	2.080
22	1.717	2.074
23	1.714	2.069
24	1.711	2.064
25	1.708	2.060
30	1.697	2.042
∞	1.645	1.960

Note: The values noted in the table correspond to those associated with α of five percent in a one-sided test. A more comprehensive table is included in Appendix A as Table A.2.

Source: Beyer, W.H. ed., *CRC Handbook of Tables for Probability and Statistics* (Cleveland, OH: Chemical Rubber Company, 1966). Reprinted with permission.

Note that Gosset's original derivation tested only whether a sample differed from its population mean μ or, alternatively, allowed the specification of confidence limits on the estimate of the mean.

Now consider an alternative aspect of the t-distribution. The t-test allows calculations of the confidence levels for the mean, knowing the estimate S of the standard deviation but not the population standard deviation σ. If σ is known and we assume a Gaussian distribution for the data, 95 percent confidence limits for μ are given by the relation:

$$\bar{x} - \frac{1.96\,\sigma}{\sqrt{n}} \leq \mu \leq \bar{x} + \frac{1.96\,\sigma}{\sqrt{n}} \qquad [8.2]$$

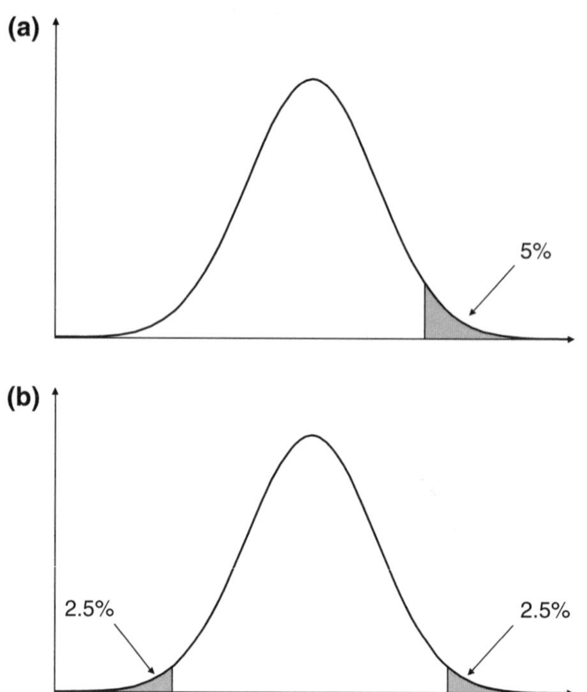

FIGURE 8.3 (a) A one-sided test of significance (b) A two-sided test of significance

EXAMPLE 8.4

A lengthy history of monitoring records from a wastewater treatment facility has established that the long-term population mean for BOD_5 is 12 mg/L. Current operational characteristics over the last four weeks of sampling have resulted in the following BOD_5 concentrations in the effluent stream: 14.6, 12.8, 13.7, and 15.4. Do the current operating characteristics suggest an operational problem in the wastewater treatment facility?

The mean of the sample concentrations is $\overline{x} = 14.1$ mg/L and standard deviation $S = 1.12$ mg/L. The t-statistic from Eq. 8.1 is then calculated as

$$t^* = \frac{|\overline{x} - \mu|}{S/\sqrt{n}}$$

$$t^* = \frac{14.1 - 12}{1.12/\sqrt{4}} = 3.75$$

The four samples indicate $4 - 1 = 3$ degrees of freedom and, for a five percent level of significance (from Table 8.1), $t_c = 2.353$ and $\alpha = 5$ percent. The result is that $t^* > t_c$. In words, this means that in all likelihood the results indicate a statistically significant difference in the mean operating characteristics of the treatment facility.

When σ is replaced by S, the only change needed is to replace the number 1.96 in Eq. [8.2] by a quantity represented as $t_{\alpha/2}$ (where $\alpha/2$ is used to indicate a two-sided test, above and below the mean as further described below). Thus, Eq. [8.2] becomes,

$$\bar{x} - \frac{t_{\alpha/2}\, S}{\sqrt{n}} \leq \mu \leq \bar{x} + \frac{t_{\alpha/2}\, S}{\sqrt{n}} \qquad [8.3]$$

The t-distribution is symmetrical about the mean (as is the normal distribution).

In other words, the t-distribution is used to establish confidence intervals and to test hypotheses when the population variance is not known and the sample size is small (as is apparent from Tables 8.1 and A.2, the t-distribution approaches the normal distribution for 30 or more samples, so less than 30 samples may be regarded as small).

When the degrees of freedom are infinite, $t_{0.05}$ for a two-sided test is 1.96. The two-sided test for $\alpha = 5$ percent means there is 2.5 percent chance of exceeding the upper bound and 2.5 percent chance of being lower than the lower bound. The t-statistic is usually written as $t_{\alpha/2}$ for a two-sided test indicating $\alpha/2$ for each side. Alternatively, for a one-sided test, when the degrees of freedom are infinite $t_{0.05} = 1.65$. As the degrees of freedom becomes large, the t-distribution becomes the normal or Gaussian distribution.

EXAMPLE 8.5

Given chloride measurements from a river—24.1, 27.9, 36.2, 26.3, 22.1, 29.2—estimate the 95 percent and 99 percent confidence limits for the mean chloride concentration. From the data provided, $\bar{x} = 27.6$ and $S = 4.92$ mg/L. For $n = 6$ measurements, the degrees of freedom are $6 - 1 = 5$ and $t_{\alpha/2} = 2.571$ for $\alpha = 5$ percent and $t_{\alpha/2} = 4.032$ for $\alpha = 1$ percent (from Table A.2). The resulting confidence bounds are

- for the 95 percent confidence limits:

$$\mu = 27.6 \pm \frac{2.571(4.92)}{\sqrt{6}}$$

$$= 27.6 \pm 5.16 \text{ or between } 22.4 \text{ and } 32.8$$

- for the 99 percent confidence limits:

$$\mu = 27.6 \pm \frac{4.032(4.92)}{\sqrt{6}}$$

$$= 27.6 \pm 8.10 \text{ or between } 19.5 \text{ and } 35.7$$

When the number of degrees of freedom are infinite, $t_{0.05/2} = 1.96$ (i.e., the t-distribution becomes the normal distribution). With 40 degrees of freedom, $t_{0.05/2}$

has increased to 2.021 (i.e., the confidence intervals on μ are larger using the t-test, because the sample size is only 41); with 20 degrees of freedom it has become 2.086, and it continues to increase steadily as the number of degrees of freedom decrease. The t_c value increases with a smaller data set and the confidence interval on the mean becomes larger, reflecting the greater uncertainty in the parameter estimates.

Equation [8.3] when manipulated becomes

$$t^* = \frac{|\bar{x} - \mu|}{S/\sqrt{n}} \qquad [8.4]$$

where the absolute value sign | | is used to incorporate the +/− signs. Thus, t^* is the deviation of the estimated mean from the population mean, measured in units of $S/\sqrt{n}$—the standard error of the estimate of the mean—and Eq. [8.1] has been obtained.

EXAMPLE 8.6

Given the chloride measurements for a river as 24.1, 27.9, 36.2, 26.3, 29.2, estimate the 95 percent confidence limits for the mean chloride concentration. From the data provided, $\bar{x} = 28.7$, $S = 4.58$, and $n = 5$. For $n = 5$, the degrees of freedom are $5 - 1 = 4$, and $t_{\alpha/2} = 2.776$ for $\alpha = 5$ percent. The resulting confidence bounds are

$$\mu = 28.7 \pm \frac{2.776(4.58)}{\sqrt{5}}$$

$$= 28.7 \pm 5.7 \text{ mg/L}$$

Now, when additional data are collected (27.3, 29.1, 28.7, 32.3) estimate the new confidence bounds on the mean concentration. From the entire data set, $\bar{x} = 29.0$, $S = 3.50$, and $n = 9$. The value of $t_{\alpha/2} = 2.262$ for $\alpha = 5$ percent. The resulting confidence bounds are

$$\mu = 29.0 \pm \frac{2.262(3.50)}{\sqrt{9}}$$

$$= 29.0 \pm 2.6$$

(II) ONE-SIDED VERSUS TWO-SIDED TESTS

In one-sided (one-tailed) tests of a mean or the difference between two means, the investigator pays attention only to deviations from the null hypothesis in one direction, ignoring the deviations in other directions. This happens primarily in situations where the investigator knows enough about the circumstances of the test to be certain that if μ_2 is not equal to μ_1, then μ_2 is greater than μ_1. For example, when monitoring trichloroethylene upgradient and downgradient of a landfill, the only concern is whether the concentration of trichloroethylene is higher down-

gradient than upgradient. For this circumstance, the one-tail test is appropriate. Conversely, if one is monitoring pH levels, the downgradient location might have higher or lower pH levels as a result of environmental contamination arising from the landfill. For this situation, a two-tailed test is appropriate.

(III) COMPARING TWO SAMPLES FOR SIGNIFICANCE OF DIFFERENCE

The development of the t-test revolutionized the statistics of small samples. However, it was Fisher (1925a,b) who realized that an extension of the test was possible to determine significant difference between two sets of data.

EXAMPLE 8.7

Assume that a random sample of size n_1 is collected from a population having unknown mean μ_1, the groundwater quality at location MW1 in Fig. 8.1(a). Further, a second sample of size n_2 is taken from a population with an unknown mean μ_2 at location MW2. The two sample means are $\bar{x}_1$ and $\bar{x}_2$. Does $\bar{x}_1$ differ significantly from $\bar{x}_2$? In other words, the problem is to test the hypothesis $H_0 : \mu_1 = \mu_2$. The alternative hypothesis (H_A:) may be $\mu_1 \neq \mu_2$, $\mu_1 > \mu_2$ or $\mu_1 < \mu_2$. A two-tailed test is appropriate in the first case; a one-tailed test, in the second and third.

If the variances are (approximately) equal, the t-statistic, t^*, is written as

$$t = \frac{|\bar{x}_2 - \bar{x}_1|}{SE_{\text{Diff}}}$$

where SE_{Diff} is the standard error of the difference.

One now compares the calculated t^* with tabulated one-sided $t_c = t_{df,\alpha}$ where the degrees of freedom df is defined as $df = n_1 + n_2 - 2$ and the level of significance is α. This is a more complex determination of the number of degrees of freedom than in the analysis in Subsec. (i), since now two parameters are being estimated instead of one. See Ex. 8.8.

(IV) ASSUMPTIONS IMPLICIT IN THE t-TEST

The t-test has built-in assumptions, some of which have not yet been acknowledged lest they unnecessarily complicate the issues. If the following assumptions are violated, then the t-test cannot be utilized (at least not directly):

a. independence between observations;
b. equal variances in both groups; and
c. normally distributed population(s).

The question of independence between samples is a difficult and subtle one when dealing with environmental quality measurements. Independence is probably seldom achieved. For example, as reported in McBean and Rovers (1985), a single groundwater sample split into subsamples or aliquots for individual analysis showed a considerably smaller variance than analyses of successive groundwater samples. Clearly, aliquots are not independent observations.

EXAMPLE 8.8

The monitoring record for BOD_5 (all values in mg/L) in the outfall for a wastewater treatment facility lists the following results:

October 1994	12, 11, 14, 7
November 1994	8, 9, 6, 13

Are concentrations in October larger (statistically significant) than average concentrations for November at five percent level of significance?

The statistics for the records are:

	October	November
Mean $\overline{x}$	11	9
Standard Deviation S	2.94	2.94

$$t^* = \frac{11 - 9}{2.94/\sqrt{4}} = 1.36$$

Degrees of freedom $df = 4 + 4 - 2 = 6$

t-statistic is $t_c = 1.94$

$t^* < t_c$

The test indicates insufficient evidence for a statistically significant difference between the means for October and November.

Additional discussions of the assumptions implicit in the t-test are in Draper and Smith (1981), Milliken and Johnson (1984), Kerlinger and Pedhazur (1973), Koch and Link (1970), Krumbein and Graybill (1965), and Kufs (1992).

In summary, under the conditions for which the test holds, the t-test is a powerful tool. Nevertheless, a significant t-value is evidenced only in relation to a specified level of significance. It does not mean that every member of one population is larger than every member of a second population. In other words, some overlap of the distributions is likely, as indicated on Figure 8.4.

Common sense suggests that the larger the difference between $\overline{x}_1$ and $\overline{x}_2$, the less likely the null hypothesis H_0 (that the means are the same or not different).

Take the situation of two independent samples with means $\overline{x}_1$ and $\overline{x}_2$ (estimates of their respective means, μ_1 and μ_2). The t-test becomes

$$t^* = \frac{(\overline{x}_1 - \overline{x}_2) - (\mu_1 - \mu_2)}{S_{\overline{x}_1 - \overline{x}_2}} \qquad [8.5]$$

The denominator in [8.5] is a sample estimate of the standard error of the difference between the means, $(\overline{x}_1 - \overline{x}_2)$, as

$$\sigma^2_{\overline{x}_1 - \overline{x}_2} = \sigma^2_{\overline{x}_1} + \sigma^2_{\overline{x}_2} \qquad [8.6]$$

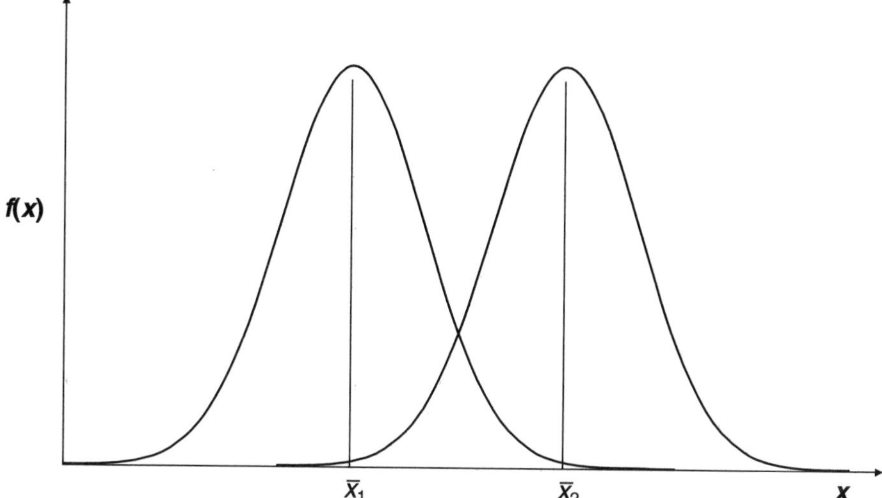

FIGURE 8.4 Schematic of probability distributions with some overlap

Note that the variance of a difference is determined from the sum of the variances. If the two means $\bar{x}_1$ and $\bar{x}_2$ are drawn from populations with variance σ^2, then Eq. [8.6] becomes

$$\sigma^2_{\bar{x}_1 - \bar{x}_2} = \frac{\sigma^2}{n} + \frac{\sigma^2}{n} = \frac{2\sigma^2}{n} \quad [8.7]$$

In most applications, the value of σ^2 is not known. However, each sample furnishes an estimate of σ^2, namely $S_1{}^2$ and $S_2{}^2$. With samples of the same size n, the best combined estimate is their pooled average

$$S^2_{\bar{x}_1 - \bar{x}_2} = \frac{S_1{}^2 + S_2{}^2}{2} \quad [8.8]$$

The statistical analysis for groups of unequal sizes follows almost exactly the same pattern as that for groups of equal sizes. We assume the variance is the same in both populations:

$$\begin{aligned} \sigma^2_{\bar{x}_1 - \bar{x}_2} &= \frac{\sigma^2}{n_1} + \frac{\sigma^2}{n_2} = \frac{2\sigma^2}{n} \\ &= \sigma^2\left(\frac{1}{n_1} + \frac{1}{n_2}\right) \quad [8.9] \\ &= \sigma^2\left(\frac{n_1 + n_2}{n_1 n_2}\right) \end{aligned}$$

The above equations assume the standard deviation is known. In most situations in environmental engineering, the standard deviation can only be estimated using data from the two sampling locations. Thus we estimate

$$S_1^2 = \frac{\sum_{i=1}^{n} (x_{1i} - \overline{x}_1)^2}{n_1 - 1} \qquad [8.10]$$

and

$$S_2^2 = \frac{\sum_{i=1}^{n} (x_{2i} - \overline{x}_2)^2}{n_2 - 1} \qquad [8.11]$$

The unbiased estimate of σ^2 is then

$$S^2_{\overline{x}_1 - \overline{x}_2} = \frac{(n_1 - 1)\, S_1^2 + (n_2 - 1)\, S_2^2}{n_1 + n_2 - 2} \qquad [8.12]$$

If the null hypothesis is true, it can be shown that the *t*-statistic follows a distribution with $(n_1 + n_2 - 2)$ degrees of freedom.

8.4.2 Effect of Unequal Variances

As evidenced in the derivation of the *t*-test, there is a built-in assumption that stipulates the variances should be the same (at least approximately). As noted earlier, a key to the valid use of the *t*-test is the assumption that the distributions of the two random variables x_1 and x_2 have the same variance. Therefore, before using the *t*-test to investigate the difference between the sample means, a prior test must be done. The intent of the prior test is to investigate whether the sample variances are sufficiently alike to warrant the assumption that they are independent estimates of the sample population variance.

Suppose we have a normal distribution with variance σ^2. Two random samples, sizes n_1 and n_2, are drawn from this population, and the two sample variances S_1^2 and S_2^2 are calculated in the usual way. As S_1^2 and S_2^2 are both estimates of the same quantity σ^2 we expect the ratio S_1^2/S_2^2 to be close to unity.

Let S_1^2 and S_2^2 be independent estimates of σ_1^2 and σ_2^2 and assume that the observations in the two samples are normally distributed. Then we are often interested in testing the hypothesis that S_1^2 and S_2^2 are both estimates of the same variance σ^2. In other words, we want to test the null hypothesis H_0 as

$$H_0 : \sigma_1^2 = \sigma_2^2 = \sigma^2$$

For alternative hypotheses of the form

$$H_1 : \sigma_1^2 > \sigma_2^2 \text{ or } H_1 : \sigma_1^2 < \sigma_2^2$$

a one-tailed test would be appropriate. If the ratio S_1^2/S_2^2 (or S_2^2/S_1^2) is much greater than unity, we would be inclined to reject H_0.

The significance of the difference between sample variances can be tested using the variance ratio test. The F-distribution is the sampling distribution of the ratio of two independent, unbiased estimates of the variance of a normal distribution; it has widespread application in the analysis of variance (to be considered in Chap. 9). The variance ratio is defined as:

$$F = \frac{\text{greater estimate of the variance of the population}}{\text{lesser estimate of the variance of the population}} \quad [8.13]$$

It follows that the larger the value of F, the less likely that the null hypothesis (that the two samples are drawn from the same population) is valid. However, with small sample sizes (n_1 and n_2 small), F may be different from unity. If the null hypothesis is true, large samples should give a value of F that differs little from unity. Tables 8.2 and 8.3 show the value of F that will be exceeded with a given degree of probability for various sample sizes of five percent and one percent, respectively. Note that the number of degrees of freedom for a sample of n items is equal to (n–1).

The results of the analyses can be of two forms:

1. pass the F-test; or
2. fail the F-test. A failure of the F-test means that in the present form the t-test cannot be used because the assumption of equal variance within the test is violated.

In situations without prior reason to anticipate inequality of variances, the alternative to the null hypothesis is a two-sided test ($\sigma_1 \neq \sigma_2$). On occasion, where one variance is known to be larger, a one-tailed t-test is warranted.

When n_1 and n_2 are large, an alternative to the F-test is to test the significance of a difference in standard deviations using the standard error of difference methodology and the fact that the variance for the distribution of sample standard deviation is $\sigma^2/(2n)$ (Note: it is the standard deviation and not the mean, hence the 2 in the denominator), as well as the fact that the variance of a difference is equal to the sum of the variances.

TABLE 8.2 5% level of variance ratio

		Number of degrees of freedom in the greater variance estimate							
		1	**2**	**3**	**4**	**5**	**10**	**20**	**∞**
	1	161	200	216	225	230	242	248	254
Number of	2	18.5	19.0	19.2	19.2	19.3	19.4	19.4	19.5
degrees of	3	10.1	9.6	9.3	9.1	9.0	8.8	8.7	8.5
freedom	4	7.7	6.9	6.6	6.4	6.3	6.0	5.8	5.6
in lesser	5	6.6	5.8	5.4	5.2	5.0	4.7	4.6	4.4
variance	10	5.0	4.1	3.7	3.5	3.3	3.0	2.8	2.5
estimate	20	4.3	3.5	3.1	2.9	2.7	2.3	2.1	1.8
	∞	3.8	3.0	2.6	2.4	2.2	1.8	1.6	1.0

Source: Fisher, R. A., and F. Yates, *Statistical Tables for Biological, Agricultural and Medical Research*, 6th ed. (Edinburgh: Oliver and Boyd, 1963), p. 53.

TABLE 8.3 1% level of variance ratio

		Number of degrees of freedom in the greater variance estimate							
		1	**2**	**3**	**4**	**5**	**10**	**20**	**∞**
	1	4100	5000	5400	5600	5800	6000	6200	6400
Number of	2	98	99	99	99	99	99	99	99
degrees of	3	34	31	29	29	28	27	27	26
freedom	4	21	18	17	16	16	15	14	13
in lesser	5	16	13	12	11	11	10	9.6	9.0
variance	10	10	7.6	6.6	6.0	5.6	4.8	4.4	3.9
estimate	20	8.1	5.8	4.9	4.4	4.1	3.4	2.9	2.4
	∞	6.6	4.6	3.8	3.3	3.0	2.3	1.9	1.0

Source: Fisher, R. A., and F. Yates, *Statistical Tables for Biological, Agricultural and Medical Research, 6th* Ed. (Edinburgh: Oliver and Boyd, 1963), p. 55.

EXAMPLE 8.9

Benzene concentrations were measured at two monitoring well locations (MW1 and MW2) as listed in the table below. Are the variances of the measurements at the two monitoring locations significantly different at the five percent level of significance?

	MW1	**MW2**
	1.6	1.8
	1.9	1.9
	2.3	2.5
	1.7	2.1
	2.4	2.2
	1.8	1.9
	1.4	1.3
	2.7	2.8
	2.1	2.3
	1.9	2.0
	2.3	2.4
	1.8	1.8
Standard Deviations	0.375	0.400

$$F\text{-ratio} = \frac{(.400)^2}{(.375)^2}\ \frac{.160}{.141} = 1.13$$

Degrees of freedom of larger variance estimate = 12 – 1 = 11
Degrees of freedom of smaller variance estimate = 12 – 1 = 11

From Table 8.2, then the *F*-test = 2.9

$$F\text{-ratio} < F\text{-test}$$

There is not a statistically significant difference between the variances, and the t-test can be utilized.

(*Note*: If we had found the F-ratio = ${S_1}^2/{S_2}^2 > F_{0.05,df_1,df_2}$ where ${S_1}^2$ and ${S_2}^2$ are based on degrees of freedom df_1 and df_2, respectively, then the result is significant at the five percent level and we have reasonable evidence that H_0 is untrue.

POOLED VARIANCE

Assume two variances from the two monitoring locations are not the same magnitude but assume that the F-test did not indicate rejection of the hypothesis indicating the two samples are not significantly different. Each sample furnishes an estimate, $\hat{S}^2$ of the population variance, σ^2. Assuming the variances have not proven significantly different, the pooled variance is estimated from

$$\hat{S}^2 = \frac{(n_1 - 1){S_1}^2 + (n_2 - 1){S_2}^2}{n_1 + n_2 - 2} \qquad [8.14]$$

where n_1 = the number of samples in the first data set and n_2 = the number of samples in the second data set.

The two variances are pooled to get the best estimate of a population variance that the t-test assumes is the same for both populations.

The standard error of difference of the two means is then

$$S_m = \sqrt{\hat{S}^2\left(\frac{1}{n_1} + \frac{1}{n_2}\right)} \qquad [8.15]$$

and the t-test is then calculated as

$$t = \frac{\overline{x}_1 - \overline{x}_2}{\sqrt{\frac{n_1 + n_2}{n_1 n_2}\left(\frac{(n_1 - 1){S_1}^2 + (n_2 - 1){S_2}^2}{n_1 + n_2 - 2}\right)}} \qquad [8.16]$$

with the number of degrees of freedom calculated as $df = n_1 + n_2 - 2$. See Ex. 8.10 and 8.11.

8.4.3 Effect of Nonnormality on the Hypothesis Test

The t-test is premised on the assumption that the data are normally distributed. The t-test is fairly robust to deviations from this assumption, but the appropriateness of the assumption should be assessed. A simple indication of normality (although only approximate) involves the utilization of the coefficient of variation test (COV from Chap. 3, Sec. 3.3). See Ex. 8.12 on p. 181.

EXAMPLE 8.10

Using the benzene concentration information in Ex. 8.9, determine whether the concentrations measured at MW2 are statistically significantly larger than those measured at MW1 at a five percent level of significance

	Mean (μg/L)	Standard deviation (μg/L)
MW1	1.99	.375
MW2	2.08	.400

The F-test in Ex. 8.9 has already demonstrated no statistically significant difference in the variance. Thus the pooled variance is

$$\hat{S}^2 = \frac{11(.375)^2 + 11(.400)^2}{22} = .150$$

The standard error of the difference is

$$S_m = \sqrt{.150\left(\frac{1}{12} + \frac{1}{12}\right)} = .158$$

The t-statistic

$$t^* = \frac{2.08 - 1.99}{.158} = .57$$

The degrees of freedom $df = 12 + 12 - 2 = 22$ and $\alpha = 5$ percent gives $t_c = 1.72$ and $t^* < t_c$, indicating there is no evidence of a difference in the benzene concentrations at MW1 and MW2.

EXAMPLE 8.11

Monitoring results for pH at two locations in the wastewater treatment plant are indicated below:

Monitoring point	pH values
MW1	7.2, 7.5, 7.1, 7.4, 7.7
MW2	6.7, 7.3, 6.7, 7.1

Is there a significant difference between the pH monitoring results at the two monitoring locations?

Since pH values can be significantly larger or smaller, a two-sided test for statistical significance is needed.

Monitoring point	Mean	Standard deviation	Sample size
MW1	7.38	.24	5
MW2	6.95	.30	4

From Table 8.2, F-test = 6.6. Comparing this to the calculated F-ratio

$$F\text{-ratio} = \frac{(.30)^2}{(.24)^2} = 1.56 < 6.6$$

The pooled variance is

$$\hat{S}^2 = \frac{4(.24)^2 + 3(.30)^2}{5 + 4 - 2} = .071$$

The standard error of difference of the two means is

$$S_m = \sqrt{.071\left(\frac{1}{5} + \frac{1}{4}\right)} = 2.40$$

$$\text{The } t\text{-test statistic } t^* = \frac{|7.38 - 6.95|}{.179} = 2.40$$

The degrees of freedom = 5 + 4 − 2 = 7

For α = 5 percent and a two-sided test, $t_c = 2.37$. Thus $t^* > t_c$ or 2.40 > 2.37, which indicates sufficient difference to reject the null hypothesis. We conclude that there is a statistically significant difference in the pH levels at the two monitoring locations.

In summary, the t-test for pH and similar monitoring parameters is constructed in the same manner except we utilize the absolute value in the numerator and use the two-tailed table for determination of t_c.

EXAMPLE 8.12

Benzene concentrations as measured at two monitoring locations are as listed below:

Monitoring location	Benzene concentrations (μg/L)
MW1	1.6, 1.9, 2.3, 1.7, 2.4, 1.8, 1.4, 2.7, 2.1, 1.9, 2.3, 1.8
MW2	1.8, 1.9, 2.5, 22.2, 2.2, 1.9, 1.3, 2.8, 2.3, 2.0, 2.4, 1.7

Is there a statistically significant difference in the concentrations at the two monitoring locations?

Sample statistics are as follows:

Monitoring location	Mean (μg/L)	Standard deviation (μg/L)	Coefficient of variation
MW1	1.99	.38	.19
MW2	3.75	5.82	1.55

Noteworthy is the sizable magnitude of the fourth monitored value at MW2 (22.2), which results in a substantial increase in the mean and the standard deviation. The coefficient of variation of the results at MW2 is 5.82/3.75 = 1.55, indicating a failure to pass the test of normal distribution assumption of normality.

Alternatives for testing the assumption of normality are indicated in Chap. 3, Sec. 3.3.

In the event the data deviate from the assumption of normality, the usual practice is to transform the data so it can be described by the normal distribution. The transformations may include the log transformation, square root transformation, and the arcsine transformation. Alternatively, if an analyst cannot identify an appropriate transformation, an alternative approach is to use a distribution-free procedure, examples of which are described in Chap. 11.

8.4.4 Assumption of Independence

The t-test model assumes that observations are the result of a random sampling and that each observation is an independent random sample from the parent population. Problems can occur, for example, when the collection of a single groundwater sample is divided into aliquots for laboratory analyses, because the laboratory results are not independent of one another.

EXAMPLE 8.13

The air quality as quantified by 16 samples from one area of a city showed a mean of 107 and a standard deviation of 10, while the air quality of 14 samples from another area of the city showed a mean of 112 and a standard deviation of 8. Is there a significant difference between the air quality at the two locations at significance levels of (a) 0.01 and (b) 0.05?

If μ_1 and μ_2 denote the population means of air quality from the two areas, respectively, we have to decide between two hypotheses:

$H_0 : \mu_1 = \mu_2$ and there is essentially no difference between the air quality of the areas, or
$H_1 : \mu_1 \neq \mu_2$ and there is a significant difference between the air quality

Under hypothesis H_0,

$$t^* = \frac{\bar{x}_1 - \bar{x}_2}{S\sqrt{\dfrac{1}{n_1} + \dfrac{1}{n_2}}} \qquad \text{where} \qquad \hat{S} = \sqrt{\frac{n_1 S_1^{\,2} + n_2 S_2^{\,2}}{n_1 + n_2 - 2}} \qquad [8.17]$$

Thus,

$$\hat{S} = \sqrt{\frac{16(10)^2 + 14(8)^2}{16 + 14 - 2}} = 9.44 \text{ and}$$

$$t^* = \frac{112 - 107}{9.44\sqrt{\frac{1}{16} + \frac{1}{4}}}$$

$$= 1.45$$

a. Using a two-tailed test at the 0.01 significance level, we would reject H_0 if t^* were outside the range $-t_{.995}$ to $t_{.995}$ which for $(n_1 + n_2 - 2) = 16 + 14 - 2 = 28$ degrees of freedom is the range −2.76 to 2.76. Thus, we cannot reject H_0 at the 0.01 significance level.

b. Using a two-tailed test at the 0.05 significance level, we would reject H_0 if t^* were outside the range $-t_{.975}$ to $t_{.975}$, which for 28 degrees of freedom is the range −2.05 to 2.05. Thus, we cannot reject H_0 at the 0.05 significance level.

We conclude that there is no significant difference between the air quality of the two locations.

8.4.5 Examples of *t*-Test Applications

Consider three sets of data A_1, A_2, and A_3. The individual values are listed in Table 8.4.

(I) A Standard *t*-Test Evaluation

Comparison of A_1 and A_2

a. The coefficient of variation is less than 1.0, which indicates that the assumption of the normal distribution is probably reasonable.

TABLE 8.4 Sample monitoring data

	Monitoring Locations		
	A1	**A2**	**A3**
Monitoring Data	4.8	5.9	6.6
	6.8	7.3	6.3
	4.2	7.2	21.9
	4.7	7.5	6.0
	7.0	8.1	5.8
	6.5	2.2	7.2
	3.9	5.6	5.7
	5.1	6.2	2.8
	2.1	5.8	5.2
	5.6	7.1	5.4
	4.1	6.0	6.1

b. F-test

$$F = \frac{(1.58)^2}{(1.45)^2} = 1.19$$

From Table 8.2, for $n_1 = 11$ and $n_2 = 11$ and thus, degrees of freedom of 10 and 10,

$$F_{\text{crit}} = 3.0$$

1.19 < 3.0 so the two variances may be considered the same.

c. Pooled variance is

$$\hat{S}^2 = \frac{10(1.45)^2 + 10(1.58)^2}{20} = 2.30$$

and the standard error of difference is

$$S_m = \sqrt{2.30\left(\frac{1}{11} + \frac{1}{11}\right)} = .98$$

d. $t^* = \dfrac{6.26 - 4.98}{.98} = 1.30$

and degrees of freedom = 11 + 11 − 2 = 20 and five percent level of significance, and

$$t_c = 1.812$$

A comparison of t^* and t_c gives

$$1.30 < 1.812$$

which indicates insufficient evidence for a statistically significant difference.

(II) t-TEST WHERE A DATA TRANSFORMATION IS NEEDED

Consider now a comparison of A_1 and A_3. The sampling results at A_3 are similar in many respects to those at location A_2, with the exception of the third reported value which is much larger. Assuming that the large value is real, and not a laboratory or sampling error, the analytical procedure is carried out as follows:

a. As with Subsec. 8.4.5(i)(a) above, the coefficient of variation test is passed.

b. The F-test gives

$$F = \frac{(5.01)^2}{(1.45)^2} = 11.9$$

From Table 8.2, 11.9 > 3.0 and thus the variances cannot be considered to be samples from the same population. One approach to the problem involves transforming the data by taking logarithms. Note that the average and standard deviations of the transformed data are as indicated below (i.e., the mean of the log-transformed data is not the same as the log of the mean).

	Statistics of the log-transformed data	
Sampling location	**Mean**	**Standard deviation**
A_1	1.56	.34
A_2	1.84	.48

The F-ratio now becomes

$$F = \frac{(.48)^2}{(.34)^2} = 1.99$$

Since $F < F_{crit}$; that is, 1.99 < 3.0, the two variances may be considered the same.

c. The pooled variance $\hat{S}^2 = .173$, the standard error of difference $S_m = .177$, and the t-test becomes

$$t^* = \frac{1.84 - 1.56}{.177} = 1.57$$

Since $t_c = 1.58$ from Subsec. 8.4.5(i)c

$$t^* < t_c$$

indicating there is insufficient evidence to reject the null hypothesis, that is, insufficient evidence to indicate a significant difference between the means.

8.5 ACCEPTANCE AND REJECTION REGIONS

In conducting statistical inference tests, incorrect conclusions sometimes occur. An acceptance rate is assumed, which really means that an error occurs a specified percentage of the time.

In reality, because the population means are estimated from small samples, the comparison of these means involves the risk of making one of two possible errors, either a type I or a type II error. We attempt to achieve a proper balance between two conditions: (1) the risk that the procedures will falsely indicate, for example, that a landfill is causing background values or concentration limits to be exceeded, when in fact the landfill is not causing a problem (false positive), and (2) the risk that the procedures will fail to indicate that background values or concentration limits are being impacted when the landfill is in fact contaminating the groundwater (false negative).

		Truth	
		H_0	H_A
Decision	Accept H_0	No error Probability = $1 - \alpha$	Type II error Probability = β
	Accept H_A	Type I error Probability = α	No error Probability = $1 - \beta$ = power

FIGURE 8.5 Types of errors in statistical testing

Consider Fig. 8.5 and the following discussion as a demonstration:

Null hypothesis H_0: no contamination at the compliance well
Alternative hypothesis H_A: contamination has occurred

The four possibilities are indicated in Fig. 8.5, where the left side of the diagram represents the decision and along the top is truth or reality.

1. We can accept hypothesis H_0, and the hypothesis H_0 is true, and no error is made;
2. We can reject H_0 (accept H_A, the alternative hypothesis) but, in fact, H_0 is true. This is a Type I error, denoted by α, which represents the rejection of the null hypothesis when it is true. This error is also called a false positive (e.g., a source is identified as impacting the groundwater when, in fact, it is not).
3. We can reject H_0, and H_A is true, and thus no error is made.
4. We can accept H_0 when H_A is true. This is a type II error, denoted by β, which represents the acceptance of the null hypothesis when it is not true. We accept H_0 when in fact H_0 should be rejected. This error is also called a false negative (e.g., a source is identified as nonpolluting when it is in fact polluting—saying something is not present when it really is). As a result of this type of error, a remedial measure may not be implemented on a contaminating source that is actually degrading the groundwater quality.

Since the type II error has environmental consequences, we frequently tend to be rather conservative.

In quantifying the likelihood of these errors, the level of significance α represents the probability of making a type I error. By specifying $\alpha = 0.05$, we indicate that the type I error will be made five percent of the time over the long run. However, the two types of errors, α and β, are not independent. By making α smaller (e.g., $\alpha = 0.01$), we make the false positive error less often but we also increase the probability of making the false negative or type II error. The smaller we make the significance level α the more confidence we shall have that the null hypothesis is really false on the occasion when we reject it. The price we pay for this increased assurance is that the smaller we make α, the more difficult it becomes to disprove the hypothesis that is false.

For a given scenario, α and β are linked; they cannot be minimized simultaneously. To examine this tradeoff in greater detail, consider the following exam-

ple: After carrying out a significance test, we have evidence whether or not to reject the null hypothesis. Assume $\alpha = 5$ percent, $\sigma = 150$ mg/L of chloride, sample size $n = 25$, and

$$H_0 : \mu = 1250 \text{ mg/L and}$$
$$H_A : \mu \neq 1250 \text{ mg/L}$$

Then

$$t^* = \frac{\bar{x} - 1250}{150/\sqrt{25}}$$

If H_0 is true, this corresponds to

$$\text{Probability } (t^* > 1.64) = 0.05$$

Thus, H_0 is rejected at the five percent level if the observed value of t^* is greater than 1.64. Therefore, 1.64 is the critical value of t^*. This is equivalent to rejecting H_0 if $\bar{x} > 1250 + 1.64(150/\sqrt{25}) = 1299$. We have chosen the critical value in such a way that the probability of a Type I error or false positive rate is five percent.

In general, the probability of getting a significant result when H_0 is true is denoted by α. In words, there is only a five percent probability of getting a mean chloride concentration greater than 1299 mg/L.

8.6 POWER OF THE DISCRIMINATION TESTS

If we wanted to be more certain, we could make $\alpha = 0.01$ or one percent (this would correspond with making the Type I error smaller, the situation in which we would find that the source is impacting the groundwater when it is actually doing so). Then, if H_0 is true, we have

$$\text{Probability} \left(\frac{\bar{x} - 1250}{150/\sqrt{25}} < 2.33 \right) = 0.99 \qquad [8.18]$$

so that the critical value of t^* is 2.33, i.e.,

$$\bar{x} < 1250 + \frac{2.33(150)}{\sqrt{25}} = 1319.9$$

As expected, we need to have a larger mean (1320), to reject the null hypothesis when $\alpha = 1$ percent than when $\alpha = 5$ percent (1299).

If we have a specific alternative hypothesis and a given value of α, we can also find the probability of an error of type II, or β.

If H_0 is tested with $\alpha = 0.05$, then the critical value of $\bar{x}$ is 1299. If the observed value of $\bar{x}$ exceeds the critical value, then the null hypothesis will be correctly rejected. However, a type II error will occur if the observed value of $\bar{x}$ is less than critical value, so that there is a probability that the null hypothesis H_0 is accepted when it is false. If the alternative hypothesis is true, we have

$$\text{Probability}\ (\bar{x} < 1299) = \text{prob}\left[\frac{\bar{x} - 1319.9}{150/\sqrt{25}} < \frac{1299 - 1319.9}{150/\sqrt{25}}\right] \quad [8.19]$$

$$= \text{prob}[t^* < -.70]$$

$$= 1 - .755 = 0.245 \quad [8.20]$$

and this is the probability of an error of type II.

Alternatively, if $\alpha = 0.01$ or one percent, the critical value is 1250 + 2.33 (150) = 1319.9. If the alternative hypothesis is actually true, the probability of getting a nonsignificant result is

$$\text{Probability}\ (\bar{x} < 1319.9) = \text{prob}\left[\frac{\bar{x} - 1319.9}{150/\sqrt{25}} < \frac{1319.9 - 1319.9}{150/\sqrt{25}}\right]$$

$$= \text{prob}[t^* < 0]$$

$$= 0.5 \quad [8.21]$$

Thus, with the lower level of significance (0.05 decreased to 0.01), there is much higher chance of making a type II error: The value 0.25 from Eq. [8.20] increased to 0.5 in Eq. [8.21].

The foregoing example shows that the two types of errors are dependent on one another. If the critical value is increased in order to reduce the probability of a type I error, then the probability of a type II error will increase.

For a specific alternative hypothesis, the power of a significance test is obtained by calculating the probability that H_0 is rejected when it is false (that is, the correct decision is made). To quantify this power,

$$\text{Power} = 1 - \text{probability (error of type II)}$$

$$= 1 - \beta \quad [8.22]$$

or, in summary form,

α	β	**Power**
0.05	0.245	0.755
0.01	0.500	0.500

Figure 8.6 summarizes the alternative configurations for the α and β errors. The power of a statistical test is defined as the probability $1 - \beta$ that the test will

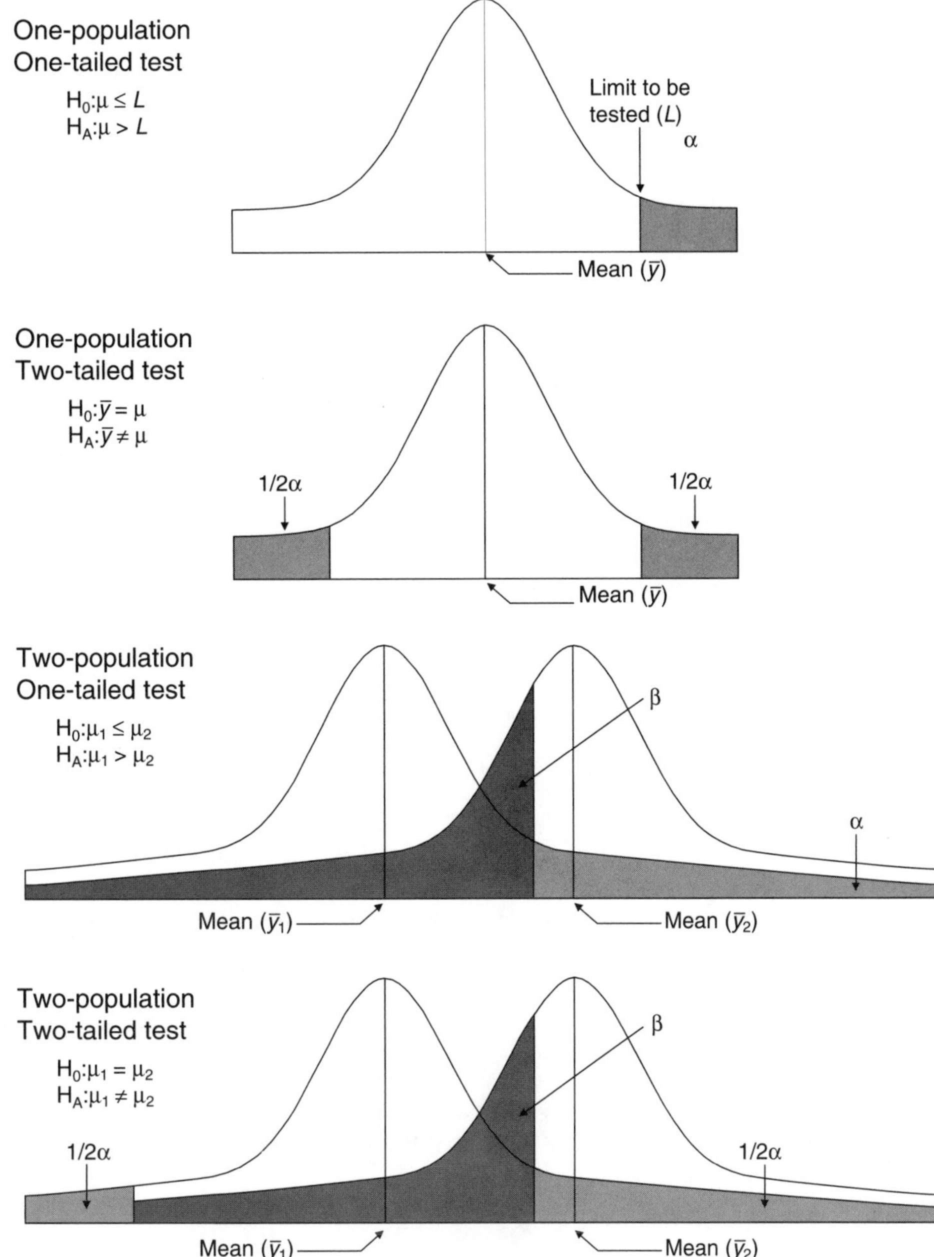

FIGURE 8.6 Probabilities of errors in statistical testing

successfully detect a statistically significant difference. Alternatively, it is the probability of rejecting a null hypothesis when it is in fact false and should be rejected.

A major impact with strict regulations is that they tend to minimize the false negative rates (i.e., the probability that you conclude that there is no contamination when there is) by increasing the false positive rates (i.e., the probability that you conclude that there is contamination when there is none). Simply selecting a test based on its low false negative rate is a problem, however, because a test with a low false negative rate will necessarily have a high false positive rate. In the extreme case in which the false negative rate is 0 percent, the false positive rate is 100 percent.

A reasonable tradeoff between type I and type II errors is desirable. We can develop power curves to examine the tradeoffs. For a given sample power curves allow selection of an optimal test in the sense that it achieves its intended false positive (e.g., five percent) while simultaneously achieving a reasonable false negative rate (e.g., five percent).

In summary, the power of a test may be increased in one or more of the following ways:

1. by reducing the variance of the measurements to be tested;
2. by increasing the size of the difference to be detected;
3. by increasing the number of samples; and
4. by using a higher significance level. For example, a test that uses a significance level of 0.10 will have a higher power (i.e., a lower chance of a Type II error) but also a higher probability of a Type I error, than will a test using the 0.05 level of significance. The power may be increased by raising the significance level, but only at the risk of increasing the probability of incorrectly rejecting the null hypothesis.

Approximate power may be found early for three differences in the two population means (U.S. EPA, 1994):

a. at a difference of zero, the power is α, the significance level;
b. when the difference in mean concentration equals the critical value of the t-statistic (from the table) times the standard error of the difference (the denominator of the form of the t-statistic that is being used) the power is approximately 50 percent; and
c. when the difference in mean concentration is twice that in (b), the power is approximately $100(1 - \alpha)$ percent.

The power of the statistical test for detecting differences tends to increase as the alternative is farther from the null hypothesis.

POWER OF THE t-TEST

For given sample sizes n_1 and n_2, significance level (α), and hypothesized difference between population means d, the power of the t-test may be estimated (after Zar, 1982) by computing

$$t_{\beta(1)} = \frac{d}{\sqrt{2\hat{S}^2/n}} - t_{\alpha/2} \qquad [8.23]$$

where $\hat{S}^2 =$ the pooled variance, $t_{\alpha(2)}$ is the critical value of t (for $n_1 + n_2 - 2$ degrees of freedom) and n is the size of each sample. If the two samples are of different sizes, n_1, and n_2, then

$$n = \frac{2n_1 n_2}{n_1 + n_2} \tag{8.24}$$

Finally, $t_{\beta(1)}$ denotes a "one-tailed" t, the value of t that delineates the proportion β in one tail of the distribution.

EXAMPLE 8.14

If the difference between the population means is 0.41 mg/L, what is the probability that the t-test would conclude the null hypothesis, when $n_1 = 12$, $n_2 = 17$, $\hat{S}^2 = 0.31$, and $t_{\alpha(2)} = 2.052$.

Then

$$n = \frac{2(12)(17)}{12 + 17} = 14.07$$

and

$$t_{\beta_{(1)}} = \frac{0.41}{\sqrt{\frac{2(0.31)}{14.07}}} - 2.052 = -0.10$$

To estimate β, consider what proportion of a normal curve is ≥ -0.10. The probability of $t_{\beta(1)} \geq -0.10$ (estimated from Table A.1) is 0.54; therefore, the power of the test is estimated as $1 - \beta = 1 - 0.54 = 0.46$. This indicates to the analyst that there is only a 46 percent chance of detecting a difference of 0.45 mg/L between the two population means. Increased power in testing may be achieved by increasing the sample size n and/or increasing α, the probability of a type I error.

8.7 EXTENSIONS OF THE *t*-TEST

The limitations/assumptions implicit within the t-test preclude utilization of the test on a number of environmental quality problems. As alternatives, a number of modified t-tests have been developed. These include:

- Satterthwaite's (1964) modified t-test;
- Cochran's approximation to the Behrens-Fisher t-test (Behrens, 1929; Cochran, 1964); and
- the paired sample t-test.

These alternative tests have focused on relaxation of the assumptions implicit in the standard t-test, but caution in their application must be exercised since the tests are approximate, which means they have corresponding limitations.

8.7.1 SATTERTHWAITE'S MODIFIED t-TEST

Satterthwaite's test represents a modified t-test that allows unequal variance between the samples compared (after Iman and Conover, 1983). The remaining assumptions necessary for utilization of the t-test must still be satisfied, namely independence between observations and the assumptions that the populations are normally distributed.

When the variances are unequal, $S_1^2 \neq S_2^2$, the variances cannot be pooled as per Eq. [8.14]. However, the formula for the variance of $\bar{x}_1 - \bar{x}_2$ in independent samples still holds as:

$$S^2_{\bar{x}_1 - \bar{x}_2} = \frac{S_1^2}{n_1} + \frac{S_2^2}{n_2} \qquad [8.25]$$

Satterthwaite's test then computes the t-statistic from

$$t^* = \frac{|\bar{x}_1 - \bar{x}_2|}{\sqrt{\dfrac{S_1^2}{n_1} + \dfrac{S_2^2}{n_2}}} \qquad [8.26]$$

However, t^* does not follow the t-distribution when $\mu_1 = \mu_2$. Two different forms of the distributions of t^* have been identified after Behrens, 1929, and Fisher and Yates, 1957. In Satterthwaite's model (1964), an approximate number of degrees of freedom is assigned to t^* so that the ordinary t-tables may be used. The approximate number of degrees of freedom is determined from

$$df = \frac{\dfrac{S_1^2}{n_1} + \dfrac{S_2^2}{n_2}}{\dfrac{\left(\dfrac{S_1^2}{n_1}\right)^2}{n_1 - 1} + \dfrac{\left(\dfrac{S_2^2}{n_2}\right)^2}{n_2 - 1}} \qquad [8.27]$$

which is, in turn, used to obtain the critical t-value t_c from the standard t-tables for a selected level of significance α. The value of df to use in the ordinary t-tables is rounded down (e.g., 16.8 becomes 16). As apparent from the procedure, Satterthwaite's approximation works by adjusting the degrees of freedom.

8.7.2 COCHRAN'S APPROXIMATION TO THE BEHRENS-FISHER t-TEST

An alternative to Satterthwaite's approximate t-test, to examine whether there is a statistically significant difference between the environmental quality at two locations when the variances are not equal, is Cochran's approximation to the Behrens-Fisher test (CABF). The CABF test functions by relaxing the requirement that the variance of the data at both sampling locations be of the same magnitude by adjusting the degrees of freedom. Cochran's test computes the t-statistic from

$$t = \frac{|\bar{x}_1 - \bar{x}_2|}{\sqrt{\frac{S_1^2}{n_1} + \frac{S_2^2}{n_2}}} \qquad [8.28]$$

The critical t-statistic t_c is computed from

$$t_c = \frac{\frac{S_1^2}{n_1}t_1 + \frac{S_2^2}{n_2}t_2}{\frac{S_1^2}{n_1} + \frac{S_2^2}{n_2}} \qquad [8.29]$$

where

$$S_1 = \sqrt{\frac{\Sigma(x_{i1} - \bar{x}_1)^2}{n_1 - 1}}$$

$$S_2 = \sqrt{\frac{\Sigma(x_{i2} - \bar{x}_2)^2}{n_2 - 1}}$$

where t_1 and t_2 are determined from t-tables for df of $n_1 - 1$ and $n_2 - 1$ respectively, for a specified level of significance α. See Ex. 8.16.

The difference between Cochran's test and Satterthwaite's test is in the calculation of the degrees of freedom df. Cochran's test weighs the individual t-values for each sample while Satterthwaite's test calculates a new degree of freedom and then uses the t-tables.

As noted above, CABF allows relaxation of the equality of variances requirement. However, there are situations in which the assumptions implicit within the basic t-test are *not* violated, and the utilization of the approximate test in such a situation will imply different findings from those intended. This concern was described in McBean et al. (1988). For example, when $n_1 = 3$ (the downgradient location has three sample results) and $n_2 = 4$ (the upgradient has four samples), with the variance ratio $S_1^2/S_2^2 = 4$, the t-test allows a deviation of $|\bar{x}_1 - \bar{x}_2|$ of 2.28 before a significant difference is identified ($\alpha = 5$ percent if a one-tailed test). Translation of the allowable deviation by Cochran's test into an equivalent deviation corresponds to $|\bar{x}_1 - \bar{x}_2| = 3.56$ which in turn gives a significance level of 1.3 percent. In other words, if Cochran's test is used instead of the t-test, the level of significance is not 0.05 but 0.013.

EXAMPLE 8.16

Consider the analysis in Section 8.4.5(ii) where a data transformation was utilized because the variances could not be considered to be samples from the same population. As an alternative to the data transformation, the CABF test can be employed.

Specifically, using Eq. [8.28],

$$t^* = \frac{7.10 - 4.98}{\sqrt{\frac{(4.53)^2}{2} + \frac{(1.45)^2}{11}}} = \frac{2.12}{1.43} = 1.48$$

The critical t-statistic is

$$t_c = \frac{\frac{(4.53)^2}{11}(1.81) + \frac{(1.45)^2}{11}(1.81)}{\frac{(4.53)^2}{11} + \frac{(1.45)^2}{11}} = 1.81$$

Since $t^* < t_c$ (1.48 < 1.81) there is insufficient evidence to indicate a significant difference between the sampling results at A_1 and A_3.

Cochran's test and Satterthwaite's test are both approximate formulae for the t-test. Conditions exist in which both approximate tests give substantially different results from the t-test. The following recommendations are made on the approximate t-tests (after McBean et al., 1988):

- when the number of samples at the two locations being compared are similar in magnitude, the difference between the tests is minimal.
- when the variances are determined not to be different in the statistically significant sense (using the F-test), the t-test should be utilized.

Cochran's test is more stringent than the t-test when

$$n_1 > n_2 \text{ for } \frac{S_1^2}{S_2^2} > 1 \text{ and less stringent when } n_1 < n_2 \text{ for } \frac{S_1^2}{S_2^2} > 1$$

Cochran's test is more stringent than the t-test when

$$n_1 < n_2 \text{ for } \frac{S_1^2}{S_2^2} < 1 \text{ and less stringent when } n_1 > n_2 \text{ for } \frac{S_1^2}{S_2^2} < 1.$$

When the variances are determined to be different (using the F-test), Cochran's test should be used since the t-test assumptions are violated.

8.7.3. Paired t-Test

The paired t-test is a statistical test to determine if data from two separate treatments or environmental media fall within the same population or are "equivalent." For example, Anderson et al. (1993) used two pilot plants to examine features such as chlorine versus ozone and alum dosage versus silicate dosage as alternatives for water treatment. If the two sides of the treatment plant are equivalent in terms of impact on the treatment of water, the expected difference between two data pairs would be zero.

Paired sample t-tests are used when the sample populations are not independent. This may occur when successive sampling of the same water samples takes place upstream and downstream of a contaminant source, as depicted in Fig. 8.2(a). A single element of water may be considered to move from an upstream point through a zone of possible contamination to a point downstream, such that it is virtually the same sample and is therefore not independent. The paired t-test represents a variation on the standard t-test.

For some types of correlated data, it is possible to define the population being sampled in order to remove or minimize the effect on the test outcome (e.g., cyclical variations in both background and monitored data are not included in the calculation of the standard error). This is done by pairing the individual observations from the correlated populations. Once the differences in the pairs are calculated, the differences are treated as a single, random, independent sample. With the pairwise test there is one-half the degrees of freedom, but the standard error is smaller.

The aim of pairing is to make the comparison more accurate by having the numbers of any pair as similar as possible except in the "treatment" difference. For example, the variable upon which the pairing takes place should predict accurately the criterion by which the effects of the "treatments" are to be judged.

Pairwise testing is also appropriate when there is a sinusoidal variation in the input. For sinusoidally varying data, the variance will be large and the t-value low. However, the variance is large because of a recurring feature that is predictable, not random. Seasonal fluctuations in water chemistry, for example, are particularly apparent in shallow aquifers where groundwater quality may vary with recharge cycles (e.g., Pettyjohn, 1976; Spalding, 1984; Pekny et al., 1989). For these types of situations, the data must be combined to yield a new population.

When comparing two treatment methods, as when two alternative procedures are used to analyze the same samples, it may happen that experiments are carried out in pairs. The difference between each pair of measurements is of interest.

This difference can be tested when we test whether both samples are equally affected by the external effects

$$t^* = \frac{\overline{D} - \delta}{\frac{S_D}{\sqrt{n}}} \qquad [8.30]$$

where δ may be taken as $\delta = 0$ or $\delta \neq 0$.

The paired t-test involves the derivation of a new population. The differences D_i, for each of the individual pairs are assumed to be distributed about a mean, μ_D. The deviations $D_i - \mu_D$ may be due to various causes. In the analysis, the deviations $D_i - \mu_D$ are assumed to be normally and independently distributed with population mean zero. When these assumptions hold, the sample mean difference $\overline{D}$ is normally distributed about μ_D with a standard error $\sigma_D/\sqrt{n}$ where σ_D is the standard deviation of the population of differences. Since the value σ_D is seldom known, the sample estimate is utilized as

$$S_D = \sqrt{\frac{\Sigma(D_i - \overline{D})^2}{n - 1}}$$

Hence, $S_{\overline{D}} = S_D / \sqrt{n}$ is an estimate of $\sigma_{\overline{D}}$ based on $(n - 1)$ degrees of freedom

$$t^* = \frac{|\overline{D} - \mu_D|}{S_{\overline{D}}} \qquad [8.31]$$

If the H_0 is true, the distribution of t^* will be a t-distribution with $n - 1$ degrees of freedom. The t-distribution may be used to test the null hypothesis that $\mu_D = 0$.

Similarly, the confidence interval for the mean is $\overline{D} \pm t_\alpha S_{\overline{D}}$. If the observation (difference) falls inside the tolerance interval, one concludes that no significant shift in level has occurred.

EXAMPLE 8.17

Determine if there is a statistically significant difference in the arsenic concentrations between the quality at the two monitoring locations (twice/year of sampling). To examine this question, subtract the concentration at the two location to end up with a single set of concentration changes. The test may then be conducted as an examination of whether the changes are significantly different from zero.

Date of Sampling	Sample Number	Monitoring Results (μg/L) Station I	Station 2	Difference in Concentration (μg/L)
July	1	37	43	6
Jan.	2	28	33	5
July	3	33	37	4
Jan.	4	26	25	–1
July	5	41	45	4
Jan.	6	22	23	1
July	7	33	32	–1
Jan.	8	24	27	3
July	9	31	32	1
Jan.	10	23	21	–2
Mean		$\bar{x} = 29.8$	31.8	2.0
Standard Deviation		$S = 6.30$	8.11	2.79

a. Consider first the test as a standard t-test determination. We begin by calculating the ratio of the variances as

$$F = \frac{(8.11)^2}{(6.30)^2} = 1.52$$

According to Table 8.2, there is not a statistically significant difference between the variances.

The pooled variance estimate is

$$\hat{S} = \frac{9(6.30)^2 + 9(8.11)^2}{18} = 5.00$$

The standard error of the mean

$$S_m = \sqrt{50.0\left(\frac{1}{10} + \frac{1}{10}\right)} = 3.25$$

and the t-value

$$t^* = \frac{31.8 - 29.8}{3.25} = .61$$

and $t_c = 1.73$ where t_c is determined using 18 degrees of freedom.

Since $t^* < t_c$, the analyses indicate no significant difference between the two monitoring locations. To leave it at this point would clearly be incorrect. The reason that the test fails to detect a significant difference is the seasonality (the values are considerably higher during the July period, as opposed to the January period). However, when the seasonality is retained in the data set, the variance is high and the t^* value is low.

b. Consider now the test as one of difference between the two monitoring locations.

The result is

$$t^* = 2.0/(2.79/\sqrt{10}) = 2.27$$

$$t_c = 1.83$$

where t_c is determined using $10 - 1 = 9$ df and $\alpha = 5$ percent.

Since $t^* > t_c$, these findings indicate that there is a significant difference.

Note that with pairing we have fewer degrees of freedom (the t-value used in the test of significance or in computing confidence limits would be smaller with independent samples).

EXAMPLE 8.18

The surface water quality monitoring results obtained from measuring both upstream and downstream of a potential nonpoint source, are included as Columns II and III in the table below. Some remedial technologies were implemented in October–November 1980, and the water quality monitoring data are as characterized by Column V (McBean, 1986), as measured in 1981–1982. Of interest are two questions:

a. Is the source contributing significantly to the river?

b. Did the remedial technologies significantly improve the water quality?

Upstream and downstream water quality monitoring records

	Preremedial records		Postremedial records	
Date of sampling	**Upstream measurements (mg/L)**	**Downstream measurements (mg/L)**	**Date of sampling (mg/L)**	**Downstream measurements**
I	**II**	**III**	**IV**	**V**
10/79	.29	4.3	10/81	.53
11/79	12	16	11/81	1.5
12/79	.32	6.1	12/81	1.3
1/80	—	—	1/82	—
2/80	—	—	2/82	—
3/80	.49	2.66	3.82	2.1
4/80	.14	3.0	4/82	1.1
5/80	1.58	4.42	5/82	—
6/80	1.77	5.74	6/82	1.8
7/80	1.07	1.40	7/82	1.2
8/80	.07	1.49	8/82	.64
9/80	.14	2.3	9/82	1.1
Mean	1.69	4.74		1.25
Std.Deviation	3.67	4.28		.50

Note: "—" means no measurement available.

Using Satterthwaite's Approximate test, an examination of the upstream and downstream concentration finds,

$$\bar{x}_1 = 1.69 \qquad \bar{x}_2 = 4.74$$

$$n_1 = 10,\ df_1 = 9 \qquad n_2 = 10,\ df_2 = 9$$

$$\frac{S_1^2}{n_1} = 1.35 \qquad \frac{S_2^2}{n_2} = 1.83$$

$$t^* = 1.71$$

$$df = 17.5, \text{ which is taken as } 17$$

Finally, for a one-sided test (from the standard t-tables)

$$t_{c,0.05} = 1.74$$

Since $t_{c,0.05} > t^*$, these results indicate that a statistically significant change has not been identified at the 95 percent level. However, a visual comparison of the upstream/downstream data in Columns I and II clearly demonstrates that the downstream water is lower quality.

For the type of correlation existing between upstream and downstream points, pairing individual observations and then examining only the differences between the observations is appropriate. Once the differences in the pairs are calculated, they are treated as a single, random, independent sample.

Using the paired test

$$\overline{D} = 2.95$$

$$S_D = 1.55$$

and

$$t^* = 6.0$$

For 9 degrees of freedom, $t_c = 1.83$ for five percent and a statistically significant impact ($t^* \geq t_c$), indicating the nonpoint source has impacted the river.

Impact of the Remedial Technology

Using Satterthwaite's approximation, in comparing pre- and postremedial measurements gives

$$t^* = 2.56$$

$$t_c = 1.7$$

which indicates the remedial measures have had a statistically significant impact.

EXAMPLE 8.19

Consider the problem of comparing two methods for finding the percentage of iron in a compound. Six different samples were analyzed by both methods and the results are as given below:

Sample number	Method A	Method B
1	13.3	13.4
2	17.6	17.9
3	4.1	4.1
4	5.7	5.1
5	7.6	7.0
6	12.9	12.8

Is there a significant difference between the two methods of analysis? Note that it would be incorrect to calculate the average percentage for method A and method B and proceed in the standard t-test because the variations between compounds will overwhelm any difference there may be between the two methods. Instead, the correct approach is to proceed with comparing differences as follows:

Sample number	Difference in concentration
1	.1
2	.3
3	0
4	−.6
5	−.6
6	−.1

If the two methods gives similar results, the differences above should be a sample of six observations from a population with a mean of zero.

The statistics of the difference in concentration are:

$$\overline{D} = -0.15$$

$$S_D = 0.37$$

$$t^* = \frac{-.15}{\frac{0.37}{\sqrt{6}}} = -.99$$

$$t_c = 2.015 \text{ for } df = 6 - 1 = 5 \text{ and } \alpha = 5 \text{ percent}$$

Since $t^* < t_c$, there is not sufficient evidence to indicate a difference between the two methods of analysis.

8.7.4 Summary of Alternative Tests

A series of alternative statistical significance tests exist, each of which involves a series of assumptions. In utilizing a particular test for a specific application, it is important to be cognizant of the limitations/approximations implicit in the individual procedures. All of the tests described in this chapter are parametric tests, meaning the value of concentration is utilized. Later chapters will deal with nonparametric procedures, where the nonparametric procedures use relative rankings and not the concentrations. Less stringent assumptions will be necessary for some of these procedures.

Both Cochran's and Satterthwaite's tests are approximate formulas for the t-test that allow relaxation of the equality of variances requirement. However, these are situations in which the assumptions implicit in the basic t-test are violated, and the utilization of the approximation tests in such situations may generate findings different from those intended (McBean & Rovers, 1990).

8.8 REFERENCES

ANDERSON, W., I. DOUGLAS, J. VAN DEN OEVER, S. JASIM, J. FRASER, and P. HUCK. "Experimental Techniques for Pilot Plant Evaluation," presented at AWWA Water Quality Technology Conference, Miami, FL, November 7–11, 1993.

BEHRENS, W.V. *Landwirtschaftliche Jahrbucher,* 68 (1929), 807.

COCHRAN, W. "Approximate Significance Levels of the Behrens-Fisher Test," *Biometrics* (March, 1964), 191–195.

DRAPER, N.R., and H. SMITH. *Applied Regression Analysis.* New York, NY: John Wiley and Sons, 1981.

FISHER, R.A. *Statistical Methods for Research Workers.* Edinburgh: Oliver & Boyd, 1925a.

FISHER, R.A. "Applications of Student's Distribution." *Metron* 5 (1925b), 90–104.

FISHER, R.A., and F. YATES. *Statistical Tables for Biological, Agricultural and Medical Research,* 6th ed. Edinburgh: Oliver and Boyd, 1963.

IMAN, R.A., and W.J. CONOVER. *A Modern Approach to Statistics.* New York, NY: John Wiley and Sons, 1983.

KERLINGER, F.N., and E.J. PEDHAZUR. *Multiple Regression in Behavioral Research.* New York, NY: Holt, Rinehart and Winston, 1973.

KOCH, G.S., and R.F. LINK. *Statistical Analysis of Geologic Data.* New York, NY: John Wiley and Sons, 1970.

KRUMBEIN, W.C., and F.A. GRAYBILL. *An Introduction to Statistical Models in Geology.* New York, NY: McGraw-Hill Inc., 1965.

KUFS, C.T. "Statistical Modeling of Hydrogeologic Data—Part I: Regression and ANOVA Models," *Ground Water Monitoring Review* (Spring, 1992), 120–130.

McBEAN, E. "Alternatives for Identifying Statistically Significant Differences," in *Statistical Aspects of Water Quality Monitoring,* eds. A. El-Shaarawi and R. Kwiatkowski, vol. 27 of *Developments in Water Science.* Amsterdam: Elsevier, 1986.

McBEAN, E., M. KOMPTER, and F. ROVERS. "A Critical Examination of Approximations Implicit in Cochran's Procedure," *Ground Water Monitoring Review* (Winter, 1988), 83–87.

McBEAN, E., and F. ROVERS. "Flexible Selection of Statistical Discrimination Tests for Field-Monitored Data," eds. D. Nielsen and A. Johnson, *Ground Water and Vadose Zone Monitoring,* ASTM STP 1053. Philadelphia, PA: American Society for Testing and Materials, pp. 256–265.

MILLIKEN, G.A., and D.E. JOHNSON. *Analysis of Messy Data.* New York, NY: Van Nostrand Reinhold Co., 1984.

MORONEY, M.J. *Facts From Figures,* 3rd ed. Middlesex, England: Pelican Books, 1965.

PEKNY, V., J. SKOREPA, and J. VRBA. "Impact of Nitrogen Fertilizers on Ground-Water Quality—Some Examples from Czechoslovakia," *Journal of Contaminant Hydrology,* 4 (1989), 51–67.

PETTYJOHN, W.A. "Monitoring Cyclic Fluctuations in Ground-Water Quality," *Ground Water,* 14, no. 6 (1976), 472–479.

SATTERTHWAITE, F. *Biometric Bulletin*, 2 (1964), 110.

SPALDING, M.E. "Investigation of Temporal Variations and Vertical Stratification of Ground Water Nitrate-Nitrogen in the Hall County Special Use Area," Project G854-06, Nebraska Water Resource Center, 33 pp.

U.S. EPA, *Statistical Training Course for Groundwater Monitoring Data Analysis*, 530-R-93-003 (April 1994), Environmental Protection Agency.

ZAR, J.H. "Power and Statistical Significance Impact Evaluation," *Groundwater Monitoring Review* (Summer, 1982), 33–35.

8.9 PROBLEMS

8.1. We flip a coin 100 times and for each situation, the result is heads.

- **a.** What is the chance of obtaining a head on the next toss?
- **b.** If we say the coin is fair, what evidence can we use to support the claim?

8.2. Two locations on a stream separated by 1 km were monitored, yielding the results in the table below. Without transforming the data (i.e., use the data as is), determine whether any incidental (i.e., in between) loading has created a significant change in concentration.

Upstream location (mg/l)	Downstream location (mg/l)
.35	1.2
2.1	2.8
.37	1.6
.45	.50
.74	.50
.61	.50
.36	.85

8.3. The background measurements from ten monitoring wells are 0.8, 1.5, 1.7, 0.6, 1.1, 1.8, 1.0, .9. Set 95 percent limits on the mean of the background concentrations, using this random set.

8.4. From a sampling program, we summarize the results obtained from two monitoring programs below.

Location A: 2.6, 3.7, 3.0, 4.1, 2.9, 3.6

Location B: 3.1, 3.6, 3.1, 4.2, 3.6, 4.1

- **a.** Is there a statistically significant difference between the results as monitored at Locations A and B? Assume $\alpha = 5$ percent.
- **b.** If we use the mean and variances calculated from the above data to predict results of additional samples at the same locations, how many samples would we need in order to show a statistically significant difference?

8.5. The following data have been obtained upstream and downstream from an agricultural field with significant seasonality in the runoff pattern of nitrogen concentration:

Sample	Location A	Location B
1	.06	.07
2	.08	.09
3	.13	.15
4	.14	.15
5	.09	.12
6	.04	.07
7	.06	.06
8	.09	.08
9	.10	.11
10	.14	.16

Use the paired t-test to determine if there is a statistically significant difference between the sampling results at the two monitoring locations.

CHAPTER 9

Multiple Comparisons Using Parametric Analyses

> *"Testing is often the only way to answer our questions, but it doesn't produce unassailable, universal truths that should be carved on stone tablets. Instead, testing produces statistics which must be interpreted."*
>
> *Robert Hooke*

9.1 INTRODUCTION

The focus in Chap. 8 was on single comparisons using various forms of the *t*-test to test significance of the difference between monitoring records. However, for situations such as that depicted in Fig. 9.1, in which a number of comparisons are to be made (for the situation illustrated, comparisons of a number of upgradient and downgradient wells), we are interested in not one statistical comparison as per Chap. 8 but many comparisons. One approach involves testing whether concentrations at one or more monitoring locations are statistically significantly different from other monitoring locations (e.g., do the concentrations at the compliance well significantly exceed the mean background concentration level at the upgradient well?).

An appropriate statistical analysis procedure to simultaneously analyze whether mean concentration differences between monitoring locations, or among groups of monitoring locations, are statistically significant is through use of the analysis of variance or ANOVA—for example, to examine whether there is significant contamination at one or more downgradient monitoring locations (1 through 4 in Fig. 9.1). ANOVA is a method by which the total variation in a set of data may be reduced to components associated with possible sources of variability, and whose relative importance are of interest.

ANOVA is the name given to a wide variety of flexible and powerful statistical procedures. All of these procedures compare the means of different groups of observations to determine whether there are any significant differences among the groups, and if differences are identified, additional procedures may be used to determine where the differences lie.

EXAMPLE 9.1

The types of questions to which ANOVA can be applied include:

a. considering the data for one time period at several wells to determine if there are significant differences between different groupings of wells; or,

b. considering the monitoring results from several wells over several time periods, assuming no seasonality is present, to determine if changes are occurring over time.

Note that in the examples as described above, the data are assumed to not exhibit seasonality or trends, which allows an analysis by the so-called "one-way ANOVA". If the data show evidence of seasonality (observed, for example, by plotting the data over time) a trend analysis or perhaps a two-way ANOVA may be appropriate. The ramifications of a one-way ANOVA versus a two-way ANOVA will be considered in Sec. 9.2.3.

ANOVA represents a technique closely related to regression. In ANOVA, the independent variables represent group membership (in contrast to discriminate analysis in which the dependent variable represents group membership). While regression techniques are generally used to evaluate data trends, the purpose of the ANOVA model is to detect differences in the average values of the dependent values of the dependent variables in the various groups.

Two general types of ANOVA exist, namely, parametric and nonparametric. Both methods of ANOVA are appropriate for application to a wide array of environmental quality problems. However, parametric ANOVA is the focus of this chapter and nonparametric ANOVA is dealt with in Chap. 11. Of the two, the

FIGURE 9.1 Plan view of upgradient and downgradient monitoring well locations

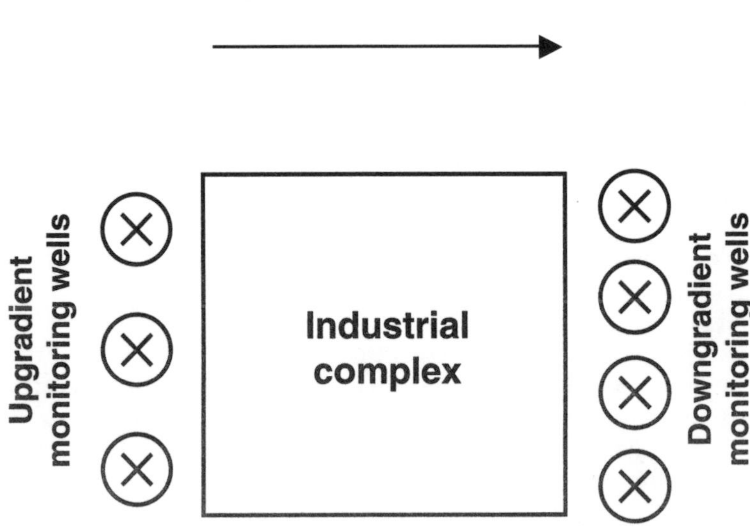

parametric ANOVA is the preferred procedure, but if the percentage of non-detects is greater than approximately 15 percent, then nonparametric one-way ANOVA must be utilized (where ranks are used as opposed to parametric values, e.g., concentrations).

The basis of the ANOVA concept was first developed by R. Fisher and the *F*-test (which is the key to ANOVA) was named in his honor. The *F*-test makes a comparison of variances to determine whether there is a significant difference.

Since this chapter represents an extension of Chap. 8, from the single comparison test to the multiple comparison test, a reasonably good understanding of Chap. 8 is presumed. As well, for ease of explanation, the problem context for describing the specifics of the calculations will be directed to comparison of upgradient and downgradient monitoring well locations as depicted in Fig. 9.1.

9.2 ANALYSIS OF VARIANCE (ANOVA)

9.2.1 DEVELOPMENT OF THE NULL HYPOTHESIS

The mathematical procedures involved in ANOVA are best demonstrated by example. The null hypothesis is made that the means (e.g., the mean water quality for two sets of wells for a particular constituent) are equal and are therefore all simultaneously estimates of the same population. Rejecting the null hypothesis means that a statistically significant difference is identified.

9.2.2 MULTIPLE COMPARISONS AND STATISTICAL POWER

Applying the same statistical significance test to each comparison is acceptable if the number of comparisons is small, but when the number of comparisons is moderate to large, the false positive rate can be quite high. This means that if a sufficient number of tests are run, there will be a significant chance that at least one test will indicate a significant difference (i.e., that contamination has occurred, even if it has not).

To avoid this problem, one could lower the type I error but generally, for a given statistical test, a lower false positive rate reduces the power of the test to detect a real difference (e.g., contamination at the downgradient well). If the statistical power drops too much, real contamination will not be identified when it occurs.

To establish a recommended standard for the statistical power of a testing strategy, it must be understood that the power is not a single number, but rather a function of the level of contamination actually present. For more tests, the higher the level of contamination, the higher the statistical power; likewise, the lower the contamination level, the lower the power. For example, when increasingly contaminated groundwater passes a particular well, it becomes easier for the statistical test to distinguish background levels from the contamination—the power is a function of the contamination level.

The power of any testing method can be increased merely by relaxing the false positive rate, or letting α become larger than five percent.

9.2.3 One-Way ANOVA and Two-Way Tests of ANOVA

The one-way ANOVA uses a row to represent each sample observation and a separate column to represent each grouping to be compared (e.g., groups of groundwater monitoring wells). The two-way ANOVA involves the addition of a third factor, the blocking variable, which is used to associate some of the variation in the sample results. Thus, as an approach, we classify the monitoring locations into P mutually exclusive groups (where $P \geq 3$). If there are only two groups of monitoring wells, the two-sample tests from Chap. 8 are used. The testing procedures described in this chapter need at least three to four samples per group. Each of the P groups should have at least three observations and the total sample size n should be large enough so that $n - P \geq 5$.

Assumptions implicit in the ANOVA procedure include that the data have normally distributed residuals with equal variances across P groups of wells, where the residual is defined as:

$$\text{Residual} = \text{observed value} - \text{predicted value} \qquad [9.1]$$

The hypothesis tests with parametric analysis of variance usually assume that the errors (residuals) are normally distributed with constant variance. These assumptions can be checked by saving the residuals (the difference between the observations and the values predicted by the ANOVA model). Therefore, the procedure involves the calculation of the residuals and then testing the normality of the residuals. The testing of normality involves pooling all residuals together and checking the distribution using, for example, the Shapiro-Wilk test (see Sec. 3.3.6). The ANOVA procedure then tests equality of variances across the groups. Assuming equality of variance, we then compute for each source of variability in turn (a) the sum of squares, and (b) the number of degrees of freedom.

The total variation of the observations is partitioned into two components, one measuring the variability between the group means, $\bar{x}_1, \bar{x}_2, \ldots, \bar{x}_n$ and the other measuring the variation within each group.

9.3 TESTING FOR HOMOGENEITY OF VARIANCE

Assumptions within the parametric ANOVA procedures include the assumption that the different groups (e.g., different wells) have approximately the same variance. If not, the power of the F-test (its ability to detect differences among the group means) is reduced. Mild differences in variance are acceptable but the effect of differences in the variances becomes noticeable when the largest and smallest group variances differ by a ratio in excess of four and become severe when the ratio is ten or more (Milliken and Johnson, 1984a).

9.3.1 Box Plots

Probably the simplest procedure to check the variances across the P groups is to use box plots (described in Sec. 2.4.2). A box plot is constructed for each well group and the boxes are compared to see if the assumption of equal variances is reasonable. Three horizontal lines are drawn for each well grouping—one line each at the lower and upper quartiles and another at the median concentration. If the box length for each group is less than three times the length of the shortest box, the sample variances are probably sufficiently close to assume equal group variances. If the group variances appear to be different, the data should be further checked using Levene's test. If Levene's test is significant (see Sec. 9.3.2), the data may need transformation or a nonparametric procedure should be considered.

9.3.2 Levene's Test

Levene's test is a more formal procedure than box plots for testing the homogeneity of variance. Levene's test is not sensitive to nonnormality in the data and has been shown to have power nearly as great as Bartlett's test (Sec. 9.3.3) for normally distributed data and to have power superior to Bartlett's for nonnormal data (Milliken and Johnson, 1984a).

To conduct Levene's test, first compute the new variable.

$$Z_{ij} = | x_{ij} - \overline{x}_i | \qquad [9.2]$$

where x_{ij} represents the jth value for the ith well and $\overline{x}_i$ is the ith well mean. The values Z_{ij} represent the absolute values of the residuals. A standard one-way ANOVA is then computed on the variables Z_{ij}. If the F-test is significant, the hypothesis of equal group variances is rejected.

EXAMPLE 9.2

Apply Levene's test for homogeneity of variance to the monitoring data collected at four individual wells.

	Sulfide Concentration (mg/L)			
Sampling time	**Monitoring well 1**	**Monitoring well 2**	**Monitoring well 3**	**Monitoring well 4**
1	29.4	16.1	7.8	27.9
2	12.6	4.7	12.4	21.9
3	18.2	36.1	3.6	12.6
4	31.3	32.1	18.4	36.7
Group Mean $\overline{x}_i$	22.9	22.3	10.6	24.8

Step 1. Calculate the group mean for each well ($\overline{x}_i$)

Step 2. Compute the absolute residuals $Z_{ij} = | x_{ij} - \overline{x}_i |$ in each well and the well means of the residuals ($\overline{Z}_i$)

Sampling time	Absolute residuals			
	Well 1	Well 2	Well 3	Well 4
1	6.5	6.2	2.8	3.1
2	10.3	17.6	1.8	2.9
3	4.7	13.8	7.0	12.2
4	8.4	9.8	7.8	11.9
Well mean ($\overline{Z}_i$) =	7.48	11.85	4.85	7.53
Overall mean ($\overline{Z}$) =	7.93	Variance of the residuals S_z^2 =		20.0

Step 3. Compute the sums of squares for the absolute residuals

$$SS_{\text{Total}} = (n - 1)\, S_z^2 = (16 - 1)\,(20.0) = 300.0$$

$$SS_{\text{Wells}} = \sum n_i \overline{Z}_i^2 - n\overline{Z}^2 = 4(7.48)^2 + 4(11.85)^2 + 4(4.85)^2 + 4(7.53)^2 - 16(7.93)^2$$

$$= 223.8 + 561.7 + 94.1 + 226.8 - 1005.5$$

$$= 100.9$$

$$SS_{\text{Error}} = SS_{\text{Total}} - SS_{\text{Wells}} = 300.0 - 100.9 = 199.1$$

Step 4. Construct an analysis of variance table to calculate the F-statistic. The degrees of freedom df are equal to (number of groups – 1) = 4 – 1 = 3 df, and (number of samples – number of groups) = 16 – 4 = 12 df.

Source	Sum of squares	*df*	Mean square	*F*-ratio
Between wells	100.9	3	33.6	2.02
Error	199.1	12	16.6	
Total	300.0	15		

Step 5. The tabulated value of $F_{.05} = 3.49$ with 3 and 12 df is larger than the F-ratio of 2.02 and so the assumption of equal variances cannot be rejected.

(*Note:* If the F-ratio had been larger than the F-statistic, the assumption of equal variances would be rejected. In this situation, the most likely next step would be to log transform the data and retest the data).

9.3.3 Bartlett's Test

Another means of testing whether a number of population variances of normal distributions are equal involves use of Bartlett's test. Recall that homogeneity of variances is an assumption made in the analysis of variance when comparing concentrations of constituents between background and compliance monitoring locations.

Barlett's test examines whether the sample data provide evidence that the well groups have different variances. However, Bartlett's test is sensitive to nonnormality in the data and may give misleading results unless one knows in advance that the data are approximately normal (Milliken and Johnson, 1984a). With long-tailed distributions, the test too often rejects equality (homogeneity) of the variances.

Test Procedure—assume that data from k monitoring well groupings are available and that there are n_i data points for monitoring group i.

Step 1. Compute the k sample variances

$$S_i^2 = \frac{\sum_{j=1}^{n_i} (x_{ij} - \overline{x}_i)^2}{n_i - 1} \text{ for all } i \text{ groups} \qquad [9.3]$$

Each variance has associated with it $f_i = n_i - 1$ degrees of freedom.

Take the natural logarithms of each variance as $\ln(S_1^2), \ldots, \ln(S_k^2)$

Step 2. Compute the test statistic

$$\chi^2 = f \ln(S_p^2) = \sum_{i=1}^{k} f_i \ln(S_i^2) \qquad [9.4]$$

where

$$f = \sum_{i=1}^{k} f_i = \left(\sum_{i=1}^{k} n_i\right) - k \qquad [9.5]$$

In words, f is the total sample size minus the number of well groups. Also,

$$S_P^2 = \frac{1}{f} \sum_{i=1}^{k} f_i S_i^2 \qquad [9.6]$$

is the pooled variance (weighted variance across wells groups).

Step 3. Using the chi-square table, find the critical value for χ^2 with k–1 degrees of freedom at a predetermined significance level (e.g., five percent). If the calculated value χ^2 is larger than the tabulated value, then conclude that the variances are not equal at that significance level.

EXAMPLE 9.2

Utilize Bartlett's Test to determine if there are statistically significant differences in the variances for the six monitoring well records for iron.

Sample number	Iron concentrations (μg/L) MW1	MW2	MW3	MW4	MW5	MW6
1	812	712	675	556	623	472
2	694	695	490	503	591	499
3	774	—	526	—	—	526
4	858	—	499	—	—	403
n_i	4	2	4	2	2	4
$f_i = n_i - 1$	3	1	3	1	1	3
S_i	69.4	12.0	86.4	37.5	22.6	52.8
S_i^2	4819.7	144.5	7459	1404	512	2790
$f_iS_i^2$	14459	144.5	22377	1404	512	8370
$\ln(S_i^2)$	8.48	4.97	8.92	7.25	6.24	7.93
$f_i\ln(S_i^2)$	25.44	4.97	26.75	7.25	6.24	23.79

$$f = \sum f_i = 3 + 1 + 3 + 1 + 1 + 3 = 12$$

$$S_P^2 = \frac{1}{f}\sum_{i=1}^{k} f_i S_i^2 = \frac{1}{12}(14459.1 + 144.5 + 22377 + 1404 + 512 + 8370)$$

$$= 3938.9$$

$$\ln(S_P^2) = 8.28$$

$$\chi^2 = f \ln (S_P^2) - \sum f_i \ln (S_i^2)$$

$$= 12(8.28) - (25.44 + 4.97 + 26.75 + 7.25 + 6.24 + 23.79)$$

$$= 99.34 - 93.37 = 5.97$$

The critical χ_c^2 value with $6 - 1 = 5$ degrees of freedom at five percent level is 11.1. Since $5.97 < 11.1$ we cannot reject the null hypothesis; that is, the test is not significant at the five percent level, suggesting that the variances are not significantly different.

It is noteworthy that the differences in the variances between the various monitoring groups is quite large and yet the null hypothesis is not rejected. With very small data sets, the differences in the variances have to be very large for the differences to be considered statistically significant. As a demonstration of this, consider the redefinition of the example where n_i for each of the six monitoring wells is taken to be five times as large, but the remaining values are as indicated previously. Then

n_i	20	10	20	10	10	20
f_i	19	9	19	9	9	19
S_i	69.4	12.0	86.4	37.5	22.6	52.8
S_i^2	4819.7	144.5	7459	1404	512	2790
$f_iS_i^2$	91574	1300.5	141721	12636	4608	53010
$\ln(S_i^2)$	8.48	4.97	8.92	7.25	6.24	7.93
$f_i\ln(S_i^2)$	161.1	44.73	169.5	65.25	56.16	150.7

$$f = \sum_{i=1}^{6} f_i = 84$$

$$S_P^{\,2} = \frac{1}{f}\sum_{i=1}^{k} f_i\, S_i^2 = 3629$$

$$\ln(S_P^{\,2}) = 8.20$$

$$\chi^2 = 41.39$$

The critical χ_c^2 is 11.1, less than the calculated 41.39; the null hypothesis is rejected.

If the results were shown to have different (heterogeneous) variances, then a logarithmic transformation of the data should be utilized and Barlett's test redone. On the other hand, unequal variances among monitoring data could be a direct indication of well contamination, since the individual data could come from different distributions (i.e., different means and variances). The next step, given the significant difference in the variance, is to test which variances are different. In the case of two sets of data, the test of equality of variances becomes the *F*-ratio test described in Chap. 8. Bartlett's test simplifies in the case of equal sample sizes for each of the groups (i.e., for $n_i = n$, for $i = 1, \ldots, k$). The test used then is Cochran's test, which focuses on the largest variance and compares it to the sum of all the variances (U.S. EPA, 1989). Technical aids for the procedures under the assumption of equal sample sizes are given by Nelson (1987).

9.4 ANOVA PROCEDURE

The mathematical steps of the ANOVA procedure involve the following: assume there are *P* well groupings with n_i data points (concentrations) at the *i*th well. These data can be from either a single sampling period or from more than one. The statistical analyst can check for seasonality before proceeding, by plotting the data over time.

Step 1. Arrange the $N = \sum_{i=1}^{p} n_i$ monitoring results in a data table as indicated below and calculate the residuals

Well number	Monitored results				Well mean	Residuals			
1	x_{11}	x_{12}	x_{13}	$x_{1,n1}$	$\overline{x}_1$	R_{11}	R_{12}	R_{13}	$R_{1,n1}$
2	x_{21}	x_{22}	x_{23}	$x_{2,n2}$	$\overline{x}_2$	R_{21}	R_{22}	R_{23}	$R_{2,n2}$
i	x_{i1}	x_{i2}	x_{i3}	$x_{i,ni}$	$\overline{x}_i$	R_{i1}	R_{i2}	R_{i3}	$R_{i,ni}$
p	x_{p1}	x_{p2}	x_{p3}	$x_{p,np}$	$\overline{x}_p$	R_{p1}	R_{p2}	R_{p3}	$R_{p,np}$
					grand mean = $\overline{x}$				

The well means are calculated as

$$\bar{x}_i = \frac{\sum_{j=1}^{n_i} x_{ij}}{n_i} \quad [9.7]$$

The grand mean calculated as

$$\bar{x} = \frac{\sum_{i=1}^{p} \sum_{j=1}^{n_i} x_{ij}}{N} \quad [9.8]$$

where N is the total number of observations,

$$N = \sum_{i=1}^{p} n_i$$

Step 2. Compute the sum of squares of difference *between* well means and the grand mean

$$SS_{\text{wells}} = \sum_{i=1}^{p} n_i (\bar{x}_i - \bar{x})^2 \quad [9.9]$$

This sum of squares has p–1 degrees of freedom associated with it and is a measure of the variability *between* wells.

Step 3. Compute the corrected total sum of squares

$$SS_{\text{total}} = \sum_{i=1}^{p} \sum_{j=1}^{n_i} (x_{ij} - \bar{x})^2 \quad [9.10]$$

This sum of squares has N–1 degrees of freedom associated with it.

Step 4. Compute the sum of squares of differences of observations *within* wells from the well means. This is the sum of squares due to error and is obtained by subtraction

$$\begin{aligned} SS_{\text{error}} &= SS_{\text{total}} - SS_{\text{wells}} \\ &= \sum_{i=1}^{p} \sum_{j=1}^{n_i} (x_{ij} - \bar{x}_i)^2 \quad [9.11] \\ &= \sum \sum R_{ij}^2 \end{aligned}$$

It has associated with it (N–P) degrees of freedom and is a measure of the variability *within* wells.

Step 5. Set up an ANOVA table (where MS = mean square)

Source of variation	Sum of squares	Degrees of freedom	Mean squares	F
Between wells	SS_{wells}	$P-1$	MS_{wells}	$F = \dfrac{MS_{\text{wells}}}{MS_{\text{error}}}$
	$= \sum(\bar{x}_i - \bar{x})^2$		$= SS_{\text{wells}}/(P-1)$	
Error (within wells)	SS_{error}	$N-P$	MS_{error}	
(= "residual variation")	$= \sum\sum(x_{ij} - \bar{x}_i)^2$		$= SS_{\text{error}}/(N-P)$	
Total	SS_{total}	$N-1$		
	$\sum_{i=1}^{p}\sum_{j=1}^{n_i}(x_{ij} - \bar{x})^2$			

The two mean squares, MS_{wells} and MS_{error} are determined by dividing the appropriate sum of squares by the appropriate number of degrees of freedom.

If the between-sample variation is significantly greater than the within-sample variation, then the potential exists that the samples were not in fact drawn from the same population but from populations whose average values differed. If this proves to be the situation, in addition to the within-population variation, there exists also a between-population variation.

Step 6. To test the hypothesis of equal means for all P wells, compute $F = MS_{\text{wells}}/MS_{\text{error}}$. Compare this statistic to the tabulated F-statistic with $(P-1)$ and $(N-P)$ degrees of freedom. If the calculated F value exceeds the tabulated value, reject the hypothesis of equal well means. Otherwise conclude there is no significant difference between the concentrations at the P wells (groupings).

If a significant difference exists, at least one pair of well means is different. To find out which compliance well exceeds background, multiple comparisons must be calculated.

In the case of a significant F, one must determine which of the compliance well(s) are contaminated. This is done by comparing each compliance well with the background well(s). Concentration differences between a pair of background wells and compliance wells or between a compliance well and a set of background wells are called, respectively, *contrasts* in the ANOVA and multiple comparison framework (U.S. EPA, 1989).

Step 7. Determine if the significant F is due to differences between background and compliance wells (involves computation of Bonferroni t-statistics as described below). Compute the Bonferroni t-statistic (on each compliance well).

Assume that of the P wells, u are background and m are compliance wells (i.e., $u + m = P$). Then m differences (m compliance wells are each compared with the average of the background wells) need to be computed and tested for statistical significance. If there are more than five downgradient wells, the individual comparisons are done at the comparisonwise significance level of one percent, which make the experimentwise significance level greater than five percent.

If more than five comparisons are involved, perform each comparison at one percent significance level. If less than five, divide by the number of comparisons.

Obtain the total sample size of all u background wells

$$n_{up} = \sum_{i=1}^{u} n_i \qquad [9.12]$$

Compute the average concentration from the u background wells

$$\overline{x}_{up} = \frac{1}{n_{up}} \sum_{i=1}^{n} \overline{x}_i \qquad [9.13]$$

Compute the m differences between the average concentrations from each compliance well and the average background wells

$$\overline{x}_i - \overline{x}_{up} \text{ for all } i = 1, \ldots m$$

Compute the standard error of each difference as

$$SE_i = \left[MS_{\text{error}} \left(\frac{1}{n_p} + \frac{1}{n_i} \right) \right]^{0.5} \qquad [9.14]$$

where MS_{error} is determined from the ANOVA matrix and n_i is the number of observations at well i (i.e., they are all tested at once). Obtain the t-statistic $t = t_{N-P, 1-\alpha/2}$ from Bonferroni's t-table with $\alpha = 0.05$ and $df = N-P$ degrees of freedom. Then compute the m quantities $D_i = SE_i{}^*t$ for each compliance well i. If $m > 5$ use the entry for $t_{N-P, 1-0.01}$ (i.e., the entry at $m = 5$).

If the difference $\overline{x}_i - \overline{x}_{up}$ exceeds D_i then the ith compliance well is concluded to have significantly higher concentrations than the average background wells. The comparison needs to be performed for each of the m compliance wells individually. The test is designed so that the overall experimentwise error is five percent when there are not more than five compliance wells.

EXAMPLE 9.3

The sulfide concentration data from Ex. 9.2 plus data for two additional monitoring wells are listed below. If monitoring wells 1 and 2 are upgradient, determine if any of the monitoring wells 3 through 6 are significantly different.

Monitoring well	Monitoring data				Mean concentration $\overline{x}_i$	Standard deviation S_i
1	29.4	12.6	18.2	31.3	22.9	8.96
2	16.1	4.7	36.1	32.1	22.3	14.5
3	17.8	22.4	13.6	28.4	20.6	6.35

Monitoring well	Monitoring data				Mean concentration $\bar{x}_i$	Standard deviation S_i
4	27.9	21.9	12.6	36.7	24.8	10.1
5	29.8	32.1	41.0	28.9	33.0	5.53
6	39.6	46.2	42.0	39.9	41.9	3.04
					$\bar{x} = 27.6$	

Complete the between-well sum of squares

$$\begin{aligned} SS_{\text{wells}} &= 4(22.9 - 27.6)^2 + 4(22.3 - 27.6)^2 + 4(20.6 - 27.6)^2 + 4(24.8 - 27.6)^2 + \\ &\quad 4(33.0 - 27.6)^2 + 4(41.9 - 27.6)^2 \\ &= 88.4 + 112.4 + 196 + 31.4 + 116 + 818 \\ &= 1362 \end{aligned}$$

with six wells – 1 = 5 degrees of freedom total.

Compute the corrected total sum of squares.

$$\begin{aligned} SS_{\text{total}} &= (29.4 - 27.6)^2 + (12.6 - 27.6)^2 + (18.2 - 27.6)^2 + \ldots \\ &= 2793 \end{aligned}$$

with 24 observations – 1 = 23 degrees of freedom

Obtain the within-well or error sum of squares by subtraction

$$SS_{\text{error}} = SS_{\text{total}} - SS_{\text{wells}} = 2793 - 1362 = 1431$$

with 24 observations – 6 wells = 18 degrees of freedom

Source of variation	Sums of squares	Degrees of freedom	Mean squares	F
Between wells	1362	5	$\frac{1362}{5} = 272.4$	$\frac{272.4}{79.5} = 3.4$
Error (within wells)	1431	18	$\frac{1431}{18} = 79.5$	
Total	2793	23		

The calculated F-statistic is 3.4. The tabulated F value with 5 and 18 df at the $\alpha = 0.05$ level of significance is 2.77 (see Table 8.2). Since the calculated value exceeds the tabulated value, the hypothesis of equal means must be rejected. Subsequent analyses must be carried out to determine where the difference exists.

Compute the Bonferroni t-statistic. There are four downgradient or compliance wells, so $m = 4$ comparisons are involved.

$n_{up} = 8 =$ the total number of samples in the upgradient or background wells.

$$\overline{x}_{up} = 22.6 = \text{the average concentration of upgradient wells.}$$

Compute next the differences between the four compliance wells and the average of the two background wells as

$$\overline{x}_3 - \overline{x}_{up} = 20.6 - 27.6 = -7.0$$

$$\overline{x}_4 - \overline{x}_{up} = 24.8 - 27.6 = -2.8$$

$$\overline{x}_5 - \overline{x}_{up} = 33.0 - 27.6 = 5.4$$

$$\overline{x}_6 - \overline{x}_{up} = 41.9 - 27.6 = 14.3$$

Compute the standard error of each difference. Note that for this case, since the number of observations is the same for all compliance wells, the standard errors for the four differences are all the same.

$$SE_i = \left[79.5\left(\frac{1}{8} + \frac{1}{4}\right)\right]^2 = 5.46 \text{ for } i = 3, \ldots, 6$$

From Table A.15, the critical t with $(24 - 6) = 18$ degrees of freedom, $m = 4$ and $\alpha = 0.05$ gives $t = 2.4$.

Compute the quantities D_i as

$$D_i = SE_i \times t = 5.46 \times 2.4 = 13.1 \text{ for } i = 3, \ldots, 6$$

Only for $\overline{x}_6 - \overline{x}_{up} = 14.3$ exceeds the critical value of 13.1. Consequently, there is significant evidence of contamination at monitoring well 6.

9.5 TWO-WAY ANOVA

One-way ANOVA was used in the previous sections to examine differences between a number of monitoring locations. In the one-way ANOVA, the variation of the observations is partitioned into two components, one measuring the variability between the group means and the other measuring the variation within each group. Alternatively, the two-way ANOVA, Term [1] of Eq. [9.15] below, measures the variance between treatments, Term [2] measures the variation between blocks, and Term [3] measures the residual variation. The algebraic identity is

$$\sum_{i,j} (x_{ij} - \overline{x})^2 = c\sum_{i=1}^{r} (\overline{x}_i - \overline{x})^2 + r\sum_{j=1}^{c} (\overline{x}_j - \overline{x})^2 + \sum_{i,j} (x_{ij} - \overline{x}_i - \overline{x}_j + \overline{x})^2$$

[1] [2] [3] [9.15]

9.6 ITERATIONS AND DATA TRANSFORMATIONS

The ANOVA is a useful test but its use does entail a series of assumptions that sometimes do not fit a particular application. As a result, data transformations are often performed to obtain approximate normality and to stabilize the variance of ANOVA residuals across different groups. Avoidance of heteroscedasticity or "unequal variances" is of interest because approximately equal variances are required for a parametric ANOVA to give valid results. This homogeneity (equality) of variances of residuals across monitoring locations is the most important assumption in parametric ANOVA. If this assumption is not met, the power of the *F*-test—its ability to detect differences among the group means—is reduced. Checking the equality of variance is important, for the analysis of variance procedure is often more sensitive to unequal variances than to moderate departures from normality. The intent of this test is to determine whether group variances are equal or significantly different.

If the residuals from the transformed data do not meet the parametric ANOVA requirements, then nonparametric approaches to ANOVA may work (see Chap. 11 for a description of the nonparametric or Kruskal-Wallis test, which utilizes the ranks of the observations as opposed to the monitored values or magnitudes). Alternatively, use of a data transformation to stabilize or equalize the variances may be utilized. Levene's test of homogeneity of group variances is not as sensitive to departures from normality as Bartlett's test. Bartlett's test checks equality, or homogeneity, of more than two variances with unequal sample sizes. The *F*-test is a special situation when there are only two groups to be compared.

In the case of equal sample sizes but more than two variances to be compared, the user may want to use Hartley's (or maximum *F*-ratio) test (Nelson, 1987).

9.7 CONCERNS WITH MULTIPLE COMPARISONS

Although the error rate of any single test is low, if a sufficiently large number of comparisons are completed, the chance of making at least one mistake (false positive error) overall is high. Miller (1981) showed mathematically that the chance of one or more false positives when performing several statistical tests can never be greater than the sum of the chances of false positives on each of the individual tests.

$$\text{In general, Pr \{at least one false positive\}} = 1 - (1 - \alpha)^{wc}$$

where w = number of monitoring locations and c = number of constituents being tested for. For 100 tests each run at five percent type I error rate, we would expect on average

$$100 \times .05 = 5 \text{ false positives}$$

The probability of at least one false positive = $1 - (.95)^{100}$ = 99.4 percent. In other words, in an assessment for a large monitoring program, it is likely we would see at least one significant test result during statistical testing.

The Bonferroni technique (see Sec. 9.4) is conservative in that the overall significance level (risk of at least one false positive) is generally smaller than the nominal rate, although it will not be excessively so if the number of tests is small and if the tests are not highly correlated with each other.

The use of *t*-tests can be carried out with critical values adjusted to account exactly for a known correlation structure between tests. When the correlations are induced by using the same denominator and background mean for tests with means of several downgradient wells, it is appropriate to use *multiple* comparisons with control (MCC) tests, of which the best known is Dunnett's (1955) test.

9.8 SUMMARY

ANOVA models are typically used to detect differences in observed phenomena rather than to predict new values of a dependent variable. In large networks, even the parametric ANOVA has a difficult time finding the needle in a haystack. The reason for this is that the ANOVA *F*-test combines all downgradient wells simultaneously so that any clean wells are mixed together with one or more contaminated well(s).

References to ANOVA techniques include Winer (1971), Hays (1973), Kerlinger and Pedhazur (1973), Milliken and Johnson (1984b), and Kufs (1992). Multiple discriminate analysis determines statistically, by an analysis of variance, those linear combinations of variables which best discriminate among the different variables (e.g., Chappell, 1976).

9.9 REFERENCES

CHAPPELL, V.G. "Determining Watershed Sub-Areas with Principal Component Analysis," *Water Resources Bulletin*, 12, no. 6 (1976), 1133–1139.

DUNNETT, C.W. "A Multiple Comparisons Procedure for Comparing Several Treatments with a Control," *Journal of the American Statistics Association*, 50 (1955), 1096–1121.

HAYS, W.L. *Statistics for the Social Sciences*, 2nd ed. New York, NY: Holt, Rinehart and Winston, 1973.

KERLINGER, F.N., and E.J. PEDHAZUR. *Multiple Regression in Behavioral Research.* New York, NY: Holt, Rinehart and Winston, 1973.

KUFS, C.T. "Statistical Modeling of Hydrogeologic Data—Part I: Regression and ANOVA Models," *Groundwater Monitoring and Remediation* (Spring, 1992), 120–130.

MILLIKEN, G.A. and D.E. JOHNSON. *Analysis of Messy Data: Volume 1, Designed Experiments.* Belmont, CA: Lifetime Learning Publications, 1984a.

MILLIKEN, G.A. and D.E. JOHNSON. *Analysis of Messy Data.* New York, NY: Van Nostrand Reinhold, 1984b.

NELSON, L.S. "Upper 10 Percent, 5 Percent and 1 Percent Points of the Maximum F-Ratio," *Journal of Quality Technology*, 19 (1987), 165–167.

U.S. EPA, *Statistical Analysis of Ground-Water Monitoring Data at RCRA Facilities. Interim Final Guidance.* Washington, DC: Office of Solid Waste, Waste Management Division (April, 1989).

WINER, B.J. *Statistical Principles in Experimental Design.* New York, NY: McGraw-Hill Book Co., 1971.

9.10 PROBLEMS

9.1. Chloride measurements at a series of monitoring wells are listed below. The first two monitoring locations are upgradient and the remainder are downgradient. Determine whether there is evidence to indicate contamination at any of the downgradient monitoring locations.

Sample number	MW1	MW2	MW3	MW4	MW5
1	26	31	35	33	26
2	21	27	17	32	24
3	23	22	31	28	35
4	29	28	29	27	37

9.2. The air quality in the vicinity of a steel mill is monitored at a series of monitoring locations. All measurements are in $\mu g/m^3$. In September 1993, an additional pollution control device was installed. Did the pollution control device cause a significant change to the air quality?

	Monitoring Location			
Time	A	B	C	D
Aug 90	27	18	25	21
Aug 91	25	16	20	25
Aug 92	36	21	22	29
Aug 93	32	22	28	35
Aug 94	28	19	21	30
Aug 95	27	18	25	32
Aug 96	31	14	24	27
Aug 97	29	21	18	33

9.3. Use Levene's test to examine homogeneity of variance for monitoring data, using the data from the diskette data file NORM.

a. Examine the homogeneity of variances for data sets A through J.
b. Examine the homogeneity of variances for data sets K through T. How effectively does logarithmically transforming the data show homogeneity of variances?
c. Examine the homogeneity of variances for data sets F through O. Note while undertaking this problem that the first ten monitoring records (A through J) were generated from a normal distribution, while the second ten monitoring records were generated from a lognormal distribution.

9.4. Repeat Prob. 9.3, but utilize Bartlett's test to examine homogeneity of variance.

9.5.
- **a.** Repeat Prob. 9.3, but utilize only the first five sample values from each of the sampling records. Comment on how your answers to this problem differ from your answers in Prob. 9.3.
- **b.** Repeat Prob. 9.3, but utilize only the first ten sample values from each of the sampling rounds.

9.6. Using the Data file NORM from the diskette, assume that monitoring locations A through C are upgradient and D through G are downgradient. Determine whether the hazardous waste disposal facility is impacting the environment.

CHAPTER 10

Testing Differences Between Monitoring Records When Censored Data Records Exist

"The public has become used to conflicting opinion . . . many have come to feel that for every Ph.D., there is an equal and opposite Ph.D."

Tim Hammonds, Food Marketing Institute

10.1 INTRODUCTION

The public has become quite vocal about decreasing their exposures to environmental contaminants. This considerable interest, due in part to widespread media coverage of environmental concerns, has influenced regulatory action to limit exposure risks. The result has been expansions in monitoring programs to include many more chemical constituents. The stringency of environmental quality standards has increased to the point that the maximum concentrations limit (MCL) allowable is at or very near the detection capability of laboratory instrumentation for some constituents. The upshot for statistical analysis is that monitoring programs contain numerous censored data entries; that is, concentrations are reported as "less than" a method detection level (MDL).

The degree of censored environmental data varies from one sampling program to another. Different laboratory procedures have imposed different limitations on their measurement capabilities. Other sources of difficulty arise due to matrix interferences, where the presence of one constituent influences the ability to quantify the concentrations of another constituent. Further discussion of the rationale for different magnitudes of detection level is presented in Sec. 10.2; we wish to acknowledge here that the actual concentration of censored data can lie anywhere between zero and the method detection limit. In the absence of a specified magnitude, estimation of the means and standard deviations becomes difficult and calculations using the t-test are less defensible. The problem is how to deal with less-than detection level values.

One approach has been to ignore all censored values. This gives a biased overestimate of the mean and a biased underestimate of the standard deviation.

To ignore the values below the detection level by truncating the data set is to ignore information. Another simple approach is to set the less-than detection level magnitudes equal to the detection level, but the drawbacks to this approach are similar to ignoring the data.

EXAMPLE 10.1

a. Assume that the actual air quality concentrations of Benzo(a)pyrene are 0.4, 1.2, 2.6, 0.3, 1.7, 0.6 $\mu g/m^3$ but that the available instrumentation cannot measure at levels less than one. The resulting vector of concentrations would be reported as $< 1, 1.2, 2.6, < 1, 1.7, < 1$. Determine the impact on estimates of the mean and the standard deviation using the simple procedure of truncating the less-than values.

The estimated mean and standard deviation using the actual concentrations are as follows:

$$\text{For } n = 6, \bar{x} = 1.13 \text{ and } S = .89$$

Using the abbreviated data set (truncating the censored data).

$$\text{For } n = 3, \bar{x}_{\text{trun}} = 1.83 \text{ and } S = .71$$

The result is that the mean has increased and the standard deviation has decreased.

b. Redo the calculations for $\bar{x}$ and S with less thans set equal to the method detection level.

For $n = 6$ and censored data set equal to the detection limit, $\bar{x} = 1.42$ and $S = .64$. Again, the result is that the mean has increased and the standard deviation has decreased, but both statistics are lower than those calculated using the trucated data set.

Other approaches exist to deal with censored data; the maximum likelihood method and regression analyses, for example, but these approaches are more involved mathematically (e.g., see Sharma et al., 1995 a and b). Clearly some form of statistical analysis must extract information from the censored data, though its appropriateness in a particular circumstance is likely problem specific.

The objective in this chapter is to clarify the pros and cons of several alternatives to dealing with censored data, particularly those approaches applicable to brief data sets. Each will be seen to have its own advantages and disadvantages. Different alternatives associated with different circumstances will be examined to identify a consensus as to which procedure to apply and when.

The key focus in this chapter is retention of the parametric magnitudes of the data. Tests must replace less-than values by some parametric value. Chapter 11 deals with nonparametric or ranking approaches.

10.2 ALTERNATIVE TYPES OF CENSORING

Various laboratory procedures have limitations in their measurement capabilities resulting in data censoring. In current environmental monitoring records, the frequency of occurrence of censored data may be very high. Reasons include:

1. Concerns about environmental quality are becoming more prevalent. Current sampling programs involve dramatically-increased numbers of constituents over those required, for example, 10 years ago. Since many programs utilize a standard set of constituents for which analyses are required, many of the constituents are at very low levels.
2. As noted in Chap. 2, many environmental quality data sets are skewed to the left because concentration data are constrained on the low side (concentrations cannot be negative) but not on the upper side.
3. Improved analytical instrumentation now provides measurement capability of water sample constituents in the μg/L range and lower, while our awareness of the toxicological effects of various contaminants has widened.

The result is that many of the environmental quality data sets have left-censored data as depicted schematically in Fig. 10.1.

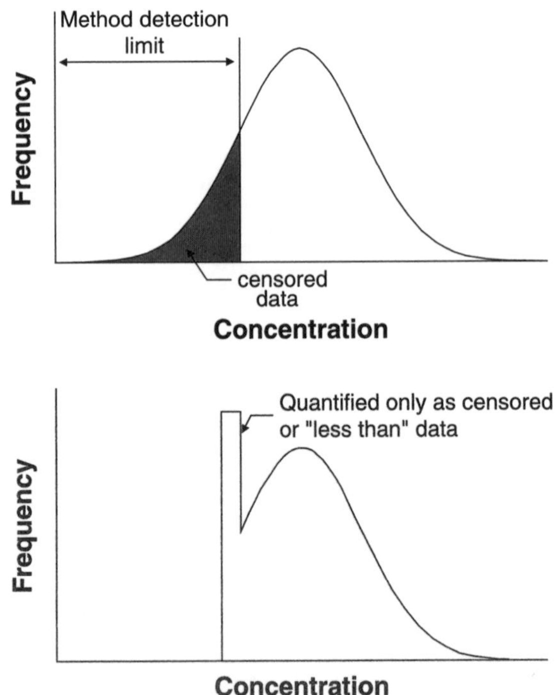

FIGURE 10.1 Influence of left-censoring of concentration data

It is important to recognize that uncertainty in reported values arises in the instrumentation for a variety of reasons:

1. Some degree of variability in test data is inevitably present in any chemical test, no matter how carefully conducted.
2. The results of detection limit calculations are less certain when the data have been generated by a number of different laboratories and/or over time while the technology for measurements has evolved. The effect of different standards, reagents, analysts, instruments, and lab protocols often results in biases among labs and variable detection limits.
3. Multiple measurements of the same sample by the same laboratory will give different results, which may lead to conflicting determinations.

All of these features relate to QA/QC where QA is defined as quality assurance, a definitive plan for laboratory operation that specifies the measures used to produce data of known precision and bias, and QC refers to quality control where a set of measures is utilized to ensure that the process is under control.

There have been numerous investigations of methods for estimating parameters for censored data drawn from assumed specified parent distributions such as the normal or lognormal (e.g., David, 1981). There have been far fewer studies of the application of these methods to environmental data for which parent distributions are unknown and sample sizes are small. When the frequency of data censoring is low the approach utilized may not make a significant difference, but as the frequency of data censoring increases, the reliability of statistical analyses that can be applied to censored data decreases.

The complexity of statistical analyses increases substantially when multiple levels of detection limits exist in environmental quality records. As an example of this situation see the values listed in Table 10.1, obtained from one of the monitoring wells at the Operating Industries Landfill site in California. The variation in detection levels apparent in these values is not unusual in monitoring data now being reported.

The analysis process normally detects and measures data below the reporting limit but the difficulty is the quantification; the presence of the constituents is known but not the magnitude. The most frequently used of these limits of measurement are the detection and quantitation limits, but others exist. There are several detection and quantitation limit acronyms commonly used by laboratories (see Table 10.2).

Further complications in the statistical analysis of data arise with the data qualifiers. Examples of data qualifiers include those listed in Table 10.3.

The detection limit is the lower limit at which you can differentiate that measurements are quantitatively meaningful (Taylor, 1987). The method detection limit (MDL) is similar and in some cases the same as the limit of detection (LOD). The only difference is that the LOD is defined with a sample blank, whereas the MDL is defined with either a sample blank or with one, or several matrix samples.

The practical quantitation limit (PQL) is usually equal to the MDL multiplied by a factor of 3 to 5 (*Federal Register,* 1987) but varies for different chemicals and different samples.

TABLE 10.1 Example of a groundwater quality monitoring record with multiple detection values

Date of sampling	Cadmium concentration (μg/L)
03/07/85	<5
05/27/85	<3.6
08/13/85	<1.9
09/09/86	6
02/05/87	<0.2
08/13/87	<2.4
10/14/87	<4
10/14/87	5.4
05/12/88	12.6
09/28/88	<3.1
08/09/89	<4
11/07/89	<5
02/05/90	<2.4
04/30/90	<5
06/20/91	<3
05/15/92	<3.7
07/19/93	<1.5

Source: CH2M-Hill, "Remedial Investigation Report for Operating Industries, Inc. Landfill, Monterey Park, California," report to U.S. EPA, San Francisco, CA, October 25, 1994.

The most important interpretive features of data qualifiers are that

1. below the MDL, the results do not demonstrate the presence of the chemical;
2. between the MDL and PQL, the constituent is present but we are unsure of the magnitude; and
3. above the PQL, take the concentration as given.

10.3 ALTERNATIVE PROCEDURES FOR STATISTICAL ANALYSIS OF ENVIRONMENTAL QUALITY DATA SETS

A number of procedures for examining data records with varying degrees of censorship have been published. The majority of these procedures assume the data are singly-censored (e.g., < 5) as opposed to multiply censored data sets (e.g., the monitoring record consists of concentrations reported as < 5 as well as < 15). Gilbert and Kinnison (1981), Gilliom and Helsel (1986), Gleit (1985), Helsel and Gilliom (1986), Kushner (1976), and Owen and deRouen (1980) all examine procedures for estimating parameters based on singly censored data sets.

TABLE 10.2 Indication of detection and quantitation limits

Limit of detection (LOD)	The lowest concentration that can be determined to be statistically different from a blank.
Method detection limit (MDL)	The minimum concentration of an analyte that can be determined with 99 percent confidence that the true value is greater than zero. The measure is suitable only for detecting that a pollutant is actually present. The MDL is estimated on the basis of ideal laboratory conditions with ideal analyte samples and does not account for matrix or other inferences encountered when analyzing specific, actual field samples.
Instrument detection limit (IDL)	Smallest signal above background noise that an instrument can detect reliably. This measure is based on a signal-to-noise ratio and takes into account no chemical or other preparatory steps involved in an analysis.
Limit of quantitation (LOQ)	The concentration above which quantitative results can be obtained with a specific degree of confidence.
Method quantification limit (MQL)	The minimum concentration of a substance that can be measured and reported.
Practical quantitation limits (PQL)	The lowest level that can be reliably achieved within specified limits of precision and accuracy during routine laboratory operating conditions. The PQL is the concentration at which laboratories should be able routinely to determine quantitatively how much of a pollutant is present. For this reason, the PQL should be taken as the most reasonable upper bound for nondetect concentrations. PQL is time and laboratory independent for regulatory purposes, e.g., numerous current NPDES permits and cleanup levels for U.S. sites include limits for PCBs at or even below the PQL. Compliance with such limits may be the result of random chance as much as environmental diligence (Scroggin, 1994).

Source: Adapted from Anne, D.C., "Know Your Detection and Quantitation Limits," *Pollution Engineering* (May 1, 1992), pp. 901–908. Also Scroggin, D.G., "Detection Limits and Variability in Testing Methods for Environmental Pollutants: Misuse May Produce Significant Liabilities," *Hazardous Waste and Hazardous Materials*, 11, no. 1 (1994), 1–4. Reprinted with permission.

10.3.1 Simple Substitution Methods

Consider the difficulties that arise when using the *t*-test for data with censored data. From Chap. 8, recall that a simple form of the *t*-test evaluation for the purposes of explanation is written as

$$t^* = \frac{\overline{x}_1 - \overline{x}_2}{\dfrac{S}{\sqrt{n}}} \qquad [10.1]$$

where $\overline{x}_1$ and $\overline{x}_2$ are the means, S is the standard deviation, and n is the number of samples. When the data are constant or nearly so, this leads to a zero or near-zero standard deviation. When the data set contains a high percentage of censored values and these censored values are transformed to some constant value, the standard deviation in Eq. [10.1] approaches zero. As S approaches zero, the calculation of t^* in Eq. [10.1] either requires a division by zero, or leads to ill-conditioning, a situation in which very small changes in the value of a single observation produce enormous changes in the value of t^*.

TABLE 10.3 Explanation of some of the data qualifiers commonly used in relation to monitoring data

Data qualifier	
U	This qualifier indicates that the compound was analyzed for, but not detected at, the reported concentration, and is commonly referred to as "not detected." This means that the reported concentration is the quantitation limit (or reporting limit). The signal characteristic for the constituent being analyzed could not be observed or distinguished from background noise during laboratory analyses. For example, 3U μg/L of cadmium means that the sample contains less than 3 μg/L of cadmium.
B	This qualifier means the chemical was identified in the sample and in the method blank analyses with the sample, which means that all or part of the reported concentration may be due to laboratory contamination. For example, consider bis(2-ethylhexyl) phthalate (or bis for the sake of simplicity) reported as 2B μg/L. Bis is a common contaminant associated with sample collection apparatus and an analysis of a blank sample (one that is collected for checking of the lab QA/QC) is 1.5 μg/L. One should suspect that the bis in the sample is due to laboratory contamination, not environmental contamination. It is left to the judgment of the data user, not the laboratory.
J	This flag indicates that the reported concentration is an estimated value. It may be referred to as "estimated" for one of a number of reasons, including that the concentration is a tentatively identified compound, the quantitation is below the reporting limit, or the sample exceeded the holding time. For example, if the sample quantitation limit is 10 μg/L but a concentration of 3 μg/L is calculated, the value may be reported 3 J.

The development of this problem, a result of a simple transformation to some constant value, may occur as indicated in Ex. 10.2. Extension of the situations in which the less thans are set equal to, are the *simple substitution methods.* These methods set all censored observations equal to zero, or to some fraction of the detection limit (e.g., the MDL, 1/2 MDL, or zero). See Ex. 10.3.

If a small proportion of the observations are not detected, one approach is to replace the censored data with a small number (e.g., MDL/2). The U.S. EPA recommends the MDL/2 approach if less than 15 percent of all samples are nondetect, but these simple substitution methods tend to perform poorly in statistical tests when the nondetect percentage is substantial (Gilliom and Helsel, 1986). Any of these simple substitution procedures provide a degree of quantification but may seriously affect subsequent utilization of the parameters in the t-test. Hashimoto and Trussell (1983) compared several estimators of the mean for cen-

EXAMPLE 10.2

Assume the true concentrations are as follows: 3.6, 4.2, 1.7, 5.0, 4.1, but that the instrumentation has an MDL of 5.0. The resulting reported concentrations would be < 5, < 5, < 5, 5.0, < 5. Compute the standard deviations for both situations.

For actual data, $S = 1.24$.

For censored data, $S = 0.0$ by the simple substitution rule of replacing the less thans with values equal to the MDL.

EXAMPLE 10.3

If the monitoring record consists of the following eight values, compute the mean and standard deviation,

$$< 5, 7, 8, 12, < 5, < 5, 6, 9$$

using the simple substitution rule.

For the values reported as < MDL use both zero and MDL to calculate a mean and standard deviation, and report the value of mean as a range

For values reported as < MDL taken as zero

$$\bar{x} = 4.75 \text{ and } S = 4.43$$

For < MDL taken as equal to the MDL

$$\bar{x} = 7.13 \text{ and } S = 2.47$$

The mean is between the interval 4.75 and 7.13 and the standard deviation is in the range 2.47 to 4.43.

sored water quality data. Their examples illustrate the bias caused by three commonly used methods: discarding censored observations, setting all censored observation equal to zero, or assigning the detection limit to all censored observations.

10.3.2 Test of Proportions

The test of proportions is a procedure that does not account for potentially different magnitudes among the concentrations of detected values. Rather, each sample is treated as a one or zero depending on whether the measured concentration is above or below the detection limit. The test of proportions relies on a normal probability approximation to the binomial distribution of zeros and ones and assumes that the sample size is reasonably large. Generally, if the proportion of detected values is denoted by P and the sample size is n, then the normal approximation is adequate, provided that nP and $n(1 - P)$ both are greater than or equal to 5.

The test of proportions is appropriate for use when the proportion of quantified values is small to moderate (e.g., 10 to 50 percent) or if more than 50 percent of the data are below detection but at least 10 percent of the observations are quantified. If very few quantified values are reported (< 10 percent), a method based on the Poisson distribution (see Sec. 10.3.8) should be used.

A test of proportions might be used, for example, if none of the background well observations are above the MDL, but if all of the compliance well observations are above the detection limit, one would suspect contamination.

Assuming the findings at two monitoring locations are being examined for significant difference, the calculation steps involved in the procedure are as follows:

Step 1. Determine w, the number of background well samples in which the compound was detected. Let n be the total number of samples obtained from the background wells (u is the number of background or upgradient samples). Compute the proportion of detects as:

$$P_u = w/n \qquad [10.2]$$

Step 2. Determine v the number of downgradient well samples in which the compound was detected, out of m total samples:

$$P_d = v/m \qquad [10.3]$$

Step 3. Compute the standard error of the difference in proportions:

$$S_D = \left\{\left[\frac{w+v}{n+m}\right]\left[1 - \frac{w+v}{n+w}\right]\left[\frac{1}{n} + \frac{1}{m}\right]\right\}^{0.5} \qquad [10.4]$$

and form the statistic

$$Z = \frac{P_u - P_d}{S_D} \qquad [10.5]$$

Step 4. Compare the absolute value of Z from Eq. [10.5] to a value (e.g., 97.5 percent) from the standard normal distribution. For the 97.5 percent, Z_c = 1.96. If the absolute value of Z exceeds Z_c = 1.96, this provides statistically significant evidence at the five percent significance level (i.e., a two-sided test) that the proportion of compliance well samples where the compound was detected exceeds the proportion of background well samples where the compound was detected.

If the percentage of nondetects is very high and/or the sample size is fairly small, the test of proportions may lead to inaccurate results.

EXAMPLE 10.3 (CONTINUED)

Five samples out of 26 upgradient samples showed detected levels whereas 8 samples out of 33 downgradient showed detection. Is there a statistically significant difference between the upgradient and downgradient monitoring results?

$$P_u = 5/26 = .192$$

$$P_d = 8/33 = .242$$

$$S_D = \left\{\left[\frac{5+8}{26+33}\right]\left[1-\frac{5+8}{26+33}\right]\left[\frac{1}{26}+\frac{1}{33}\right]\right\}^{0.5}$$

$$= .017$$

$$Z = \frac{.192 - .242}{.017} = -2.94$$

$$|Z| = 2.94 > 1.96 = Z_c$$

Therefore, there is a statistically significant difference between the results at the upgradient and downgradient monitoring locations.

10.3.3 Plotting Position Procedure

Use of probability paper provides a simple means of estimating the mean and standard deviation in the presence of censored data (McBean and Rovers, 1984). The limitation of this method arises in that sufficient detected data must exist to convince the analyst that the correct type of probability paper is being utilized and that a reasonable line can be fit to the plotted data. This procedure involves fitting the probability distribution to the data above the detection limit and then extrapolating the line in turn to estimate the mean and standard deviation, as depicted schematically in Fig. 10.2. Potential candidate distributions include the normal (see Chap. 3) and lognormal (see Chap. 4), since the probability plotting papers are readily available and the distributions are generally appropriate to environmental quality data. For alternative distributions, see the discussion in Chap. 5.

The distributional assumptions for parametric procedures can be rather difficult to check when a substantial fraction of nondetects exist. To utilize this procedure, follow these steps:

Step 1. Rank order the detect data and, utilizing a plotting position formula such as the Weibull formula, plot the data on probability paper.

Rank	Plotting Position
(m)	$m/(n+1)$

where m = the rank of the sample and n = the total number of samples.

Step 2. Extrapolate the fitted line and read the mean and standard deviation from the graph (the mean concentration corresponds to the 50th percentile and the standard deviation corresponds to the difference between the 84th and the 50th percentile, since the standard deviation for the normal distribution corresponds to approximately 34 percent of the area under the normal distribution curve).

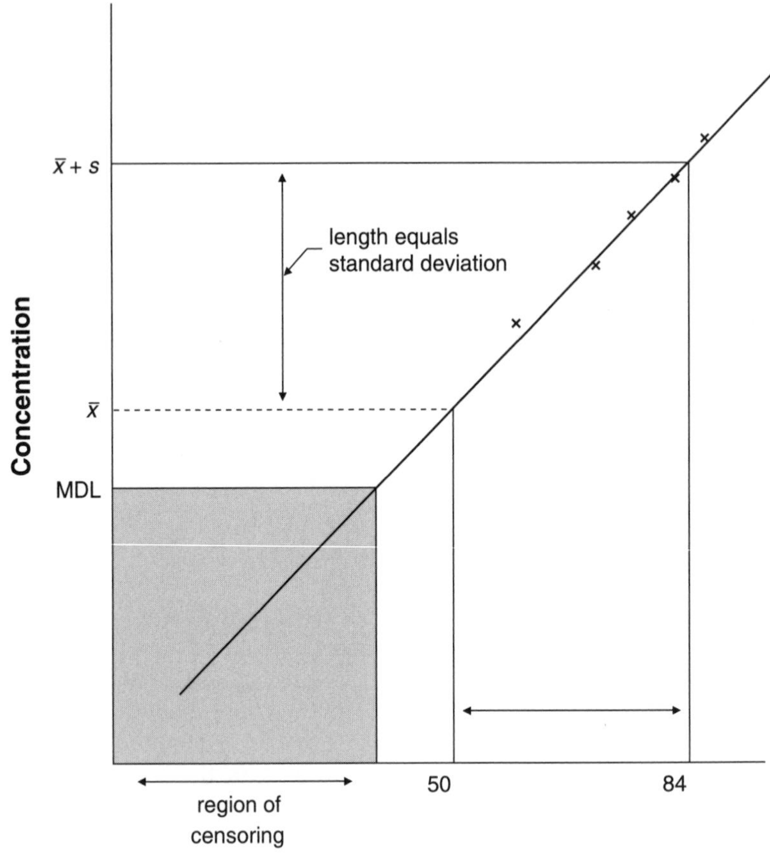

FIGURE 10.2 Extrapolation of fitted line from probability paper to estimate the mean and standard deviation

To be successful, a sufficient portion of the data must exceed the detection limit to establish that the data are reasonably described by the assumed probability distribution. As well, there must be sufficient quantities of information from which the fitted line can be drawn. For example, with a brief data set, one large value can substantially bias the resulting line (McBean and Rovers, 1984). Mage (1982) describes several techniques for plotting ranked observations on probability paper and drawing lines.

A significant advantage of the probability plotting procedure is that it gives a visual assessment of the data. The analyst can downplay the effect of an individual data point if there is a valid rationale for presuming the particular point is a data outlier. The disadvantages of probability plots are that there is no commonly accepted rule for plotting the data and drawing lines by eye; nor is there a

EXAMPLE 10.4

Consider the monitoring values summarized below. Use the plotting position procedure to estimate the mean and standard deviation.

I Monitoring data	II Rank	III Rank-ordered data	IV Plotting position[a]
<5	1	10.2	.077
10.2	2	9.2	.153
9.2	3	7.8	.231
<5	4	7.2	.308
<5	5	5.9	.385
7.2	6	5.4	.462
5.3	7	5.3	.538
<5	8	<5	.615
7.8	9	<5	.692
<5	10	<5	.769
5.4	11	<5	.846
5.9	12	<5	.923

[a] $m/(n+1)$

The plot of the data in Cols. III and IV is illustrated on Fig. 10.3. From the dashed line, the mean is estimated to be 5.4 μg/L. The 84th percentile is 8.8 with the result that the standard deviation is estimated as $S = 8.8 - 5.4 = 3.4$ μg/L.

simple objective way to judge how well the data points conform a straight line (Mage, 1982).

10.3.4 Cohen's Test

Another procedure for estimating the mean and variance of a censored normal distribution was presented by Cohen (1961). Cohen's test may be applied up to the situation of 90 percent nondetects. This approach assumes the observed data (detects and nondetects) come from the same normal or lognormal population but that the nondetect values have been censored at their detection limits. The premise of the technique is based on the censored probability model where we estimate new or adjusted mean and standard deviations.

Cohen's method adjusts the sample mean and sample standard deviation to account for data below the detection limit, assuming the data are normally distributed and the detection limit remains the same.

Step 1. Compute the sample mean $\overline{x}_d$ from the data above the detection limit as:

$$\overline{x}_d = \frac{1}{m}\sum_{i=1}^{m} x_i \qquad [10.6]$$

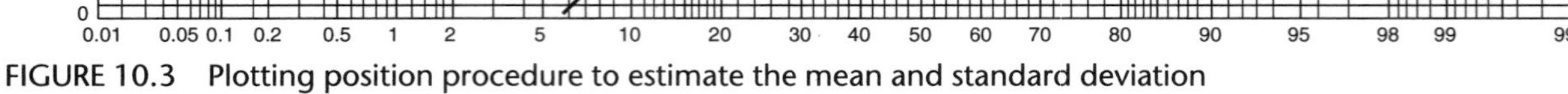

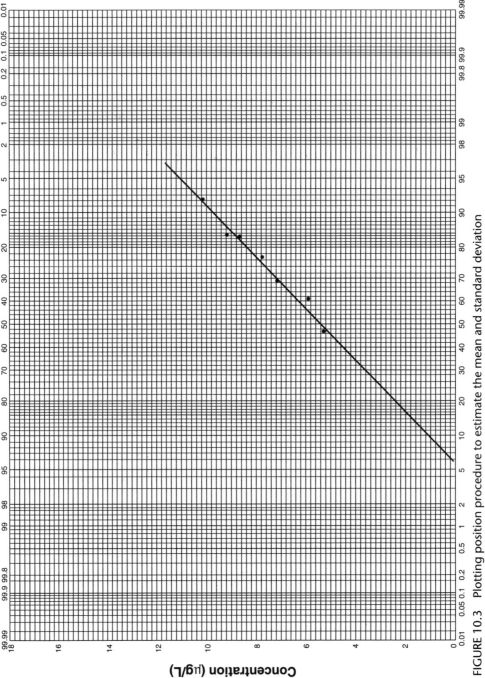

FIGURE 10.3 Plotting position procedure to estimate the mean and standard deviation

Step 2. Compute the sample variance S_d^2 from the data above the detection limit as follows:

$$S_d^2 = \frac{\sum_{i=1}^{m}(x_i - \overline{x})^2}{(m - 1)} \quad [10.7]$$

Step 3. Compute the two parameters h and γ as:

$$h = \frac{n - m}{n} \quad [10.8]$$

$$\gamma = \frac{S_d^2}{(\overline{x} - DL)^2} \quad [10.9]$$

where h is the proportion of nondetects and DL = detection limit.

Use these to determine the value of the parameter $\overline{\lambda}$ from the Table A.11.

Step 4. Estimate the corrected sample mean, which accounts for the data below the detection limit as:

$$\overline{x} = \overline{x}_d - \overline{\lambda}\,(\overline{x}_d - DL) \quad [10.10]$$

and

$$S = (S_d^2 + \overline{\lambda}\,(\overline{x}_d - DL)^2)^{0.5} \quad [10.11]$$

EXAMPLE 10.5

The monitoring record for benzene consists of the following:

$$10.1, < 5, < 5, 8.9, 6.2, 5.6, < 5\ \mu g/L$$

Use Cohen's method to estimate the sample mean and standard deviation.

Step 1. Compute the sample mean $\overline{x}_d$ for the data above the detection limit as:

$$\overline{x}_d = \frac{10.1 + 8.9 + 6.2 + 5.6}{4} = 7.7$$

Step 2. Compute the sample standard deviation for data above the detection limit:

$$S_d^2 = 4.62$$

Step 3. Compute $h = \frac{7 - 4}{7} = .429$, and

Compute $\gamma = \frac{4.62}{(7.7 - 5)^2} = .633$, thus

$$\bar{\lambda} = .817 \text{ (interpolating from Table A.11)}$$

Step 4. Estimate the corrected sample mean

$$\bar{x} = 7.7 - .817\,(7.7 - 5) = 5.49$$

Correct the standard deviation as

$$S = [4.62 + .817\,(7.7 - 5)^2]^{0.5} = 3.25$$

Cohen's method provides maximum likelihood estimates of the mean and variance of a censored normal distribution. Cohen's adjustment may not give valid results if the proportion of nondetects exceeds 50 percent. McNichols and Davis (1988) found that the false positive rate associated with the use of t-tests based on Cohen's method rose substantially when the fraction of nondetects was greater than 50 percent. This occurred because the adjusted estimates of the mean and standard deviation are more highly correlated as the percentage of nondetects increases, leading to less reliable statistical tests.

10.3.5 Aitchison's Method

When at least ten percent of the groundwater samples have a measurable (detected) value of a particular constituent, the mean and variance of the distribution can be approximated using a method developed by Aitchison (1955). Like Cohen's method, the assumption of a particular probability model for the data leads to adjusted estimates of $\bar{x}$ and S. However, Aitchison's adjustment is constructed on the assumption that the nondetect samples are free of contamination, so that all nondetects may be regarded as zero concentrations. To compute Aitchison's adjustment (Aitchison, 1955), it is assumed that the detected samples follow an underlying normal distribution or, if the detects are found to be lognormally distributed, the entire computation can be carried out using the logarithms of the data instead. Let d = the number of nondetects and n = total number of samples (detects and nondetects). Then if $\bar{x}_d$ and S_d denote respectively (where these are calculated as per Eqs. [10.6] and [10.7]), the same sample mean and standard deviation of the detected values, the adjusted overall mean $\hat{x}$ (where the ^ indicates that the sample mean is an estimate) is

$$\hat{x} = (1 - d/n)\,\bar{x}_d$$

and the adjusted overall standard deviation $\hat{S}$ is

$$\hat{S} = \left[\left(\frac{n - (d + 1)}{n - 1}\right)(S_d)^2 + \frac{d\,(n - d)\,(\overline{x}_d)^2}{n\,(n - 1)}\right]^{0.5}$$

EXAMPLE 10.6

Utilizing the same data as in Ex. 10.5, estimate the sample mean and standard deviation employing Aitchison's method.

$$d = 3$$

$$n = 7$$

$$\overline{x}_d = 7.7$$

$$S_d = \sqrt{4.62} = 2.15$$

$$\hat{x} = (1 - 3/7)\,7.7 = 4.40$$

$$\hat{S} = \left[\frac{7 - 4}{6}(2.15)^2 + \frac{3}{7}\left(\frac{4}{6}\right)(7.7)^2\right]^{0.5} = 4.39$$

It is interesting to compare these adjusted statistics with those determined using Cohen's method.

	Cohen's method	Aitchison's method
Mean	5.49	4.40
Standard deviation	3.25	4.39

The lower mean of Aitchison's method in comparison with Cohen's method is due to the assumption within Aitchison's method that nondetect samples are free of contamination.

EXAMPLE 10.7

Utilize Aitchison's method to adjust the estimate of the mean and standard deviation of the data listed below:

$$< 5, 5, < 5, 7.2, 12, 66, 6.1, 5.9\ \mu\text{g/L}$$

and then compute the 95 percent upper tolerance level.

Apparent from the detection data listed is that the data distribution is skewed (the value 66). Therefore, the data should be log transformed prior to analysis.

Reported value	Log-transformed (natural logarithm)
< 5	—
6	1.79
< 5	—
< 5	—
7.2	1.97
12	2.48
66	4.19
6.1	1.81
5.9	1.77

$$\bar{x}_d = 2.34$$

$$S_d = .946$$

$$\hat{x} = (1 - 3/9)\,(2.34) = 1.56$$

Reversing the transformation gives $e^{1.56} = 4.76$.

$$\hat{S} = \left[\frac{9-4}{8}(.946)^2 + \frac{3}{9}\left(\frac{6}{8}\right)(2.34)^2\right]^{0.5} = 1.39$$

The upper 95 percent tolerance interval is calculated as follows, using $n = 9$ and therefore degrees of freedom = 9 – 1 = 8.

First, using the log transformation

$$TI = 1.56 \pm 2.31\,(1.39) = 11.95$$

In arithmetic terms

$$TI = e^{11.95} = 154.8\ \mu g/L$$

10.3.6 Maximum Likelihood Procedure

The maximum likelihood estimator (MLE) procedure as a way of incorporating the effect of less thans in censored data sets is based on an assumption of the underlying distribution of the entire data set. MLE estimators of distributional parameters can then be derived from the uncensored observations. This procedure is useful for data with censored values but is considerably more complicated than the preceding alternatives and is best accomplished when there are large data sets. For small sample sizes, MLE are not necessarily minimum variance, unbiased estimates.

Owen and deRouen (1981) addressed the problem of estimating a mean from censored air contaminant data. They used Monte Carlo techniques to evaluate the performance of MLE methods derived for the lognormal distributions. The technical literature also includes Holland and Fitz-Simmons (1982) who describe a computer program for fitting statistical distributions to air pollution data using the maximum likelihood estimation. Sharma et al. (1995a and b) utilize the MLE for parameter estimation in air pathways migration models.

There are several methods for handling regression with censored data. These include the iterative least squares (ILS) method (Schmee and Hahn, 1979), the linear unbiased estimate method (Nelson and Hahn, 1973), and the MLE method (Dempster et al., 1977; Haas and Jacangelo, 1993). The ILS method is simpler to implement than the MLE method and has good statistical convergence properties (Schmee and Hahn, 1979). The ILS method was developed for parameter estimation of a linear model with right-censored data that arise from life tests.

10.3.7 Poisson Model

If more than 50 percent but less than 90 percent of the monitoring records are nondetects, or if the assumptions of Cohen's and Aitchison's methods are not met, parametric statistical intervals should generally be abandoned in favor of nonparametric procedures as described in Chap. 11. However, there is an additional parametric-based procedure, namely the Poisson model, which may be appropriate for application.

Specifically, when 90 percent or more of the data values are nondetect, the detected samples may be modeled as rare events by using the Poisson distribution. Recall from Chap. 5, the Poisson model describes the behavior of a series of independent events over a large number of trials, where the probability of occurrence is low but remains constant from trial to trial. The Poisson model is similar to the binomial model in that both models represent counting processes. In the binomial case, nondetects are counted as misses or zeros and detects are counted (regardless of contamination level) as hits or ones. In the case of the Poisson model each particle or molecule of contamination (as per the implied assumption implicit in the Poisson distribution model) is counted separately but cumulatively, so that the counts for detected samples with high concentrations are larger than counts for samples with smaller concentrations.

For a detect with concentration of 50 ppb, the Poisson count would be 50. Counts for nondetects can be taken as zero or perhaps equal to one-half the detection level (e.g., if detection level were 10 ppb, the Poisson count for that sample would be 5). Therefore, unlike the Binomial model, the Poisson model has the ability to utilize the magnitudes of detected concentrations in statistical tests.

10.3.8 Use of an Indicator Parameter

If the monitoring data record contains a number of constituents that are highly correlated, the potential exists to utilize the value of one constituent to estimate a censored constituent (McBean and Rovers, 1984).

10.4 MULTIPLE DETECTION LIMITS

There has been very little reported research in the environmental quality field using techniques to handle multiply censored data sets (Millard and Deveral, 1988). However, the continuing evolution of analytical chemistry techniques has enhanced detectability of increasingly smaller concentrations of chemicals in the environment so that now multiple detection limits are fairly common.

There are at least three possible causes of multiple detection limits, including:

1. The limit of detection of a particular analyte depends on the method used to measure it. There may be more than one method available, and each method may be optimal (have the smallest percent measurement error) in a certain range of analyte concentration. For example, the protocol may call for method 1 to be used if the specific conductance is above a certain threshold, and method 2 if a specific conductance is below a threshold value.
2. A second cause of multiple detection limits involves the process of dilution, resulting in varying detection limits.
3. Detection limits decrease over time as the measurement techniques improve.

Utilization of standard parametric statistical methods such as *t*-tests and multiple regression becomes difficult with multiply censored data sets.

Statisticians in the field of survival analysis and life testing have developed numerous techniques for analyzing multiply censored data sets (e.g., Kalbfleisch and Prentice, 1980). Multiple correlation procedures may be useful to infill or estimate some of the data.

10.5 REFERENCES

AITCHISON, J. "On the Distribution of a Positive Random Variable Having a Discrete Probability Mass at the Origin," *Journal of the American Statistical Association*, 50 (1955), 901–908.

ANNE, D.C. "Know Your Detection and Quantitation Limits," *Pollution Engineering* (May 1, 1992), 68–71.

CH2M-Hill, "Remedial Investigation Report for Operating Industries, Inc. Landfill, Monterey Park, California," report to U.S. EPA, San Francisco, CA, October 25, 1994.

COHEN, A.C. "Simplified Estimators for the Normal Distribution when Samples are Singly-Censored or Truncated," *Technometrics*, 1 (1959), 217–237.

COHEN, A. "Tables for Maximum Likelihood Estimates from Singly Truncated and Singly-Censored Samples," *Technometrics*, 3 (1961), 535–541.

DEMPSTER, A.P., N.M. LAIRD, and D.B. RUBIN. "Maximum Likelihood from Incomplete Data via the EM Algorithm," *Journal of the Royal Statistical Society, Series B-Methodological* 39 (1977), 1–22.

Federal Register, 52, no. 130 (July 8, 1987).

GILBERT, R.O., and R.R. KINNESON. "Statistical Methods for Estimating the Mean and Variance from Radionuclide Data Sets Containing Negative, Unreported or Less-than Values," *Health Physics*, 40 (1981), 377–390.

GILLIOM, R.J., and D.R. HELSEL. "Estimation of Distributional Parameters for Censored Trace Level Quality Data 1: Estimation Techniques," *Water Resources Research*, 22 (1986), 135–146.

GLEIT, A. "Estimation of Small Normal Data Sets with Detection Limits," *Environmental Science and Technology*, 19 (1985), 1201–1206.

HAAS, C.N., and J.G. JACANGELO. "Development of Regression Models with Below-Detection Data," *Journal of Environmental Engineering*, 119, no. 2 (1993), 214–230.

HASHIMOTO, L.K., and R.R. TRUSSEL. "Evaluating Water Quality Data Near the Detection Limit," paper presented at the Proceedings of the American Water Works Association Advanced Technology Conference, June 5–9, 1983, Las Vegas, Nevada.

HELSEL, D.R., and R.J. GILLIOM. "Estimation of Distributional Parameters for Censored Trace Level Water Quality Data 2. Verification and Applications," *Water Resources Research*, 22 (February, 1986), 147–155.

HOLLAND, D., and T. FITZ-SIMMONS. "Fitting Statistical Distributions to Air Quality Data by the Maximum Likelihood Method," *Atmospheric Environment*, 16, no. 5 (1982), 1071–1076.

KALBFLEISCH, J.D., and R.L. PRENTICE *The Statistical Analysis of Failure Time Data*. New York, NY: John Wiley and Sons, 1980.

KEITH, L.H. *Environmental Sampling and Analysis: A Practical Guide*. Chelsea, MI: Lewis Publishers, 1991.

KEITH, L.H., W. CRUMMETT, J. DEEGAN, Jr., R.A. LIBBY, J.K. TAYLOR, and G. WENTLER. "Principles of Environmental Analysis," *Analytical Chemistry*, 55, no. 14 (1983), 2210–2218.

KIRCHMER, C.J. "Estimation of Detection Limits for Environmental Analytical Procedures—a Tutorial," in *Detection in Analytical Chemistry—Importance, Theory and Practice*, ed. L.A. Currie, Washington, DC: American Chemical Society, 1988.

KLOTZ, J. "Small Sample Power and Efficiency for the One Sample Wilcoxon and Normal Scores Tests," *Annals of Mathematics Statistics*, 34 (1963), 624–632.

KUSHNER, E.J. "On Determining the Statistical Parameters for Pollution Concentration from a Truncated Data Set," *Atmospheric Environment*, 10 (1976), 975–979.

LONG, G.L., and J.D. WINEFORDNER. "Limit of Detection A Closer Look at the IUPAC Definition," *Analytical Chemistry*, 55, no. 7 (1983), 712–724.

MAGE, D.T. "An Objective Graphical Method for Testing Normal Distributional Assumptions Using Probability Plots," *The American Statistician*, 36, no. 2 (1982), 116–120.

MANTEL, N. "Calculation of Scores for a Wilcoxon Generalization Applicable to Data Subject to Arbitrary Right Censorship," *The American Statistician*, 35, no. 4 (1981), 244–247.

McBEAN, E.A., and F.A. ROVERS. "Alternatives for Handling Detection Limit Data in Impact Assessments," *Ground Water Monitoring Review* (Spring, 1984), 42–44.

McBEAN, E.A., and F.A. ROVERS. "Analysis of Variances as Determined from Replicate Versus Successive Samplings," *Ground Water Monitoring Review* (Summer, 1985), 61–64.

McNICHOLS, R.J., and C.B. DAVIS. "Statistical Issues and Problems in Ground Water Detection Monitoring at Hazardous Waste Facilities," *Ground Water Monitoring Review* (Fall, 1988), 135–150.

MILLARD, S.P., and S.J. DEVERAL. "Nonparametric Statistical Methods for Comparing Two Sites Based on Data with Multiple Nondetect Limits," *Water Resources Research,* 24, no. 12 (1988), 2087–2098.

NELSON, W., and G.J. HAHN. "Linear Estimation of a Regression Relationship from Censored Data—Part II: Best Linear Unbiased Estimation and Theory," *Technometrics,* 15, no. 1 (1973), 133–150.

OWEN, W.J., and T.A. DeROUEN. "Estimation of the mean for Lognormal Data Containing Zeroes and Left-censored Values, with Applications to the Measurement of Worker Exposure to Air Contaminants," *Biometrics,* 36 (1980), 707–719.

SCHMEE, J., and G.J. HAHN. "A Simple Method for Regression Analysis with Censored Data," *Technometrics* 21 (1979), 417–432.

SCROGGIN, D.G. "Detection Limits and Variability in Testing Methods for Environmental Pollutants: Misuse May Produce Significant Liabilities," *Hazardous Waste and Hazardous Materials,* 11, no. 1 (1994), 1–4.

SHARMA, M., N.R. THOMSON, and E.A. McBEAN. "Linear Regression Analyses with Censored Data: Estimation of PAH Washout Ratios and Dry Deposition Velocities to a Snow Surface," *Canadian Journal of Civil Engineering,* 22, no. 4 (1995a), 819–833.

SHARMA, M., E.A. McBEAN, and N.R. THOMSON. "Maximum Likelihood Method for Parameter Estimation with Below-Detection Data," *ASCE—Journal of the Environmental Engineering Division,* 11 (1995b), 776–784.

TAYLOR, J.K. *Quality Assurance of Chemical Measurements.* Chelsea, MI: Lewis Publishers Inc., 1987.

U.S. EPA. *Statistical Analysis of Ground-Water Monitoring Data at RCRA Facilities* (April 1989) Washington, D.C.: Office of Solid Waste.

10.6 PROBLEMS

10.1. For the following water quality data, estimate the mean and standard deviation using the procedures named below

$$8.2, 5.3, < 5, < 5, 10.1, 9.3, 7.6, < 5$$

- **a.** The probability plot procedure.
- **b.** The test of proportions.
- **c.** Cohen's test procedure.
- **d.** Aitchison's method.

10.2. For the following skewed data, estimate the mean and standard deviation using the procedures named below

$$1.7, < 5, 63, 8.3, < 5, 5.5, < 5, 6.7$$

- **a.** The probability plot procedure.
- **b.** The test of proportions.
- **c.** Cohen's test procedure.
- **d.** Aitchison's method.

CHAPTER 11

NONPARAMETRIC PROCEDURES

11.1 INTRODUCTION

Previous chapters have focused on the available procedures for addressing hypothesis testing questions and the identification of trends. However, as noted in Chap. 8, there are significant difficulties with these procedures when a substantial percentage of censored data exist within the data set and/or when there is a failure to adhere to assumptions implicit within the parametric tests. The result is that many existing environmental quality databases are not amenable to analyses by the parametric methods described previously. Many techniques are not suitable because of missing data, censored data, and changing laboratory techniques. As well, many of these procedures cannot be applied to data which are clearly nonnormal in distribution.

A group of tests have been devised in which no assumptions are made about the distribution of the observations. For this reason, the tests are referred to as distribution free. An array of these alternative tests exists, collectively referred to as *nonparametric* tests. The fundamental characteristic of these procedures is that the ranks of the data are utilized instead of data values. The analyses of data on ranks are a direct parallel of the more traditional parametric methods. For data that do not fulfill the necessary assumptions for the parametric analyses, the nonparametric methods are as powerful or more powerful than the equivalent parametric tests.

A large body of technical literature exists based on analyses of the ranks of the data. The mathematics involved in nonparametric tests are generally very straightforward. As well, unlike the parametric procedures examined in Chaps. 8 through 10, the nonparametric procedures require practically no knowledge about the distribution of the population. Thus, in the event that one or more of the assumptions implicit in the *t*-test are violated (e.g., there is a significant difference between the variances), an alternative class of procedures involving nonparametric analyses may be utilized.

The nonparametric methods usually have the additional advantage that they require less burdensome calculations because they are generally not related specifically to the parameters of a given distribution. The major disadvantage of nonparametric methods is that they may be wasteful of information and usually they have a smaller efficiency than the corresponding parametric methods, provided that the assumptions of the standard (parametric) methods can be met. Hamby (1994) discusses many of the features of the nonparametric tests. In large normally distributed samples, the nonparametric tests have an efficiency of approximately 95 percent relative to Student's *t*-test (Hodges and Lehman, 1946) and in small, normally distributed samples, the signed-rank test (see Sec. 11.2.2) has been shown to have an efficiency slightly higher than this (Klotz, 1963). With nonnormally distributed data the efficiency of the nonparametric tests relative to the *t*-test never falls below 86 percent in large samples and may be greater than 100 percent for distributions that have long tails (Hodges and Lehman, 1946).

Rank or nonparametric tests are generally tests that do not require any assumptions other than independent samples from continuous populations. During the last 40 years, the role played by nonparametric tests has undergone considerable change. In the 1950s and 1960s it was quite common to characterize nonparametric methods as "rough and ready." This attitude has changed until now practically every introductory statistics text contains a chapter concerning nonparametric or distribution-free or rank methods (Noether, 1981).

The nonparametric tests include the Wilcoxon-Mann-Whitney test, the Kruskal-Wallis test, the Wilcoxon signed ranks test, the Friedman test, Spearman's rho, and others. Since the tests are relatively quickly completed, the rank tests are highly useful for the investigator who is doubtful that the data can be regarded as normal.

EXAMPLE 11.1

The concentrations measured at a groundwater monitoring location are as follows: 13.2, 12.1, 2.1, 2.0 and 1.9. Assign the ranks for these five measurements.

The assignments of the ranks are as follows:

Concentration	Rank
13.2	1
12.1	2
2.1	3
2.0	4
1.9	5

All subsequent statistical analyses are now done using the rank information, implying equal spacing between successive values. Note that the relative proximity information (e.g., the relative grouping of the data into two high and three relatively low values) is lost. This loss of information content is, however, the rationale for subsequently not having to attend to the probability distribution of the concentration data.

If two or more observations are tied, each of the observations are assigned the mean of the ranks they would jointly occupy.

11.2 SINGLE-COMPARISON PROCEDURES

The single-comparison procedures are for testing the null hypothesis that two samples come from identical continuous populations against the alternative that the populations have different means. A highly efficient class of nonparametric tests of this and similar hypotheses is based on rank sums; that is, the observations are assigned ranks according to their order of magnitude and the tests are performed on the basis of various sums of these ranks.

11.2.1 MANN-WHITNEY TEST

The Mann-Whitney test or U test is a highly efficient nonparametric test based on ranked sums to compare the means of two independent samples. The Wilcoxon rank sum test for two groups is essentially the same as the Mann-Whitney test. The Mann-Whitney *U*-statistic is a linear transformation of the Wilcoxon test and therefore an equivalent statistic. Probability theory does not depend on the distribution type when operating on ranks (Conover, 1980). The data in each of the two samples are first ordered from lowest to highest values and then the combined data set is ranked. The ranks are then summed for each group.

The null hypothesis is that the two samples are taken from a common population so that there should be no consistent difference between the two sets of rankings. Denoting the respective sample sizes by n and m and the sum of the ranks occupied by the first sample by R_1, it can be shown that the mean and the variance of the sampling distribution of the statistic are as follows:

$$U = nm + \frac{n(n+1)}{2} - R_1 \qquad [11.1]$$

are given by

$$\overline{U} = \frac{nm}{2} \qquad [11.2]$$

and

$$S_U^2 = \frac{nm\,(n+m+1)}{2} \qquad [11.3]$$

If both n and m are greater than 8, the distribution of the U-statistic can be approximated closely by a normal distribution; hence the test can be based on the statistic.

$$Z = \frac{U - \overline{U}}{S_U} \qquad [11.4]$$

With knowledge of Z, conclusions can then be made as to whether the null hypothesis of identical populations can or cannot be rejected.

If one sum of the single-sample ranks is sufficiently larger than the other, then the two sample means are determined to be significantly different.

EXAMPLE 11.2

Concentrations at two locations were monitored as tabulated below:

MW1	1.3	0.9	0.8	0.2	0.4	0.6	0.1	5.1	0.2	
MW2	1.7	3.5	7.8	0.9	0.7	2.6	0.2	1.5	15.3	0.7

Utilize the Mann-Whitney test to determine whether there is a statistically significant difference between the monitoring results at the two locations.

First, arrange the 19 observations according to size, retaining the sample identity of each observation. Then, we assign the ranks 1, 2, 3, . . ., and 19 to the observations, as shown below:

Monitoring well	MW1	MW1	MW1	MW2	MW1	MW1	MW2	MW2	MW1
Observation	0.1	0.2	0.2	0.2	0.4	0.6	0.7	0.7	0.8
Rank	1	3	3	3	5	6	7.5	7.5	9

Monitoring well	MW1	MW2	MW1	MW2	MW2	MW2	MW2	MW1	MW2
Observation	0.9	0.9	1.3	1.5	1.7	2.6	3.5	5.1	7.8
Rank	10.5	10.5	12	13	14	15	16	17	18

Monitoring well	MW2
Observation	15.3
Rank	19

Note that if two or more observations are tied in rank, we assign each of the observations the mean of the rank they jointly occupy.

If we denote the respective sample sizes by n_1, and n_2 and the sum of the ranks occupied by the first sample by R_1, it can be shown that the mean and the variance of the sampling distribution of the statistic

$$U = n_1 n_2 + \frac{n_1(n_1 + 1)}{2} - R_1 = 9 \times 10 + \frac{9 \times 10}{2} - 66.5 = 68.5$$

are given by

$$\overline{U} = \frac{n_1 \times n_2}{2} = \frac{9 \times 10}{2} = 45.0$$

and

$$S_U^2 = \frac{n_1 n_2 (n_1 + n_2 + 1)}{12} = \frac{9 \times 10 \times 20}{12} = 150$$

Thus

$$Z = \frac{68.5 - 45.0}{\sqrt{150}} = 1.93$$

Since this value falls between –1.96 and 1.96, the critical values for a two-sided alternative (95 percent) and $\alpha = 0.05$ (see Table A.1), we conclude that the null hypothesis of identical populations cannot be rejected.

USE OF THE MANN-WHITNEY TEST TO TEST EQUALITY OF VARIANCE

If the ranks of the data are assigned in a different way from that outlined above, the same U-statistic can also be used to test the null hypothesis of identical populations against the alternative that the populations have unequal dispersions or variances. For this type of test, the ranks are assigned from both ends toward the middle by assigning Rank 1 to the smallest observation, Rank 2 and 3 to the largest and second largest observation, and so on. All other aspects of this test for unequal dispersions are identical with those of the Mann-Whitney U-test.

11.2.2 SPEARMAN'S RANK CORRELATION COEFFICIENT

Spearman's rank correlation coefficient procedure is a very simple ranking technique. Consider the situation where two laboratories are completing parallel analyses on groundwater quality. The results from analyses of ten different tests (repetitions for the same constituent) are shown in Table 11.1, where the entries indicate the relative quality (as determined by each laboratory) for each of the samples. Therefore, for example, Laboratory 1 identified Sample 2 as having the worst water quality, Sample 1 as the second worst, and so on. Do the laboratories

TABLE 11.1 Ranking in accordance with Spearman's rank correlation coefficient

Sample	**1**	**2**	**3**	**4**	**5**	**6**	**7**	**8**	**9**	**10**
Laboratory 1	2	1	3	4	6	5	8	7	10	9
Laboratory 2	3	3	2	4	6	7	5	9	10	8
Rank difference = d	1	2	–1	0	0	2	–3	2	0	–1
Square of difference = d^2	1	4	1	0	0	4	9	4	0	1

show evidence of agreement amongst themselves in regard to ranking? This can be addressed by calculating Spearman's rank correlation coefficient as

$$R = 1 - \frac{6 \sum d^2}{n^3 - n} \qquad [11.5]$$

where n = the number of samples ranked and d = the rank difference

For the values in Table 11.1

$$\sum d^2 = 24$$

so

$$R = 1 - \frac{6 \times 24}{1000 - 10} = 0.85 \qquad [11.6]$$

The rank correlation coefficient procedure has been designed so that when the two rankings are identical, R has the value +1; when the rankings are as greatly in disagreement as possible (i.e., one ranking is exactly the reverse of the other), R has the value −1.

To examine whether this measure of agreement could have arisen by chance, the significance of R may be tested (provided $n > 10$) as

$$t^* = R\sqrt{\frac{n-2}{1-R^2}} \qquad [11.7]$$

with $(n - 2)$ degrees of freedom.

In this example,

$$t^* = .85\sqrt{\frac{10-2}{1-(.85)^2}} = 4.6$$

with eight degrees of freedom so that the level $R = .85$ did not likely arise by chance. This demonstrates agreement in the findings of the laboratories.

11.2.3 Sign Test for Paired Observations

This test represents an alternative to the paired t-test. For example, if a sequence of x's and y's correspond to an observation and a model's prediction, respectively, then the individual values are represented by

$$x_1, x_2, \ldots, x_n \text{ and } y_1, y_2, \ldots, y_n$$

The sign test does not depend on the size of numerical values of the differences but only on their sign or rank order. The assumption of normality is not required in order for this test to be valid. If the difference between the values is expected to equal zero, we would expect approximately one-half of the differences to be negative. If a small proportion of the differences are either negative or positive, we suspect that there might be a real difference but the magnitude of the proportion is the question.

The true proportion of positive (or negative) signs is equal to $p = 1/2$. Hence, in n independent trials in which a positive or negative sign for each pair is determined,

$$b(x) = \frac{n!}{x!\,(n-x)!}\left(\frac{1}{2}\right)^n \qquad [11.8]$$

With paired observations, the sign test provides a simple test of the null hypothesis that x and y have a common distribution (as opposed to the alternative that x is less than y or x is greater than y).

To examine this, consider a new variable Z, defined as

$$Z_i = x_i - y_i \qquad [11.9]$$

If the null hypothesis is true, then

$$Pr[Z_i > 0 \mid Z_i \neq 0] = Pr\,[Z_i < 0 \mid Z_i \neq 0] = 0.5$$

The probabilities in [ ] are made conditional on $Z_i \neq 0$ to emphasize that pairs for which $Z_i = x_i - y_i = 0$ can be disregarded.

A reasonable test statistic is S the number of Z_is that are greater than zero. If the alternative hypothesis is true, a smaller value of S would occur than if the null hypothesis were true. If there are m Z_is not equal to zero ($m < n$), then with the null hypothesis, S has a binomial distribution and

$$\Pr\,[S < s] = \sum_{k=0}^{s} \frac{m!}{k!\,(m-k)!}\left(\frac{1}{2}\right)^m \qquad [11.10]$$

11.3 MULTIPLE-COMPARISON PROCEDURES

The Kruskal-Wallis H-test is utilized for testing whether k samples come from identical populations against the alternative that the populations have unequal means when there are three or more samples. The Kruskal-Wallis H-test represents a nonparametric alternative to the ANOVA test. The null hypothesis of the H-test is that the samples have been taken from populations with identical distributions and any differences between the samples are due to chance variation inherent in the process of random sampling. Hence, the Kruskal-Wallis H-test is utilized for testing whether k samples come from identical populations against the alternative that the populations have unequal means.

11.3.1 KRUSKAL-WALLIS TEST (OR NONPARAMETRIC ANOVA)

The Kruskal-Wallis H-test for deciding whether k independent samples come from identical populations is conducted in a manner similar to the U-test. As with the

U-test, the observations are ranked jointly, and if R_1 is the sum of the ranks occupied by the n_i observations of the ith sample, the test is based on the statistic

$$H = \frac{12}{N(N+1)}\left[\frac{R_1^2}{n_1} + \frac{R_2^2}{n_2} + \ldots + \frac{R_k^2}{n_k}\right] - 3(N+1) \qquad [11.11]$$

where $n_1, n_2, \ldots, n_k$ are the number of observations in each of the k groups.

When $n_i > 5$ for all i and the null hypothesis holds, the distribution of the H-statistic is well approximated by the chi-square distribution with $k - 1$ degrees of freedom. There should be at least three groups with a minimum sample size of three in each group.

The Kruskal-Wallis H-test is utilized when the data or the residuals have been found significantly different from normal and when a log transformation fails to adequately normalize the data. The Kruskal-Wallis test does not imply assumptions about the underlying distribution.

In a one-way nonparametric ANOVA, the assumption under the null hypothesis is that the data from each monitoring location come from the same continuous distribution and hence have the same mean concentration of a constituent. The Kruskal-Wallis test is equivalent except that the test is performed using ranks, as opposed to concentrations.

If the monitoring locations are found to differ, post hoc comparisons are needed to determine if contamination is present.

Define the number of groups by k and the number of observations in each group by n_i with N being the total number of well observations. Let x_{ij} denote the jth observation in the ith group, where j runs from one to the number of observations in the group, n_i and $i = 1, \ldots, k$ groups.

The steps of the procedure are then as follows:

Step 1. The analysis is initiated by ranking all the observations from the least to the greatest from the combined groups to be analyzed. Let R_{ij} denote the rank of the jth observation in the ith group. As a convention, denote the background monitoring location(s) as Group 1. Ties are given the average rank of the tied values.

Step 2. Compute the sum of the rank values for each group and the average rank within each group (k groups). The sum of the ranks for the ith group is defined as R_i, and the average rank for each $\overline{R}_i = R_i/n_i$.

Step 3. Compute the average rank of the overall data set.

Step 4. Calculate the Kruskal-Wallis test statistic H, given in equation [11.11].

$$H = \frac{12}{N(N+1)}\left[\frac{R_1^2}{n_1} + \frac{R_2^2}{n_2} + \ldots + \frac{R_k^2}{n_k}\right] - 3(N+1)$$

where $n_1, n_2, \ldots n_k$ are the number of observations in each group.

Step 5. Compare H with appropriate chi-square critical value χ^2, with degrees of freedom = $k - 1$ where k is the number of groups. If H is greater than χ^2, this indicates a significant difference between at least two of the well groups.

The null hypothesis is rejected if the computed value exceeds the tabulated critical value.

Step 6. If the H-statistic is significant, then individual comparisons are computed between the background (Group 1) and compliance wells. If the computed value exceeds the value from the chi-square table, compute the critical difference for well comparisons to the background, assumed to be Group 1.

Step 7. Compute the differences of average ranks from each group to the average rank of background.

Step 8. Calculate the critical difference

$$C_i = Z_{\alpha/(k-1)}\left[\frac{N(N+1)}{12}\right]^{0.5}\left[\frac{1}{n_1}+\frac{1}{n_i}\right]^{0.5} \qquad [11.12]$$

where $Z_{\alpha/(k-1)}$ is the upper $\alpha/(k-1)$ percentile from the standard normal distribution, with zero mean and unit variance, and $\alpha = 5$ percent.

The null hypothesis of the H-test is that the samples have been from populations with identical distributions. Any differences between the samples are due to variations inherent in the process of random sampling.

Step 9. Compare each rank difference to C. The differences of the average ranks for each group to the background are compared with the critical values found in Step 8 to determine which wells indicate evidence of contamination; that is, compare $R_i - R_1$ to C_1 for i, taking the values 2 through k (remembering that Group 1 is the background).

EXAMPLE 11.3

Suppose that the experiment with the results are shown below. Note that tied observations are again assigned the mean of the ranks they jointly occupy.

MW1	0.2	0.3	0.4	0.5	1.7	1.9	2.0
MW2	0.8	1.1	1.3	1.9	2.5	7.8	
MW3	0.7	0.9	8.2	12.0	12.1	15.3	
MW4	0.1	0.1	0.3	0.5	2.9	13.8	

The observations in the first sample set occupy Ranks 3, 4.5, 6, 7.5, 14, 15.5, and 17, so that $R_1 = 67.5$. Similarly, the observations in the second sample occupy Ranks 10, 12, 13, 15, 5, 18, and 20 so that $R_2 = 88.5$. Continuing, $R_3 = 111.0$ and $R_4 = 58.0$. Substituting into the formula for H we obtain.

$$H = \frac{12}{25 \times 26}\left(\frac{67.5^2}{7}+\frac{88.5^2}{6}+\frac{111^2}{6}+\frac{58^2}{6}\right) - 3(26) = 6.4$$

And if we compare this figure with 7.815 (from Table A.12, the value of $\chi^2_{0.05}$ with three degrees of freedom), we find that the null hypothesis cannot be rejected. In other words, we cannot reject the null hypothesis that the samples are from identical populations against the alternative that the population means are not all equal.

EXAMPLE 11.4

The data below indicate the benzene concentrations at three locations in a manufacturing facility. The null hypothesis is that the three samples are from the sample population.

Location 1		Location 2		Location 3	
Concentration	**Rank**	**Concentration**	**Rank**	**Concentration**	**Rank**
23	7	11	3	110	18
15	4	17	5	84	16
42	12	31	11	85	17
8	1	27	9.5	45	13
10	2	63	14	64	15
18	6	27	9.5	25	8

$R_1 = 32$ $\quad R_2 = 52$ $\quad R_3 = 87$

$n_1 = 6$ $\quad n_2 = 6$ $\quad n_3 = 6$

The data are ranked overall from lowest to highest. Identical values are given the mean of the ranking they would otherwise have received.

The sum of the ranks are noted for each of the sampling locations. It is then calculated as

$$H = \frac{12}{N(N+1)} \sum_{i=1}^{k} \frac{R_i^2}{n_i} - 3(N+1) \qquad [11.13]$$

$$= \frac{12}{18 \times 19}\left(\frac{32^2}{6} + \frac{52^2}{6} + \frac{87^2}{6}\right) - 3\,(19)$$

$$= 8.90$$

When sampling locations all contain more than five individual values, the sampling distribution of H is virtually identical to the sampling distributions of Chi-square with two degrees of freedom (Table A.12) which is 5.99. Since the calculated H is larger than the critical value, the null hypothesis is rejected. In other words, it is safe to assume that the benzene concentrations are different at least at some of the three sampling locations.

Next calculate the critical value to compare with Locations 2 and 3 with 1, using equation [11.12].

$$C_i = Z_{\alpha/(k-1)}\left[\frac{N(N+1)}{12}\right]^{1/2}\left[\frac{1}{n_1} + \frac{1}{n_i}\right]^{1/2}$$

There are two comparisons to be made. For $\alpha = 0.05$, we find the upper $0.05/2 = 0.025$ percentile of the standard normal distribution to be 1.96 (from Table A.1). The critical values for sampling locations, C_2 is

$$C_2 = 1.96\left[\frac{18(19)}{12}\right]^{1/2}\left[\frac{1}{6} + \frac{1}{6}\right]^{1/2}$$

$$= 3.45$$

The critical value for sampling locations C_3 is also 3.45. Compute the differences:

Differences		Critical value
Between 1 and 2	52 – 32 = 20	3.45
Between 1 and 3	87 – 32 = 55	3.45

Comparing the differences with the critical differences indicates that both sampling locations 2 and 3 are significantly different from Location 1.

11.3.2 Special Consideration of the Kruskal-Wallis Test

Equation [11.13] should be modified if there are any tied ranks in the data set. To allow for tied ranks the calculated value of H is divided by a correction factor

$$1 - \frac{\Sigma\,(t^3 - t)}{n^3 - n} \qquad [11.14]$$

where t is the number of individual values involved in each set of tied ranks. The summation $\Sigma\,(t^3 - t)$ must be calculated for each set of tied ranks.

Example 11.5(a)

Use the Kruskal-Wallis test to determine whether there is evidence of contamination at the site, given the monitoring records listed below.

	Concentrations					
	Background wells			Compliance wells		
Sampling time	**Well 1**	**Well 2**	**Well 3**	**Well 4**	**Well 5**	**Well 6**
1	<5	<5	8.1	<5	<5	<5
2	<5	<5	<5	12.1	13.4	<5
3	<5	6.9	<5	<5	8.6	15.6
4	7.7	<5	<5	8.0	<5	18.1

Step 1. Compute the overall percentage of nondetects. In this case, nondetects account for 15/24 = 63 percent of the data. More than 15 percent of the records are less thans and therefore parametric ANOVA should not be utilized. Use the Kruskal-Wallis test instead, pooling all three background wells into one group and treating each compliance well as a separate group.

Step 2. Compute ranks for all the data including the tied observations (e.g., nondetects) as in the following table. Note that each nondetect is given the same mid rank, equal to the average of the first 14 unique ranks.

	Concentrations					
	Background Wells			Compliance Wells		
Sampling time	**Well 1**	**Well 2**	**Well 3**	**Well 4**	**Well 5**	**Well 6**
1	8.0	8.0	19	8.0	8.0	8.0
2	8.0	8.0	8.0	21	22	8.0
3	8.0	16	8.0	8.0	20	23
4	17	8.0	8.0	18	8.0	24
Rank sum		$R_1 = 124$		$R_4 = 55$	$R_5 = 58$	$R_6 = 63$
Rank mean		$\overline{R}_1 = 10.3$		$\overline{R}_4 = 13.8$	$\overline{R}_5 = 14.5$	$\overline{R}_6 = 15.8$

Step 3. Calculate the sums of the ranks in each group (R_i) and the mean ranks in each group ($\overline{R}_i$).

Step 4. Compute the Kruskal-Wallis statistic H using the formula

$$H = \frac{12}{N(N+1)} \sum_{i=1}^{k} \frac{R_i^2}{n_i} - 3(N+1)$$

where N = the total number of samples, n_i = number of samples in the ith group and k = the number of groups.

$$N = 24, k = 4$$

$$H = \frac{12}{24(25)} \left(\frac{124^2}{12} + \frac{55^2}{4} + \frac{58^2}{4} + \frac{63^2}{4} \right) - 3(25)$$

$$= 0.20\,(1281 + 756 + 841 + 993) - 75 = 2.42$$

Step 5. Compute the adjustment for ties. There is only one group of distinct tied observations, containing 15 samples. The adjusted Kruskal-Wallis statistic is given by

$$H' = \frac{2.42}{1 - \left(\dfrac{15^3 - 15}{24^3 - 24} \right)} = 3.20$$

Step 6. Compare the calculated value of H' to the tabulated chi-square value with $(k–1)$ = (# of groups – 1) = 4 – 1 df, $\chi^2_{3,.05} = 7.81$. Since the observed value of 3.20 is less than the chi-square critical value, there is insufficient difference between the well groups. Post hoc pairwise comparisons are unnecessary.

EXAMPLE 11.5(B)

Consider now the situation as if Ex. 11.5(a) had not been carried out and the monitoring results were obtained as indicated in the modified table below, the change being the addition of results from Well 7.

Sampling time	Background wells	Well 7
1	(values as listed in Ex. 11.5 (a)	24.9
2		18.2
3		29
4		24.9

Step 1. The overall percentage of nondetects is 16/28 = 57 percent. More than 15 percent are nondetect; therefore use the Kruskal-Wallis test.

Step 2. Compute the ranks

Sampling time	1	2	3	4	5	6	7
1	8.0	8.0	19	8.0	8.0	8.0	26.5
2	8.0	8.0	8.0	21	22	8.0	25
3	8.0	16	8.0	8.0	20	23	28
4	17	8.0	8.0	18	8.0	24	26.5
Rank sum =		12.4		55	64	63	106
Rank mean =		10.3		13.8	16.0	15.6	26.5

Step 3. Sum the ranks and mean ranks as noted.

Step 4.

$$H = \left[\frac{12}{N(N+1)} \sum_{i=1}^{k} \frac{R_i^2}{n_i}\right] - 3\,(N+1)$$

$$N = 28, k = 5$$

$$H = \frac{12}{28\,(29)}\left[\frac{124^2}{12} + \frac{55^2}{4} + \frac{64^2}{4} + \frac{63^2}{4} + \frac{106^2}{4}\right] - 3(29)$$

$$= .015\,[1281 + 756 + 1024 + 992 + 2809] - 3(29)$$

$$= 102.9 - 87 = 15.9$$

Step 5.

$$H' = \frac{15.9}{1 - \left[\frac{15^3 - 15}{28^3 - 28}\right]} = \frac{7.5}{1 - \frac{3360}{21924}} = 18.8$$

Step 6. For $k - 1 = 5 - 1 = 4$ df, $\chi^2_{4,.05} = 9.49$

Since the observed value of 18.8 is greater than the chi-square critical value, there is evidence of significant differences between the well groups. Therefore, post hoc comparisons are needed.

Step 7. Calculate the critical difference for compliance well comparison to the background, when the background wells are taken as Group 1

$$C_i = Z_{\alpha/(k-1)} \left[\frac{N(N+1)}{12}\right]^{\frac{1}{2}} \left[\frac{1}{n_1} + \frac{1}{n_i}\right]^{\frac{1}{2}}$$

for $i = 2, \ldots, k$. $Z_{\alpha/k-1}$ is the upper $Z_{\alpha/k-1}$ percentile from the standard normal distribution. Since the number of samples at each compliance well is four, the same critical difference can be used for each comparison, namely

$$C_i = Z_{.05/4} \left(\frac{28\,(29)}{12}\right)^{\frac{1}{2}} \left(\frac{1}{12} + \frac{1}{4}\right)^{\frac{1}{2}} \qquad Z_{.05/4} = Z_{.0125} = 2.24 \text{ (from Table A.1)}$$

$$= 2.24\,(8.23)\,(.577)$$

$$= 10.64$$

Step 8. Form the differences between the average ranks of each compliance well and the background, and compare these differences to the critical value of 10.64.

Well 4	$\overline{R}_4 - \overline{R}_1 = 13.8 - 10.3 = 3.5$
Well 5	$\overline{R}_5 - \overline{R}_1 = 16.0 - 10.3 = 5.7$
Well 6	$\overline{R}_6 - \overline{R}_1 = 15.6 - 10.3 = 5.3$
Well 7	$\overline{R}_7 - \overline{R}_1 = 26.5 - 10.3 = 16.2$

Since the average rank difference at Well 7 exceeds the critical difference, there is significant evidence of contamination at Well 7 but not at Wells 4, 5, and 6.

Note: in normal circumstances, it would not be necessary to do the parallel steps in Ex. 11.5(a) and (b). The presentation here was done solely for the purposes of explanation.

EXAMPLE 11.6

For the values in Ex. 11.5 there is one set of tied ranks, involving two values. Calculate the correction factor

$$1 = \frac{2^3 - 2}{18^3 - 18} = 1 - \frac{6}{5814} - 0.999$$

The value of H from Ex. 11.4 from Eq. [11.13] is 8.90; corrected for ties this becomes

$$\frac{8.90}{.999} = 8.91$$

This represents a very small adjustment. Unless more than a quarter of the values are tied ranks, the effect of the correction factor is small.

When small samples are involved (n_i <6), the χ^2 approximation can no longer be used as a reliable procedure. In these situations, the sampling distribution of H must be used. Critical values of H for use with three small samples are listed in Table A.10. If more than five monitoring locations are involved, the individual comparisons should be performed using $Z_\alpha = 0.01$ rather than $Z_{\alpha/k-1}$.

11.4 REFERENCES

CONOVER, W.J. *Practical Nonparametric Statistics,* 2nd ed. New York, NY: John Wiley and Sons, 1980.

CONOVER, W.J., and R.L. IMAN. "Rank Transformations as a Bridge Between Parametric and Nonparametric Statistics," *The American Statistician,* 35, no. 3 (1980), 124–128.

HAMBY, D.M. "A Review of Techniques for Parameter Sensitivity Analysis of Environmental Models," *Environmental Monitoring and Assessment,* 32 (1994), 135–154.

NOETHER, G.E. "Comment," *The American Statistician,* 35, no. 3 (1981), 129.

QUADE, D. "On Analysis of Variance for the K-sample Problem," *Annals of Mathematical Statistics,* 37 (1966), 1747–1748.

VAN BELLE, G., and J.P. HUGHES. "Nonparametric Tests for Trend in Water Quality," *Water Resources Research,* 20, no. 1 (1984), 127–136.

11.5 PROBLEMS

11.1 **a.** Using the Wilcoxon rank sum test, compare the treatment group presented below with the control group

Control	Treatment
12	4
11	9
9	1
6	13
12	10
9	18
10	17
9	14
6	12
8	16

Are these groups different at $p = 0.05$?

b. Use the Mann-Whitney test to compare the two groups.

11.2. Background and downgradient monitoring results are summarized below. Determine if there is a significant difference between the background and monitoring results.

	Downgradient		
Background	**MWA**	**MWB**	**MWC**
12	26	13	26
10	8	18	12
8	11	17	13
14	12	15	17
16	13	18	21
14	18	21	22

11.3. Use the Kruskall-Wallis test to determine whether there is evidence of contamination at the site, given the data as listed below.

	Downgradient		
Background	**MWA**	**MWB**	**MWC**
<10	15	16	17
<10	<10	<10	13
11	<10	<10	<10
13	18	12	<10
15	26	<10	21
11	20	21	20

PART IV

RISK

CHAPTER 12

RISK ASSESSMENT AND DATA MANAGEMENT

12.1 RISK EXPOSURE FROM ENVIRONMENTAL CAUSES

Extensive societal pressures to decrease the risks to which people are exposed are being exerted. The pressures toward risk reduction are present in virtually all aspects of society, but are particularly evident in considering environmental quality issues for a number of reasons:

1. Most people do not understand measures of environmental "quality", and are generally mistrustful of any change, whether it is "bad," "neutral," or "good" for the environment. Also, the public is generally not well versed in risk language and terms, and therefore risks related to unfamiliar concerns tend to be exaggerated.
2. The lay public has trouble distinguishing whether 10^{-9}, indicating the probability of death in a single year, is greater or less than 10^{-12}; again, a communication issue.
3. Acknowledgment of past environmental incidents has made people generally wary.
4. Most people have not felt they had much control over risks of exposure to environmental contaminants, so they are more averse to accepting them once identified.
5. Since somebody else (i.e. the risk generator) will have to pay the economic costs associated with protection from environmental hazards, some people conclude that no expense for risk reduction is too large.
6. As reported in *Chemecology* (1991), most people perceive risk in one of three ways, according to Alan Bromley, then Assistant for Science and Technology

to the U.S. President: "An activity can be 100 percent safe—which in reality is impossible, or it can be 100 percent dangerous—which means it is to be avoided at all costs. Everything in between tends to be viewed as having a 50/50 chance of occurring," Bromley notes.

The size of risks may be broken into four broad categories:

1. Familiar high risks, which exact a large toll and on which we have good information (e.g., vehicular fatalities and hang-gliding accidents).
2. Risks of low probability whose consequences are so large that they must be taken seriously (e.g., a large earthquake).
3. An extension of the second category, events whose probability is so very low that they have never happened at all, yet whose consequences deserve attention.
4. A collection of risks that are hard to evaluate because they show up as increases in naturally occurring hazards (e.g., an increase in forest fires).

As a result of concern over environmental risks, there is extensive pressure to have these risks decreased to very small magnitudes but how small should the risks be made? On the one hand, as members of modern society people are exposed to a variety of risks on a daily basis. When the risks are voluntary and people understand them at least to some extent, societal members accept the risks (e.g., cigarette smoking, driving, and flying in airplanes). On the other hand, exposure to deteriorated environmental quality is in many situations involuntary; hence there are differences in attitude and unwillingness to accept exposure to environmental risks.

As Lehr (1990b) indicates, risk is often interpreted as bad, a thing to be avoided, and yet all economic and technological progress requires that human beings take risks. As Wildavsky (1988) indicates, "There can be no safety without risk." Ultimately, the question of risk reduction becomes one of risk management: What levels of environmental exposure risk are acceptable? The resolution to this question is complicated by the many forces that operate in our societies.

In any particular situation, dealing with environmental risk involves two steps:

1. *Risk assessment:* Assessment of risk associated with the environment involves the use of factual information to define the health effects of exposure of humans and/or the environment to hazardous materials and situations. For example, this involves quantification of the extent to which a chemical moves from the location of a spill via one or more exposure pathways to create an exposure scenario for a nearby resident, and a determination of the deleterious consequences, if any, of that exposure scenario.
2. *Risk management:* Management of the risk associated with the environment considers the process of weighing policy and remediation alternatives, integrating the results of risk assessment with engineering data and with social, economic, and political concerns to reach a decision involving the management of the risk.

Questions related to reaching risk management decisions are generally beyond the scope of this book. Readers interested in questions of how to manage risks are referred to other sources for further reading (e.g., Glickman and Gough, 1990; and Asante-Duah, 1993; and Louvar and Louvar, 1998). This book merely presents a general overview of the range in scale of environmental risks, and develops statistical tools and procedures for risk assessment, estimating the likelihood of adverse effects from human exposure to chemical, physical, and/or biological agents and of ecological impacts within an ecosystem. However, in pursuit of statistical answers to environmental questions, it is not wise to ignore that the estimation of risk may have considerable associated uncertainty, as noted in the preceding chapters.

Questions of uncertainty are not unique to matters of risk assessment. Uncertainty drives all of science. If there were certainty, there would be no science. Science is an endlessly changing series of approximations. Science is about testable conclusions based on often disputable facts (Lehr, 1990b).

The intent of this final chapter of the book is to describe the methodology of risk assessments that employ the various tools described in the earlier chapters relating to the statistical interpretation of data. The field of environmental risk assessment is an enormous and evolving field; we seek only to provide some sample indications of the processes involved. As a result, models are described herein only as a means of characterizing the migration pathways of the environmental constituents. The relative merits and advantages of various model types are left to others (e.g., McTernan and Kaplan, 1990).

12.2 ESTIMATION OF EXPOSURE RISKS

12.2.1 Exposure Risks in Everyday Living and Levels Now Considered Acceptable

In general, that we find so many chemicals in our water supply is less related to an expanded use and abuse of chemicals than it is to an ever-improving capability to quantify smaller and smaller concentrations. Our exposure to risk is not necessarily greater than in previous decades; we are simply better able to identify the chemicals that are present.

In utilizing the available tools for risk assessment, we have two major considerations:

1. There is a limit on the extent to which the risk management should be expected to reduce environmental risk. No agency can control all the sources of the risks to which our society is exposed. As a compromise, risk management for environmental concerns is often now interpreted as decreasing risk to one in a million, referred to in the legal profession as a "de minimis risk" or negligible risk. In society, we are willing to accept some involuntary risks, such as being struck by lightning, which corresponds to about one in a million (Lehr, 1990a). Some examples of risks that increase the chances of death by one in a million are summarized in Table 12.1. We are now reaching a point where acceptable risk is considered somewhere on the order of a one in

TABLE 12.1 Risks that increase chance of death by one in a million

- Smoking of 1.4 cigarettes
- Drinking 1/2 L of wine
- Spending 1 hour in a coal mine
- Traveling 6 minutes by canoe
- Traveling 10 miles by bicycle
- Flying 1,000 miles by jet
- Drinking City of Miami drinking water for 1 year
- Eating 40 tablespoons of peanut butter
- Eating 100 charcoal broiled steaks

Source: Wilson, R., 1979. "Analyzing the Daily Risks of Life," *Technology Review* (February 1979), 45. Reprinted with permission from *Technology Review*, published by the Association of Alumni and Alumnae of MIT, copyright 1998.

100,000 chance of dying from exposure over a lifetime (Lehr, 1990a). The acceptance of a risk of one in a million associated with, for example, a contaminated site remediation thus reflects an attempt to decrease the risk associated with site remediation to a magnitude on the order of the other risks to which members of modern society are exposed.

2. A difficulty with assessments of many environmental risks is the availability of only minimal data, and thus a major concern associated with the estimation of risk is the degree of confidence in the estimates of risk.

Thus, there are concerns with estimating exposure risks to environmental phenomena. We might utilize monitoring data to characterize the concentrations and thus estimate the risk by comparisons with environmental quality standards. This approach is only available in circumstances where we have monitoring records available at the point of interest, such as depicted schematically in Fig. 12.1 as Scenario A. Alternatively, we may need to utilize the data from monitoring wells to assist in quantifying the parameters for use in mathematical models, where the models are subsequently utilized to estimate the exposure concentrations, as depicted in Scenario B (Fig. 12.1). In doing the mathematical modeling, we must consider the array of pathways the environmental constituent may follow from the source to create an exposure risk to a receptor. In many situations, there are potentially numerous pathways that influence the accuracy of the calculated exposure risk. The methods for computing exposure risk are described in detail in volume 2 of this Environmental Management series (Louvar and Louvar, 1998).

12.2.2 Uncertainty of the Impacts Arising from Exposure Risk to Environmental Contaminants

A fundamental component of risk assessment is the determination of the point at which people become ill. Such determinations are nontrivial; the information utilized in making such assessments is available from both toxicological and epidemiological studies.

Most toxicological studies fit into the classic paradigm of experimental studies. In such an experiment, all environmental conditions are fixed to the extent possible, and only one, or at most a very few, are varied, with the expectation that

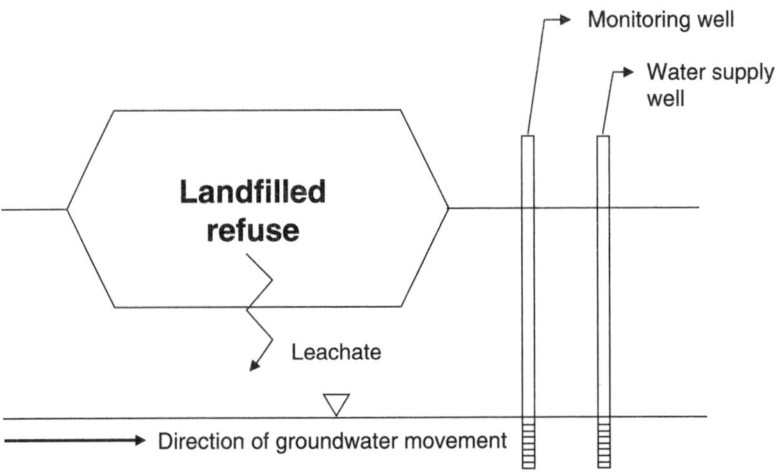

Scenario A: With monitoring data from the water supply well, the exposure risk arising from water consumption can be estimated
Scenario B: With monitoring data from the monitoring well, in combination with a mathematical model to predict exposure concentration in the water supply well

FIGURE 12.1 Schematic depiction of alternative exposure risk computational procedures

differences in outcome will result from only planned and controled modifications in exposure. Our confidence in the result is strengthened to the extent that we are aware of all causal factors and can control them in the experiment.

In contrast, most epidemiological studies of importance to risk assessment are observational. That is, there is no manipulation of the setting in which people live, but rather the study design takes advantage of "natural experiments" in which groups of people have been exposed, usually inadvertently, to different levels of potentially hazardous materials.

Economic restrictions limit the number of animals that can be included in any toxicological study. Most environmental regulations that apply to carcinogens are set at levels that are intended to ensure a lifetime cancer risk no greater than 10^{-4}, 10^{-5}, or 10^{-6}. If the experimental animals in a lifetime bioassay feeding study were exposed to a carcinogenic chemical at a level comparable to that of even relatively polluted environments, the probability of a tumor being observed would normally be less than 10^{-4}, requiring a study size of 100,000 or more animals to detect a statistically significant result. Since this is infeasible, the experiment is normally designed using relatively high concentrations of a chemical to much smaller numbers of animals to obtain a meaningful increase in adverse outcomes. The results are then extrapolated to estimate effects at the much lower exposure levels typical of the human or ecological environment.

It has been estimated that the extrapolated uncertainty at low dose can vary as much as forty-fold, depending simply on the extrapolation model (Stallones, 1988). On the other hand, epidemiological studies are often from specific occupational groups. Here the cancer risk is evaluated in, for example, working populations exposed to higher levels of contaminants than normally found in the general environment. Extrapolation of findings from high to much lower exposure levels is also required.

Even with this weakness, epidemiologic findings enjoy a major advantage over toxicological animal data for risk assessment, since there is no need to extrapolate across species lines from experimental animals to humans. Differences exist between humans and experimental species with respect to features such as size, metabolic rate, and target organs, as well as biochemical differences. Hence, even if the best estimates of health impacts arising from environmental exposure are obtained, there is still considerable uncertainty in these estimates. In response, we tend to be conservative in our extrapolations.

Data on the responses of humans exposed to some hazards exist. A considerable effort is underway to improve the utility of epidemiological data for risk estimation, but sampling errors often are very large for exposures at low concentrations in real-world situations (Rogers, 1991) and toxicological impact falls off rapidly (and not linearly) at low levels (Dinman, 1972). The safety factor approach involves first the determination of the largest experimental dose that is not significantly different from the zero-dose control group (the no-observed-adverse-effect level or NOAEL). The NOAEL represents an estimate of the threshold for a particular hazard. Then the NOAEL is modified to reflect various levels of uncertainty, typically a factor of safety to get a reference dose (RfD) or a reference concentration (RfC).

The outputs from these assessments include:

1. The reference dose (RfD) is the maximum allowable daily exposure to a non-carcinogenic contaminant for a particular body weight that is unlikely to cause adverse systemic effects during a lifetime for a particular exposure pathway (units of mg/kg-body weight/day).
2. The cancer slope factor (CSF) or potency value is a measure of the change in risk with a change in dose of a particular carcinogenic contaminant. There is presumed to be a risk of getting cancer associated with any carcinogenic dose.

Since 1990, the U.S. EPA has defined a target extra cancer risk range of 1×10^{-4} to 1×10^{-6} as a range of generally acceptable risk (NCP, 1990, U.S. EPA, 1991). From the 1×10^{-4} level and higher, the probability of cancer and the size of the population at risk should be considered unacceptable for any modeled population or exposure scenario.

12.2.3 Estimation of Risk (via Migration)

Given environmental contamination, a source of potential exposure risk (i.e., a hazard) exists. However, for risk to occur, three components must exist: (i) a source of contaminants; (ii) one or more pathways by which the contaminants may migrate; and (iii) a receptor who will be harmed if the exposure is sufficiently large. In schematic form, these three essential features are illustrated in Fig. 12.2. If any of these features (i) through (iii) are zero, there is no risk. Thus, risk

FIGURE 12.2 The three essential components of risk

assessment is really just a systematic process for making estimates of all the significant sources and exposure risk pathways that prevail over an entire range of failure modes and/or exposure scenarios. One or more processes may occur that lead to the exposure of people, animals, and the environment. The processes that create the exposure involve migration pathways via which the environmental constituents travel from source to the receptor. A depiction of some of the potential pathways is illustrated in Fig. 12.3. When a contaminant is released into the environment, it moves or partitions into the environment according to its physical properties and the properties of the environment. Developing quantitative estimates of the extent to which these migration pathways are contributing is often a considerable undertaking. Numerous coefficients exist in models. The availability of monitoring data for the purposes of calibrating the model transfer functions greatly improves the confidence in the modeling results. For example, Sharma et al. (1994) used monitoring data to ensure that the mathematical model predictions for polyaromatic hydrocarbons (PAH) depositions were accurate, as depicted in Fig. 12.4. This increases the confidence of the deposition predictions developed using the model for locations at which there are no observations.

For some situations, one migration pathway will clearly be the dominant feature. Conversely, for others, a variety of migration pathways may be relevant, which is suggestive of the degree of difficulty sometimes involved in a compre-

FIGURE 12.3 Depiction of pathways for migration of environmental constituents

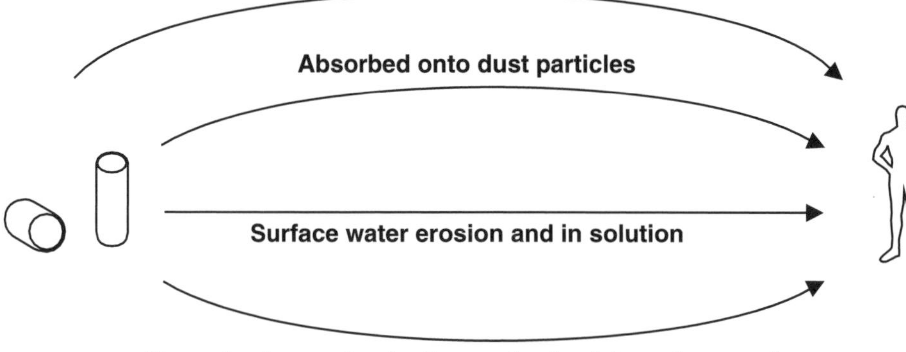

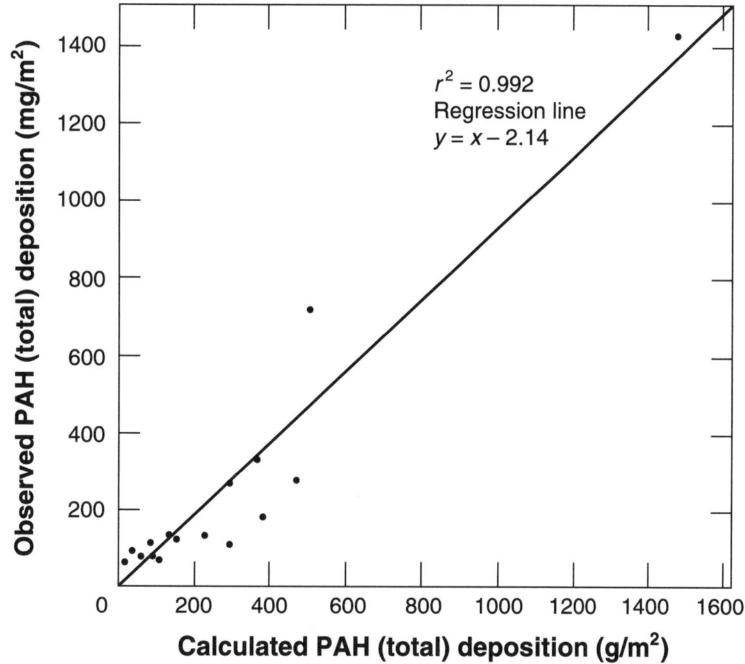

FIGURE 12.4 Regression plot of observed and computed PAH
Source: Adapted from Sharma, N., J. Marsalek, and E. McBean, "Migration Pathways and Remediation of Urban Runoff for PAH Control," *Journal of Environmental Management*, 41(1994), 325–336. Reprinted with permission.

hensive risk assessment. Examples of the array of pathways are listed in Table 12.2(a) and Table 12.2(b). The various pathways may include contaminants being transported via one or more media (including air, soils/sediments, surface water, and groundwater) to potential receptors (through, for example, inhalation, dermal contact, and ingestion). The existence of various exposure routes to alternative organs within the body (e.g., inhalation suggests that the target organ is likely the lungs, whereas ingestion may result in the target organ being the stomach) indicates that the pathways assessment and exposure risk calculations may require considerable work.

Of interest is the determination of the relative importance of the various types of exposure scenarios. The exposure assessment aspect of risk assessment

TABLE 12.2(a) Examples of mechanisms for release from the source area

- fugitive dust generation to impact air
- fugitive dust generation and deposition on other soils
- leaching of source areas to contaminated groundwater
- human uptake by direct contact, ingestion, or inhalation
- surface runoff following rain events to river
- groundwater seepage to river
- aquatic life uptake by contact

TABLE 12.2(b) Exposure pathways by medium

Medium	
Groundwater	Ingestion from drinking
	Inhalation of volatiles
	Dermal absorption from bathing
Surface water	Ingestion from drinking
	Inhalation of volatiles
	Dermal absorption from bathing
	Ingestion during swimming
	Ingestion of contaminated fish
Soil	Ingestion
	Inhalation of particulates
	Inhalation of volatiles
	Ingestion via plant uptake
	Dermal absorption from gardening

must entail the characterization of the physical and exposure setting, including contaminant distributions leading from sources of environmental contaminants to the points of exposure, the identification of the significant migration and exposure pathways, and the identification of chemical intakes for all potential receptors and significant pathways.

Mathematical models to quantify the exposure scenarios are important elements of the assessment. Numerous model classification systems exist, including (i) black box models, (ii) analytical models, and (iii) numerical models. The selection of a model is influenced by at least three conditions (after McBean, Rovers, and Schmidtke, 1990): (i) the judgment of the investigator in the specific application of the various modeling techniques, (ii) the extent of the database, and (iii) the physical system being modeled. A detailed discussion of the pros and cons of different models is left to others (e.g., for further reading, see McTernan and Kaplan, 1990).

The previous chapters have presented types of probability distributions and statistical tools as information that is utilized as part of the mathematical modeling effort. The next level of concern is to establish whether the various elements of a system combine to create significant potential exposure risk. The result is that there are four somewhat overlapping segments:

1. *Exposure assessment*—examination of the pathways identifies the substances or contaminants of concern, evaluates the contaminant concentrations in the media of interest, and examines the characteristics of the contamination and environmental settings that may affect the fate, transport, and persistence of the contaminants. The complexity of the assessment increases as we increase the number of contaminants of concern and the number of exposure pathways.
2. *Hazard identification*—used to refer to the risk level. The health risk assessment estimates the magnitudes of exposure to a compound and the increased likelihood and severity of adverse effects.

3. *Dose response*—identifies when the exposure dose is sufficient to cause an impact. The standard protocol for deriving maximum concentration limits (MCLs) addresses exposures by incorporating a default assumption that ingestion of drinking water averages 2 L/day over a lifetime. The methodology begins with the derivation of a dose calculated to protect the public from adverse health effects resulting from ingestion of a contaminant. This reference dose (RfD) is then divided by the estimated ingestion rate (2 L/day) to yield the drinking water equivalent, the level of a contaminant in water that should produce no adverse effects if all other sources of exposure are eliminated.
4. *Risk characterization*—typically the exposure is not restricted to a single constituent. This requires integration over the array of constituents to summarize the information and develop numerical expressions of risk (for carcinogens) and hazards (for noncarcinogens).

Insofar as questions relate to whether or not a chemical is likely to cause cancer, the U.S. EPA uses a conservative model to project the likely lifetime upper-limit excess cancer risk, generally assuming that no threshold is found in the dose response curve (Dourson and Jordan, 1989).

Hence the risk calculation is accomplished by summing over the migrational pathways of exposure levels and probabilities of occurrence. AERIS (1990) or API (1994) are examples of models that are utilized to complete the calculations.

NONCARCINOGENIC RISK

For noncarcinogens, hazard quotients (HQ) and hazard indices (HI) are calculated for each contaminant of concern for each exposure pathway, using the equation

$$HQ_{ij} = \frac{CDI_{ij}}{RfD_{ij}} \qquad [12.1]$$

where CDI_{ij} = chronic daily intake for exposure pathway j for chemical i

RfD_{ij} = reference dose for exposure pathway j for chemical i

HQ_{ij} = noncarcinogenic hazard quotient for chemical i via exposure pathway j

Since noncarcinogenic risks are considered additive, a hazard index is calculated for each exposure pathway j for all contaminants

$$HI_j = \sum_{i=1}^{I} HQ_{ij} \text{ over all chemicals } i = 1, \ldots, I \qquad [12.2]$$

where HI_j is the hazard index for exposure pathway j. The exposure pathway hazard indices are then summed over all pathways $j = 1, \ldots, J$, to get the total exposure hazard index.

$$HI = \sum_{j=1}^{J} HI_j \qquad [12.3]$$

If the total exposure hazard index for the site is greater than one, then risk management is required.

CARCINOGENIC RISK

The excess lifetime cancer risk (ELCR) is calculated for carcinogens as

$$ELCR_{ij} = CDI_{ij}\, CSF_{ij} \qquad [12.4]$$

where $ELCR_{ij}$ is a (unitless) probability of an individual developing cancer as a result of exposure pathway from chemical i; CDI_{ij} is the chronic daily intake averaged over a specified duration (mg/kg/day) for chemical i with exposure pathway j, and CSF_{ij} is the cancer slope factor $(\text{mg/kg/day})^{-1}$.

The total exposure excess lifetime cancer risk is calculated by adding the risk for exposure pathway. If the total exposure excess lifetime cancer risk is greater than the exposure risk considered acceptable (e.g., 10^{-5}), then risk management is required.

The risk assessment process thus results in decomposing a problem into logical pieces. However, important limitations may exist in the models utilized in characterizing the migration of the constituents from the source to the receptor, the data utilized as inputs to the models, and the scientific understanding of the reference doses. As a demonstration, consider the situation depicted in Fig. 12.5, which indicates the smokestack emissions from a steel mill, causing long-term exposure to nearby residents. The exhaust gases include a vector of environmental

FIGURE 12.5 Indications of associated features influencing the estimates of exposure risk

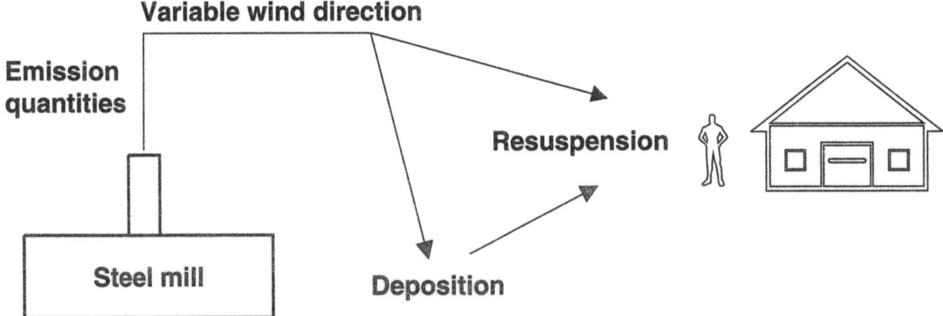

constituents, polynuclear aromatic hydrocarbons (PAHs), known to be carcinogens. Relevant concerns include:

1. To what accuracy are the emission quantities known?
2. To what degree do depositional effects influence the migration of the PAHs prior to reaching the location of the receptor?
3. To what extent does wind variability in terms of temporal variability of both direction and velocity influence the exposure of the receptor?
4. Even if the exposure concentration reaches a specific level, what is the likelihood that the individual will incur deteriorated health? Is the reference dose a magnitude in which there is confidence, or does it involve a considerable extrapolation?

Features 1 through 4 indicate some of the uncertainties associated with the estimation of the exposure risk.

Similarly, uncertainties exist in projection of concentrations in the groundwater at a point downgradient from a landfill where the migration potential of landfill leachate is a function of soil characteristics and attenuative mechanisms. We utilize a fate and transport model to integrate over the individual random variables to make projections of water quality at the downgradient location.

The likelihood of illness is a result of the exposures of the receptors. Two types of receptors are typically used, an adult and a young child. The adult and the child are assigned behavioral patterns and living conditions such that they represent the maximum cumulative intakes for individuals at the point of exposure. The adult receptor is assumed to be in the 20- to 39-year bracket, have a 70-kg weight, and a daily breathing rate of 23 m^3. The young child receptor is assumed to be a child about 2 to 3 years old, with 10 kg body weight and a daily breathing rate of 5 m^3.

The pathways to target organs within the body then relate to the pathways such as ingestion exposure, dermal exposure, and inhalation exposure.

The final step in the risk assessment process is to integrate the toxicity and exposure assessment results to characterize the risks.

12.3 ALTERNATIVE METHODS FOR GENERATING THE DISTRIBUTION OF RISK

Deterministic procedures are those for which, given a single input, there is a single output, as schematically depicted in Fig. 12.6. A simple example of this situation is one in which a doubling of the smokestack emissions from 100 $\mu g/m^3$ to 200 $\mu g/m^3$ would translate to a doubling of the downwind air quality from 1 $\mu g/m^3$ to 2 $\mu g/m^3$. However, use of a deterministic solution may be incorrect for such reasons as (a) the smokestack emissions are only quantifiable as a probability distribution (e.g., there is probability of .1 that the emissions are between 100 $\mu g/m^3$ and 200 $\mu g/m^3$); and (b) the parameters in the model that are used to characterize the translation from smokestack emissions to downwind air quality are uncertain. An emission rate of 100 $\mu g/m^3$ will produce an air quality between 1 to 1.5 $\mu g/m^3$

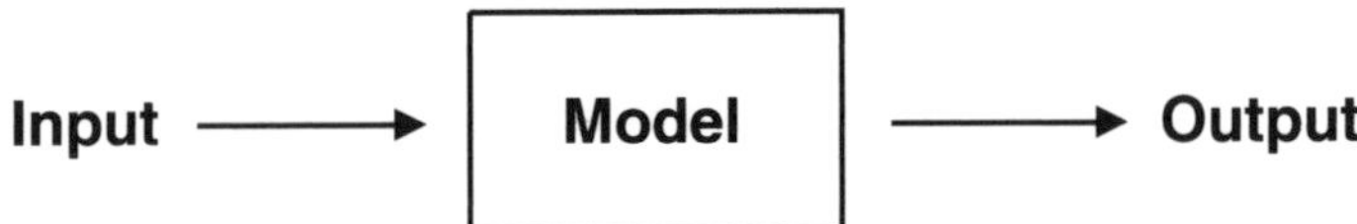

FIGURE 12.6 Schematic of input and output for a deterministic model

with a probability of 0.3. For these types of reasons, the use of deterministic models may be inappropriate for many applications.

As alternatives, probabilistic risk assessment techniques have been available for some time. Probabilistic techniques differ from deterministic algorithms by explicitly considering variability and uncertainty in parameter values and models.

The Monte Carlo method is perhaps the best known probabilistic technique to incorporate uncertainty. In a Monte Carlo simulation, distributions are specified for each parameter that account for both variability and uncertainty. A single instance of risk is computed by (i) selecting random values for each parameter from the distributions (e.g., the normal distribution) and accounting for any correlations between variables and (ii) calculating the value of risk (using deterministic models and relationships) with the set of random values for that instance. By conducting a large number of determinations, a distribution of risk values is generated from which it is possible to determine relevant statistical measures.

The stochastic approach using Monte Carlo has several advantages over the deterministic approach:

1. Input variables are defined as statistical distributions, covering the entire range of likely values.
2. Input parameters may be varied randomly and simultaneously, allowing uncertainty propagation.
3. Monte Carlo allows sensitivity analyses.

Uncertainty plays a major role in all risk assessments related to environmental quality. Uncertainty exists in the characterization of contaminant sources, the transport processes, the exposure levels, and the dose responses. It is not known for example whether a landfill will release contaminants to the groundwater and, if so, when, how much, for how long, or for what chemical compounds. For a given contaminant release, transport will occur but the parameters required to describe this movement are not known with precision. Uncertainties arise due to the variability of human exposure factors, random and systematic errors in measurements and sampling, and limitations in mathematical models.

12.3.1 Monte Carlo Analyses

Using Monte Carlo procedures, models are solved repeatedly with the parameters randomly chosen from their postulated probability distributions. The output is then used to analyze the statistical properties of the distribution of the dependent variables.

The stochastic nature of the input variables may be accounted for by use of a distribution sampling technique. The Monte Carlo method involves the generation of a statistically large number of realizations of input parameters consistent with their statistical distributions. A realization is a single simulation performed with a deterministic model using a particular set of input values. For one realization, a single value of each variable that is treated stochastically is randomly sampled from its assumed or fitted distribution function. The simulation is performed with these values in a corresponding output variable or model performance measure. By utilizing a large number of realizations, the distributions of output variables are developed. The mean and variance of the output variables are computed from the output of all realizations performed.

The Monte Carlo method is very powerful in that it requires few assumptions. It represents a simple direct means to translate the variations in input parameters into variations in output parameters (exposure concentrations). For certain types of problems, the use of the Monte Carlo modeling approach is very appropriate. For example, if we are introducing a process change in an industrial

EXAMPLE 12.1—EXAMPLE OF A RISK ASSESSMENT METHODOLOGY

A permitting agency receives a proposal to modify an existing incinerator. The proponent for the incinerator is arguing to allow the incinerator to burn organic wastes to produce energy and safely dispose of hazardous waste. Stack emissions data are then compiled as measured at similar furnaces and the fugitive emissions are estimated. Using these data and meteorological data, the concentration of air contaminants at various receptor locations can be estimated using an atmospheric dispersion model (e.g., Turner, 1969). The concern then is to identify the potential receptors. These will include within a 10-km radius a small town, a small drinking water reservoir, a small nearby bay where people do some fishing, and some backyard vegetative gardens exist.

Then we would develop a list of indicator chemicals. Perhaps three populations of receptors of concern are identified, and we look to examine the exposure pathways; inhalation, dermal contact with contaminated soil, incidental soil ingestion, intake of home-grown produce through roots and/or particulate deposition, food contamination during preparation in the kitchen, ingestion of contaminated water from the reservoir, intake of locally caught contaminated fish, and ingestion of contaminated milk.

Possibly only the latter exposure pathways, namely inhalation, dermal exposure to soil, and soil ingestion pathways are found to be relevant; other exposure pathways are approximately zero.

The receptors are identified as:

a. residents—these are adults living in the nearest residences;
b. school children—those attending school and which live in locations assumed with little or no impact; and
c. workers.

The subsequent steps of analysis then entail assessment of the concentrations to which the receptors are exposed.

process, we may be able to predict the change in the effluent stream, but the change in the air quality at some downwind point is another question and a model is used to make that projection.

The stochastic approach is a valuable tool in the decision-making process because it generates valid statistical distributions of exposure estimates. It allows sensitivity analyses that provide information on which parameters are major components in the overall probability of health and/or ecological risk.

12.3.2 Latin-Hypercube Procedure

The Latin-Hypercube sampling procedure is an alternative to the Monte Carlo procedure, but the Latin-Hypercube procedure is a more efficient method of parameter sampling. This procedure requires the division of the input parameter distribution into N segments of equal probability, $1/N$, where N is the number of realizations. An equal number of samples is drawn from each segment to provide a representation of the full range of possible input parameter values. This method ensures that the tails of the distribution are efficiently accounted for, a feature that is not ensured when using the Monte Carlo technique. The inclusion of the tail end of the distribution is important, since it is these extreme values of an input distribution that lead to values of the dependent variables that are of interest to the landfill designer. This sampling technique is performed for each input variable that has been assigned a level of uncertainty. Iman and Conover (1980) and McBean et al. (1997) provide a detailed discussion of the Latin-Hypercube sampling technique.

12.3.3 Examples of Use of the Two Procedures

A number of investigations have utilized the Monte Carlo and Latin-Hypercube methodologies. For example, Salinas (1993) examined stochastic health risk assessment using Monte Carlo simulation in estimating downward air quality near a proposed incinerator. Smith and Freeze (1979) use 300 Monte Carlo realizations in their study of a two-dimensional stochastic aquifer. Donald and McBean (1994 and 1997) used 40 Latin-Hypercube simulations in a stochastic analysis of hydraulic conductivity for a landfill liner.

Murray et al. (1995) developed a comparison of leakage rates through leachate collection systems at the bottom of a landfill. They utilized 100 Latin-Hypercube simulations in order to represent the range of possible values for the input parameters assigned to characterize the levels of uncertainty. An effective evaluation of leakage rates must include the uncertainty associated with the relevant parameters, including the clay liner hydraulic conductivity value and the spatial frequency of flexible membrane liner (FML) holes in the case of the composite liner, since knowledge of the magnitudes of these two parameters is only known in terms of a probability density function. Once the probability distribution of the leakage quantities has been estimated, an estimate of the exceedence probability (e.g., 95 percent sample from the distribution of the data) can be obtained.

EXAMPLE 12.2—RISK ASSESSMENT IN DESIGN OF A BOTTOM LINER SYSTEM FOR A LANDFILL SITUATION (ABBREVIATED FROM MURRAY ET AL. (1995))

A major component of the design for a landfill involves the leachate collection and low permeable liner system placed at the bottom of a landfill. The system includes features such as a high permeability (high transmissivity) drainage layer and drainage tiles and a low permeability liner or barrier underlying the drainage layer and tiles, as illustrated in Fig. 12.7. The low permeability barrier layer may be constructed of a single material, either using natural soils or one or more FMLs, or as a composite of natural soils and FMLs.

Given the multiple components within the leachate/liner system and the necessary care and effort required during the placement to ensure their proper functioning, the overall cost of the bottom liner system involves millions of dollars. The objective is to minimize the escape of percolating leachate from the overlying refuse which would otherwise contaminate the groundwater. We cannot make the risk of contamination of the groundwater zero. Some

FIGURE 12.7 Configuration of a leachate collection system designed to minimize percolation into the underlying groundwater

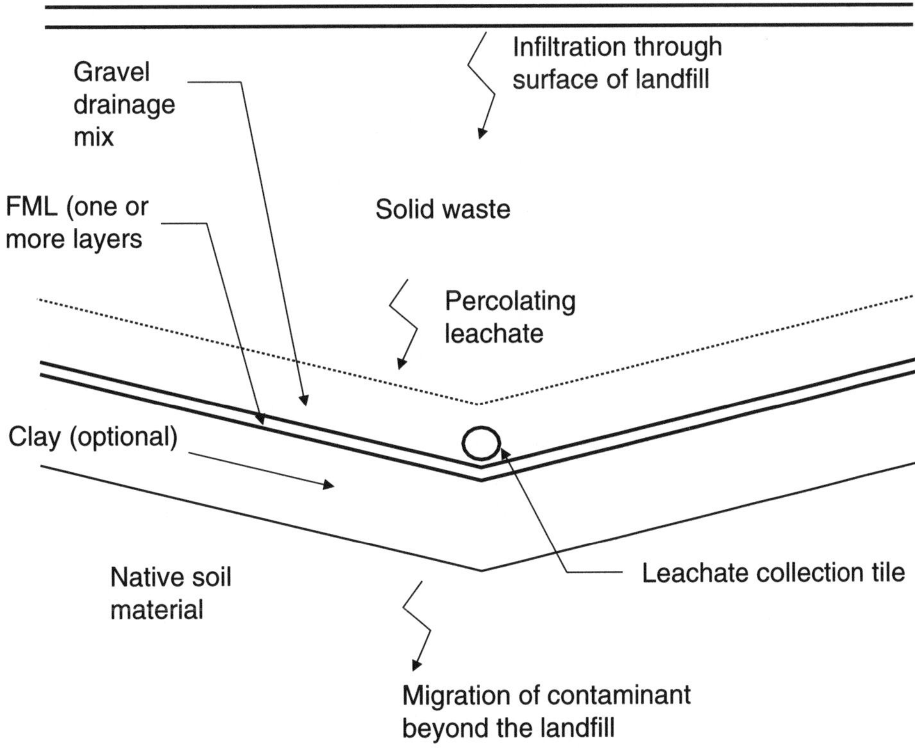

leachate will leak out. However, by investing in greater thicknesses of soils, the likelihood of significant excursion of leachate can be decreased.

Alternatively, the FMLs involve placement of a manufactured material. For each field placement, the FMLs are typically brought to the landfill in 6 to 10 m wide rolls. The individual rolls must then be seamed in the field. The integrity of the seams is critical for minimizing leachate escape. As well, there will be some defects arising from the manufacturing. A number of mathematical models have been developed to simulate the components of flow and transport of leachate and its constituents to the surrounding environment. Elements of the modeling require the characterization of

(i) leachate volumes and leachate solute concentrations generated within the landfill facility,
(ii) leachate mounting profiles overlying the low permeability barrier,
(iii) flow and solute migration through the barrier system and the underlying variably saturated zone, and
(iv) flow and solute migration through an underlying aquifer to a point of compliance.

Consider some of the uncertainty aspects of the liner. Leakage through FMLs occurs via defects due to various holes and faulty seams. Giroud and Bonaparte (1989) draw the following conclusions regarding postinstallation FML seam defect frequency based on field observations: An average of one defect per 10 m of field seam (equivalent to 90 to 150 seam defects per hectare, if 6 to 10 m wide FML panels are used) can be expected without quality assurance (QA), and without quality control (QC). An average of one defect per 300 m of field seam (equivalent to 3 to 5 defects per hectare, if 6 to 10 m wide FML panels are used) can be expected with reasonably good installation, adequate QA/QC, and repair of noted defects.

Alternatively, Laine and Miklas (1990) reported finding a typical range of 1 to 22 leaks per 4000 m^2 with an overall average of 14 leaks per 4000 m^2 during tests of 61 new or in-service FML-lined lagoons. They reported a hole frequency without quality assurance of one hole per 130 m^2 and with quality assurance, at one hole per 4000 m^2.

Saleh (1992) reports electrical leak detection data from 28 FML-lined facilities, including the total number of seam and material leaks, and in two cases differentiated between bottom and side leaks. Fifty percent of the facilities were studied as a quality assurance program and the remainder were studied as a result of known leakage problems. Results from two of the facilities where bottom and sideslope leaks were differentiated indicated that there were approximately 16 leaks per hectare in the bottom liner.

Laine and Darilek (1993) present leak frequency data from 169 FML-lined sites throughout the United States. No differentiation is made in their reporting between sites in accord with quality assurance/quality control, and no information is provided regarding the FML materials used or the thickness of the geomembranes. Figure 12.8 is a graphic representation of the reported data, and Fig. 12.9 shows the relationship between the $\log_{10}$-transformed values of leak frequency versus the standard normal deviate from the geometric mean of the $\log_{10}$-transformed distribution. The leak frequency data of Laine and Darilek (1993) have a geometric mean of 26 leaks per hectare and a log-transformed standard deviation of 0.695.

In synthesizing the above-referenced material to a form useful in predicting potential leakage rates through an FML, two features must be established, namely the frequency and size of the holes in the FML. Specifically:

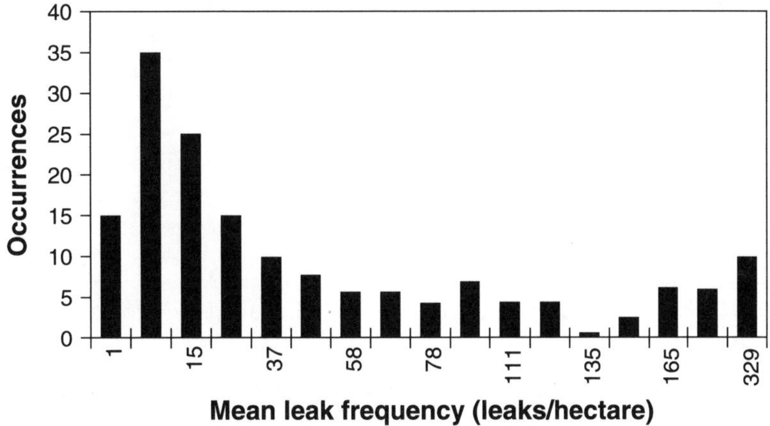

FIGURE 12.8 Leak frequency data from geomembrane-lined sites
Source: Data from Laine, D. L., and G. T. Darilek. "Detecting Leaks in Geomembranes," *Civil Engineering*, 63, no. 8 (1993), 5–53.

1. The spatial frequency of holes would be expected to follow a lognormal distribution.
2. The size of the holes for the FML seam defect size, according to Giroud and Bonaparte (1989), is expected to be on the order of 10 mm in length in the absence of quality assurance, and an average seam defect is expected to be on the order of 1 to 3 mm in length with quality assurance.

FIGURE 12.9 Lognormal probability plot of leak frequency data
Source: Data from Laine, D. L., and G. T. Darilek. "Detecting Leaks in Geomembranes," *Civil Engineering*, 63, no. 8 (1993), 5–53.

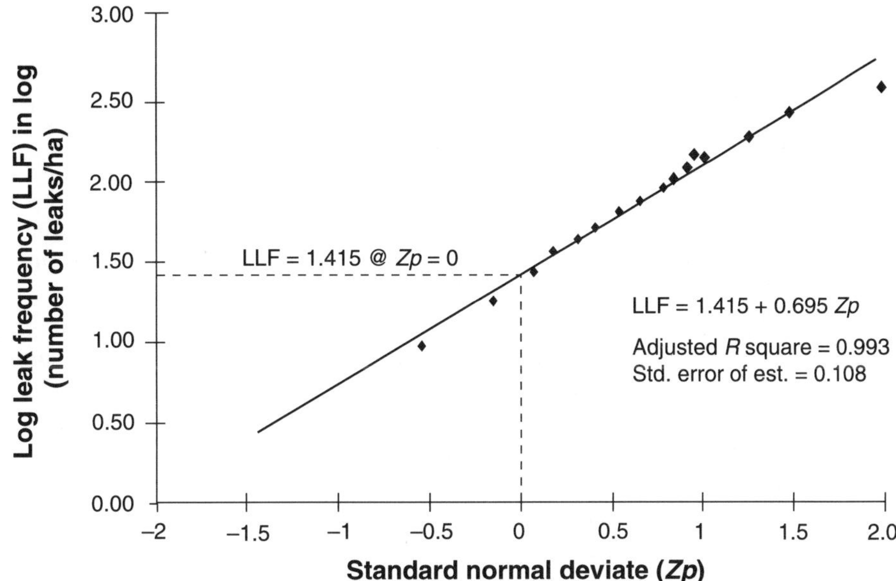

Giroud and Bonaparte (1989) propose a method for evaluating the rate of leakage due to permeation across a single FML by combining the processes of liquid permeation and vapor transmission. As well, Giroud et al. (1992) present equations for calculating leakage rates through defects in barrier systems where the FML is placed in close contact with, and on top of, a layer of permeable soil.

Murray et al. (1995) describe in detail the equations utilized in the assessment. Given the equations for estimated leakage rates, a comparison between alternative barrier system configurations is feasible. The two major types of barrier systems studies are

1. single low permeability soil liners; and
2. composite barrier systems comprising an FML placed over top and in close contact with a low permeability soil.

However, an effective evaluation of leakage rates must include the uncertainty associated with the relevant parameters. To examine the uncertainty, 100 Latin-Hypercube simulations of each barrier system were carried out. Uncertainty is assigned to the clay liner hydraulic conductivity value, and the spatial frequency of FML holes in the case of the composite liner, since knowledge of the magnitudes of these two parameters is only known in terms of a probability density function (e.g., as per Fig. 12.9 for the frequency of holes in the FML). The distribution assigned to the hole spatial frequency depends on the level of QA/QC during installation. Both variables are assigned lognormal distributions.

Donald and McBean (1994) reported close agreement between hydraulic conductivity data and field cores taken from the Keele Valley Landfill liner, using a geometric mean of 7.68×10^{-9} cm/s and $\log_{10}$ standard deviation of 0.264. These parameters are used for the first clay liner hydraulic conductivity distribution (C1) used in this analysis. This distribution produces clay liner hydraulic conductivity values ranging from 1.24×10^{-9} to 4.78×10^{-8} cm/s for ±3 standard deviations from the mean.

A second hydraulic conductivity distribution (C2) with a geometric mean of 7.68×10^{-8} cm/s and a range from 1.24×10^{-8} to 4.78×10^{-7} cm/s for ±3 standard deviations from the mean is used for the second clay liner system described below. This represents an order of magnitude shift in the hydraulic conductivity values reported in Donald and McBean (1994), with the shapes of the two distributions being otherwise identical.

In accord with the data presented previously, the geometric mean of FML hole frequency is assumed to be 3 holes per hectare for a site having undergone a good level QA/QC. A geometric mean of 125 holes per hectare is assigned to sites having a poor level of QA/QC. Both distributions are assumed to have a $\log_{10}$ standard deviation of 0.695.

Table 12.3 lists the geometric mean, maximum, and minimum values of the average leachate mounding heights over the barrier system for the various liner scenarios. A general trend exists in that larger average mounding heights are associated with the less permeable barrier systems. Composite liners exhibit a wider range of average mounding height values due to the variability of hole spatial frequency in the FMLs.

Table 12.4 displays the geometric mean, maximum, and minimum values of the liner leakage rates (as a percentage of infiltration rates) for the alternative barrier systems. Leakage rates through both clay and composite liner systems are extremely sensitive to the hydraulic conductivity of the clay liner. For both single-clay and composite barrier systems,

TABLE 12.3 Mounding height results

Liner type	Clay quality	Geomembrane quality	Average Mound Height (m) Geometric mean	Minimum	Maximum
Clay	C1	—	0.273	0.080	0.782
Composite	C1	poor QA/QC	0.274	0.058	0.696
Composite	C1	good QA/QC	0.276	0.058	0.697
Clay	C2	—	0.217	0.051	0.659
Composite	C2	poor QA/QC	0.252	0.058	0.693
Composite	C2	good QA/QC	0.275	0.058	0.700

Source: Murray, G., E. McBean, and J. F. Sykes. "Estimation of Leakage Rates Through Flexible Membrane Liners," *Ground Water Monitoring Review and Remediation* (Fall, 1995), 148–154. Reprinted with permission.

TABLE 12.4 Liner leakage rates

Liner type	Clay quality	Geomembrane quality	Liner Leakage Rates (% Infiltration) Geometric mean	Minimum	Maximum
Clay	C1	—	4.91	2.44	9.82
Composite	C1	poor QA/QC	1.86	0.02	7.53
Composite	C1	good QA/QC	0.05	0.00	4.08
Clay	C2	—	44.96	23.85	72.66
Composite	C2	poor QA/QC	11.90	0.11	60.43
Composite	C2	good QA/QC	0.27	0.00	24.49

Source: Murray, G., E. McBean, and J. F. Sykes. "Estimation of Leakage Rates Through Flexible Membrane Liners," *Ground Water Monitoring Review and Remediation* (Fall, 1995), 148–154. Reprinted with permission.

an order of magnitude reduction in the clay liner hydraulic conductivity corresponds to an order of magnitude drop in the calculated leakage values (e.g., 45 decreases to 4.9 percent of the infiltration) when the hydraulic conductivity of the clay is decreased by an order of magnitude.

Composite liners have a significant impact on reducing leakage rates through the bottom liner. Liners incorporating a minimum level of FML QA/QC reduce the average leakage rate by 62 to 73 percent when compared to clay liners alone. Liners incorporating an extensive level of geomembrane QA/QC reduce the average leakage rate by 99 percent when compared to clay liners alone. Due to the high variability in FML hole spatial frequency, composite liners exhibit larger variations from the geometric mean of the average liner leakage rate when compared to single-clay liners.

Part of the challenge in the development of the risk assessment arises because there is uncertainty in the data assignments used in the models, and there is uncertainty in the impact of different exposure levels. To demonstrate these concerns further, assume for purposes of discussion that one remediation alternative involves the excavation and treatment of contaminated soils.

The resulting uncertainties in the data assignments include features such as, (i) the meteorological conditions to be experienced during the excavation are known only to the extent to which historical data can be used to identify the probabilities of alternative meteorological conditions; (ii) the locations, quantities, and form of containment of buried wastes to be excavated are known only to a limited extent; and (iii) the resulting effectiveness of a specific remedial action is uncertain until after it has been attempted (e.g., will bioremediation of soils cleanse the soils to acceptable levels?). In addition, there is uncertainty as to what exposure level to a chemical is sufficient to initiate cancer. As a consequence, a succession of worst-case assumptions is sometimes utilized. An alternative approach, using probabilistic risk assessment, involving sampling from probability distribution as was utilized in Ex. 12.2, represents an alternative.

12.4 REFERENCES

AERIS SOFTWARE INC. *AERIS Model Version 3.0 Technical Manual,* prepared as part of Supply and Services Canada contract 09SE-DE405-6-6586, 1990.

AMERICAN PETROLEUM INSTITUTE, *Decision Support System for Exposure and Risk Assessment,* version 1.0, New York, NY: Nassau, 1994.

ASANTE-DUAH, D.K *Hazardous Waste Risk Assessment.* Boca Raton, FL: Lewis Publishers, 1993.

BROMLEY, A. *Chemecology* (February, 1991), Chemical Manufacturers Association.

DINMAN, B.D. "'Non-Concept' of 'No-Threshold': Chemicals in the Environment," *Science,* 175 (1972), 495–497.

DONALD, S.B., and E.A. McBEAN. "Statistical Analyses of Compacted Clay Landfill Liners," *Canadian Journal of Civil Engineering,* 21 (October, 1994), 872–882.

DONALD, S.B., and E.A. McBEAN. "Statistical Analyses of Compacted Soil Landfill Liners Part 2; Sensitivity Analyses," *Canadian Journal of Civil Engineering,* 22, no. 4 (1997), 658–663.

DOURSON, M.L., and W.L. JORDAN. "How 'Safe' is the Ground Water Americans Drink?" *Ground Water Monitoring and Remediation* (Fall, 1989), 73–74.

GIROUD, J.P., and R. BONAPARTE. "Leakage through Liners Constructed with Membranes, Part I: Geomembrane Liners," *Geotextiles and Geomembranes,* 8, no. 1 (1989), 27–67.

GIROUD, J.P., K. BADU-TWENEBOAH, and R. BONAPARTE. "Rate of Leakage through a Composite Liner due to Geomembrane Defects," *Geotextiles and Geomembranes,* 11 (2992), 1–28.

GLICKMAN, T.S., and M. GOUGH, eds., *Readings in Risk.* Washington, DC: Resources for the Future, 1990.

IMAN, R.L., and W.J. CONOVER. "Small Sample Sensitivity Analysis Techniques for Computer Models, with an Application to Risk Assessment," *Communications in Statistics—Series A: Theory and Methods,* 9, no. 7 (1980), 1749–1842.

LAINE, D.L., and M.P. MIKLAS. "Finding Leaks in Geomembrane Liners Using an Electrical Method: Case Histories," *Hazardous Materials Control,* 3, no. 3 (1990), 29–37.

LAINE, D.L., and G.T. DARILEK. "Detecting Leaks in Geomembranes," *Civil Engineering,* 63, no. 8 (1993), 5–53.

LEHR, J.H. "Toxicological Risk Assessment Distortions: Part I," *Ground Water,* 28, no. 1 (1990a), 2–8.

LEHR, J.H. "Toxicological Risk Assessment Distortions: Part III—A Different Look at Environmentalism," *Ground Water,* 28, no. 3 (1990b), 330–340.

McBEAN, E., K. PONNAMBALAM, and W. CURI. "Stochastic Environmental Modeling," in *Environmental Data Management,* eds. V. Singh and N.B. Harmancingh. Dordrecht, Netherlands: Kluwer Academic Publishers, 1997.

McBEAN, E., and F. ROVERS. "Utility of Risk-Time Curves in Selecting Remediation Alternatives," *Waste Management and Research,* 13, 2 (1995), 167–174.

McBEAN, E., F. ROVERS, and K. SCHMIDTKE. "Risk Assessments Using Relatively Simple Mathematical Models," in *Risk Assessment for Groundwater Pollution Control,* eds. W.F. McTernan and E. Kaplan. New York, NY: ASCE, 1990.

McTERNAN, W.F., and E. KAPLAN, eds., *Risk Assessment for Groundwater Pollution Control.* New York, NY: ASCE, 1990.

MURRAY, G., E. McBEAN, and J.F. SYKES. "Estimation of Leakage Rates Through Flexible Membrane Liners," *Ground Water Monitoring Review and Remediation* (Fall, 1995), 148–154.

MURRAY, G., E. McBEAN, and J.F. SYKES. "Risk-Based Engineering Design for a Landfill Leachate Collection and Liner System," *Ground Water Monitoring Review and Remediation* (Spring, 1996), 139–146.

NCP. "National Oil and Hazardous Substances Pollution Contingency Plan (National Contingency Plan)," 40 CFR 300 Final Rule, *Federal Register,* 55 (March, 1990), 8666–8865.

REJESKI, D.W. "Exploring Future Environmental Risks," in *Environmental Statistics, Assessment and Forecasting,* eds. C.R. Cothern and N.P. Ross. Boca Raton, FL: Lewis Publishers, 1994.

ROGERS, L.B. "Modern Analytical Chemistry and Environmental Science," *Chemical Technology,* 21, no. 4 (1991), 229–233.

SALEH, A.R.M. "Leakage Mechanism through Double Liner Systems," in Proceedings of the Environmental Engineering Session at Water Form '92, New York, NY: ASCE, 1992, 192–200.

SALINAS, J.A. "Application of Probabilistic Methods to Assess Human Health Risks of Emissions from a Hazardous Waste Incinerator," 1993 Pacific Basin Conference on Hazardous Waste, Honolulu, Hawaii, Nov. 8–12, 1993.

SHARMA, M., J. MARSALEK, and E.A. McBEAN. "Migration Pathways and Remediation of Urban Runoff for PAH Control," *Journal of Environmental Management,* 41 (1994), 325–336.

SMITH, L., and R. FREEZE. "Stochastic Analysis of Steady-State Groundwater Flow in a Bounded Domain: 2. Two-dimensional Simulations," *Water Resources Research,* 15, no. 6 (1979), 1543–1559.

STALLONES, R.A. "Epidemiology and Environmental Hazards" in *Epidemiology and Health Risk Assessment,* L. Gordis, ed., New York, NY: Oxford University Press, 1988, 3.

TURNER, D.B. "Workbook of Atmospheric Dispersion Estimates," report no. 999-AP-26, Cincinnati, OH: U.S. Environmental Protection Agency, 1969.

U.S. EPA. *Risk Assessment Guidance for Superfund: Volume 1—Human Health Evaluation Manual (Part B, Development of Risk-Based Remediation Goals) "RAGS"*, OSWER Directive, 9285-7-01B, Office of Emergency and Remedial Response, 1991.

WILDAVSKY, A. *Searching for Safety.* New Brunswick, NJ: Transaction Publishers, 1988.

APPENDIX

TABLE A.1 Areas under the normal distribution curve

$$F(z) = \int_0^z \frac{1}{\sqrt{2\pi}} e^{-\frac{1}{2}z^2} dx$$

z	0.00	0.01	0.02	0.03	0.04	0.05	0.06	0.07	0.08	0.09
0.0	0.0000	0.0040	0.0080	0.0120	0.0159	0.0199	0.0239	0.0279	0.0319	0.0359
0.1	0.0398	0.0438	0.0478	0.0517	0.0557	0.0596	0.0636	0.0675	0.0714	0.0753
0.2	0.0793	0.0832	0.0871	0.0910	0.0948	0.0987	0.1026	0.1064	0.1103	0.1141
0.3	0.1179	0.1217	0.1255	0.1293	0.1331	0.1368	0.1406	0.1443	0.1480	0.1517
0.4	0.1554	0.1591	0.1628	0.1664	0.1700	0.1736	0.1772	0.1808	0.1844	0.1879
0.5	0.1915	0.1950	0.1985	0.2019	0.2054	0.2088	0.2123	0.2157	0.2190	0.2224
0.6	0.2257	0.2291	0.2324	0.2357	0.2389	0.2422	0.2454	0.2486	0.2518	0.2549
0.7	0.2580	0.2611	0.2642	0.2673	0.2704	0.2734	0.2764	0.2794	0.2823	0.2852
0.8	0.2881	0.2910	0.2939	0.2967	0.2995	0.3023	0.3051	0.3078	0.3106	0.3133
0.9	0.3159	0.3186	0.3212	0.3238	0.3264	0.3289	0.3315	0.3340	0.3365	0.3389
1.0	0.3413	0.3438	0.3461	0.3485	0.3508	0.3531	0.3554	0.3577	0.3599	0.3621
1.1	0.3643	0.3665	0.3686	0.3708	0.3729	0.3749	0.3770	0.3790	0.3810	0.3830
1.2	0.3849	0.3869	0.3888	0.3907	0.3925	0.3944	0.3962	0.3980	0.3997	0.4015
1.3	0.4032	0.4049	0.4066	0.4082	0.4099	0.4115	0.4131	0.4147	0.4162	0.4177
1.4	0.4192	0.4207	0.4222	0.4236	0.4251	0.4265	0.4279	0.4292	0.4306	0.4319
1.5	0.4332	0.4345	0.4357	0.4370	0.4382	0.4394	0.4406	0.4418	0.4430	0.4441
1.6	0.4452	0.4463	0.4474	0.4485	0.4495	0.4505	0.4515	0.4525	0.4535	0.4545
1.7	0.4554	0.4564	0.4573	0.4582	0.4591	0.4599	0.4608	0.4616	0.4625	0.4633
1.8	0.4641	0.4649	0.4656	0.4664	0.4671	0.4678	0.4686	0.4693	0.4699	0.4706
1.9	0.4713	0.4719	0.4726	0.4732	0.4738	0.4744	0.4750	0.4756	0.4762	0.4767
2.0	0.4772	0.4778	0.4783	0.4788	0.4793	0.4798	0.4803	0.4808	0.4812	0.4817
2.1	0.4821	0.4826	0.4830	0.4834	0.4838	0.4842	0.4846	0.4850	0.4854	0.4857
2.2	0.4861	0.4865	0.4868	0.4871	0.4875	0.4878	0.4881	0.4884	0.4887	0.4890
2.3	0.4893	0.4896	0.4898	0.4901	0.4904	0.4906	0.4909	0.4911	0.4913	0.4916
2.4	0.4918	0.4920	0.4922	0.4925	0.4927	0.4929	0.4931	0.4932	0.4934	0.4936
2.5	0.4938	0.4940	0.4941	0.4943	0.4945	0.4946	0.4948	0.4949	0.4951	0.4952
2.6	0.4953	0.4955	0.4956	0.4957	0.4959	0.4960	0.4961	0.4962	0.4963	0.4964
2.7	0.4965	0.4966	0.4967	0.4968	0.4969	0.4970	0.4971	0.4972	0.4973	0.4974
2.8	0.4974	0.4975	0.4976	0.4977	0.4977	0.4978	0.4979	0.4980	0.4980	0.4981
2.9	0.4981	0.4982	0.4983	0.4983	0.4984	0.4984	0.4985	0.4985	0.4986	0.4986
3.0	0.4987	0.4987	0.4987	0.4988	0.4988	0.4989	0.4989	0.4989	0.4990	0.4990
3.1	0.4990	0.4991	0.4991	0.4991	0.4992	0.4992	0.4992	0.4992	0.4993	0.4993
3.2	0.4993	0.4993	0.4994	0.4994	0.4994	0.4994	0.4994	0.4995	0.4995	0.4995
3.3	0.4995	0.4995	0.4996	0.4996	0.4996	0.4996	0.4996	0.4996	0.4996	0.4997
3.4	0.4997	0.4997	0.4997	0.4997	0.4997	0.4997	0.4997	0.4997	0.4998	0.4998
4.0	0.499968									

Source: After Weatherburn, C.E., *A First Course in Mathematical Statistics*, 2nd ed. London: Cambridge University Press, 1957 (for z = 0 to z = 3.1); Richardson, C. H., *An Introduction to Statistical Analysis*. Orlando, FL: Harcourt Brace Jovanovich, 1944 (for z = 3.2 to z = 3.4); Bowker, A.H., and G.J. Lieberman, *Engineering Statistics*. Englewood Cliffs, NJ: Prentice-Hall, 1959 (for z = 4.0). Reprinted with permission.

TABLE A.2 Percentiles of student's *t*-distribution (*df* = degrees of freedom) (for one-sided test)

	P							
df	**0.60**	**0.75**	**0.90**	**0.95**	**0.975**	**0.99**	**0.995**	**0.9995**
1	0.325	1.000	3.078	6.314	12.706	31.821	63.657	636.619
2	0.289	0.816	1.886	2.920	4.303	6.965	9.925	31.598
3	0.277	0.765	0.633	2.353	3.182	4.541	5.841	12.941
4	0.271	0.741	1.533	2.132	2.776	3.747	4.604	8.610
5	0.267	0.727	1.476	2.015	2.571	3.365	4.032	6.859
6	0.265	0.718	1.440	1.943	2.447	3.143	3.707	5.959
7	0.263	0.711	1.415	1.895	2.365	2.998	3.499	5.405
8	0.262	0.706	2.397	1.860	2.306	2.896	3.355	5.041
9	0.261	0.703	1.383	1.833	2.262	2.821	3.250	4.781
10	0.260	0.700	1.372	1.812	2.228	2.764	3.169	4.587
11	0.260	0.697	1.363	1.796	2.201	2.718	3.106	4.437
12	0.259	0.695	1.356	1.782	2.179	2.681	3.055	4.318
13	0.259	0.694	1.350	1.771	2.160	2.650	3.012	4.221
14	0.258	0.692	1.345	1.761	2.145	2.624	2.977	4.140
15	0.258	0.691	1.341	1.753	2.131	2.602	2.947	4.073
16	0.258	0.690	1.337	1.746	2.120	2.583	2.921	4.015
17	0.257	0.689	1.333	1.740	2.110	2.567	2.898	3.965
18	0.257	0.688	1.330	1.734	2.101	2.552	2.878	3.922
19	0.257	0.688	1.328	1.729	2.093	2.539	2.861	3.883
20	0.257	0.687	1.325	1.725	2.086	2.528	2.845	3.850
21	0.257	0.686	1.323	1.721	2.080	2.518	2.831	3.819
22	0.256	0.686	1.321	1.717	2.074	2.508	2.819	3.792
23	0.256	0.685	1.319	1.714	2.069	2.500	2.807	3.767
24	0.256	0.685	1.318	1.711	2.064	2.492	2.797	3.745
25	0.256	0.684	1.316	1.708	2.060	2.485	2.787	3.725
26	0.256	0.684	1.315	1.706	2.056	2.479	2.779	3.707
27	0.256	0.684	1.314	1.703	2.052	2.473	2.771	3.690
28	0.256	0.683	1.313	1.701	2.048	2.467	2.763	3.674
29	0.256	0.683	1.311	1.699	2.045	2.462	2.756	3.659
30	0.256	0.683	1.310	1.697	2.042	2.457	2.750	3.646
40	0.255	0.681	1.303	1.684	2.021	2.423	2.704	3.551
60	0.254	0.679	1.296	1.671	2.000	2.390	2.660	3.460
120	0.254	0.677	1.289	1.658	1.980	2.358	2.617	3.373
∞	0.253	0.674	1.282	1.645	1.960	2.326	2.576	3.291

Source: Beyer, W.H. ed., *CRC Handbook of Tables for Probability and Statistics* (Cleveland, OH: Chemical Rubber Company, 1966). Reprinted with permission.

TABLE A.3 Tolerance factors (K) for one-sided normal tolerance intervals with probability level (confidence factor) $\gamma = 0.95$ and coverage $P = 95\%$

n	K	n	K
3	7.655	75	1.972
4	5.145	100	1.924
5	4.202	125	1.891
6	3.707	150	1.868
7	3.399	175	1.850
8	3.188	200	1.836
9	3.031	225	1.824
10	2.911	250	1.814
11	2.815	275	1.806
12	2.736	300	1.799
13	2.670	325	1.792
14	2.614	350	1.787
15	2.566	375	1.782
16	2.523	400	1.777
17	2.486	425	1.773
18	2.543	450	1.769
19	2.423	475	1.766
20	2.396	500	1.763
21	2.371	525	1.760
22	2.350	550	1.757
23	2.329	575	1.754
24	2.309	600	1.752
25	2.292	625	1.750
30	2.220	650	1.748
35	2.166	675	1.746
40	2.126	700	1.744
45	2.092	725	1.742
50	2.065	750	1.740
		775	1.739
		800	1.737
		825	1.736
		850	1.734
		875	1.733
		900	1.732
		925	1.731
		950	1.729
		975	1.728
		1,000	1.727

Source:

(a) for sample sizes ≤50: Lieberman, Gerald J., "Tables for One-Sided Statistical Tolerance Limits," *Industrial Quality Control,* vol. 14, no. 10 (1958), pp. 1–9. © 1958 American Society for Quality. Reprinted with permission.

(b) for sample sizes ≥50: *K*-values calculated from large sample approximation. From *Statistical Analysis of Ground-Water Monitoring Data at TLCRA Facilities, Interim Final Guidance.* Washington, DC: Office of Solid Waste, U.S. Environmental Protection Agency, April 1989.

TABLE A.4 Critical values for the *t*-distribution

	Level of Significance for a One-Tailed Test							
	0.250	0.100	0.050	0.025	0.010	0.005	0.0025	0.0005
	Level of Significance for Two-Tailed Test							
Degrees of freedom	**0.500**	**0.200**	**0.100**	**0.050**	**0.020**	**0.010**	**0.005**	**0.001**
1	1.000	3.078	6.314	12.706	31.821	63.657	27.321	536.627
2	0.816	1.886	2.920	4.303	6.965	9.925	14.089	31.599
3	0.765	1.638	2.353	3.182	4.541	5.841	7.453	12.924
4	0.741	1.533	2.132	2.776	3.747	4.604	5.598	8.610
5	0.727	1.476	2.015	2.571	3.365	4.032	4.773	6.869
6	0.718	1.440	1.943	2.447	3.143	3.707	4.317	5.959
7	0.711	1.415	1.895	2.365	2.998	3.499	4.029	5.408
8	0.706	1.397	1.860	2.306	2.896	3.355	3.833	5.041
9	0.703	1.383	1.833	2.262	2.821	3.250	3.690	4.781
10	0.700	1.372	1.812	2.228	2.764	3.169	3.581	4.587
11	0.697	1.363	1.796	2.201	2.718	3.106	3.497	4.437
12	0.695	1.356	1.782	2.179	2.681	3.055	3.428	4.318
13	0.694	1.350	1.771	2.160	2.650	3.012	3.372	4.221
14	0.692	1.345	1.761	2.145	2.624	2.977	3.326	4.140
15	0.691	1.341	1.753	2.131	2.602	2.947	3.286	4.073
16	0.690	1.337	1.746	2.120	2.583	2.921	3.252	4.015
17	0.689	1.333	1.740	2.110	2.567	2.898	3.222	3.965
18	0.688	1.330	1.734	2.101	2.552	2.878	3.197	3.922
19	0.688	1.328	1.729	2.093	2.539	2.861	3.174	3.883
20	0.687	1.325	1.725	2.086	2.528	2.845	3.153	3.850
21	0.686	1.323	1.721	2.080	2.518	2.831	3.135	3.819
22	0.686	1.321	1.717	2.074	2.508	2.819	3.119	3.792
23	0.685	1.319	1.714	2.069	2.500	2.807	3.104	3.768
24	0.685	1.318	1.711	2.064	2.492	2.797	3.091	3.745
25	0.684	1.316	1.708	2.060	2.485	2.787	3.078	3.725
26	0.684	1.315	1.706	2.056	2.479	2.779	3.067	3.707
27	0.684	1.314	1.703	2.052	2.473	2.771	3.057	3.690
28	0.683	1.313	1.701	2.048	2.467	2.763	3.047	3.674
29	0.683	1.311	1.699	2.045	2.462	2.756	3.038	3.659
30	0.683	1.310	1.697	2.042	2.457	2.750	3.030	3.646
35	0.682	1.306	1.690	2.030	2.438	2.724	2.996	3.591
40	0.681	1.303	1.684	2.021	2.423	2.704	2.971	3.551
45	0.680	1.301	1.679	2.014	2.412	2.690	2.952	3.520
50	0.679	1.299	1.676	2.009	2.403	2.678	2.937	3.496
55	0.679	1.297	1.673	2.004	2.396	2.668	2.925	3.476
60	0.679	1.296	1.671	2.000	2.390	2.600	2.915	3.460
65	0.678	1.295	1.669	1.997	2.385	2.654	2.906	3.447
70	0.678	1.294	1.667	1.994	2.381	2.648	2.899	3.435
80	0.678	1.292	1.664	1.990	2.374	2.639	2.887	3.416
90	0.677	1.291	1.662	1.987	2.368	2.632	2.878	3.402
100	0.677	1.290	1.660	1.984	2.364	2.626	2.871	3.390
125	0.676	1.288	1.657	1.979	2.357	2.616	2.858	3.370
150	0.676	1.287	1.655	1.976	2.351	2.609	2.849	3.357
200	0.676	1.286	1.653	1.972	2.345	2.601	2.839	3.340
∞ (normal)	0.6745	1.2816	1.6448	1.9600	2.3267	2.5758	2.8070	3.2905

Source: R. H. McCuen, *Statistical Methods for Engineers*, © 1985, p. 389. Reprinted by permission of Prentice-Hall, Inc., Upper Saddle River, NJ.

TABLE A.5 Critical values for T_n (one-sided test) when standard deviation is calculated from the same sample

Number of observations n	Upper 0.1% significance level	Upper 0.5% significance level	Upper 1% significance level	Upper 2.5% significance level	Upper 5% significance level	Upper 10% significance level
3	1.155	1.155	1.155	1.155	1.153	1.148
4	1.499	1.496	1.492	1.481	1.463	1.425
5	1.780	1.764	1.749	1.715	1.672	1.602
6	2.011	1.973	1.944	1.887	1.822	1.729
7	2.201	2.139	2.097	2.020	1.938	1.828
8	2.358	2.274	2.221	2.126	2.032	1.909
9	2.492	2.387	2.323	2.215	2.110	1.977
10	2.606	2.482	2.410	2.290	2.176	2.036
11	2.705	2.564	2.485	2.355	2.234	2.088
12	2.791	2.636	2.550	2.412	2.285	2.134
13	2.867	2.699	2.607	2.462	2.331	2.175
14	2.935	2.755	2.659	2.507	2.371	2.213
15	2.997	2.806	2.705	2.549	2.409	2.247
16	3.052	2.852	2.747	2.585	2.443	2.279
17	3.103	2.894	2.785	2.620	2.475	2.309
18	3.149	2.932	2.821	2.651	2.504	2.335
19	3.191	2.968	2.854	2.681	2.532	2.361
20	3.230	3.001	2.884	2.709	2.557	2.385
21	3.266	3.031	2.912	2.733	2.580	2.408
22	3.300	3.060	2.939	2.758	2.603	2.429
23	3.332	3.087	2.963	2.781	2.624	2.448
24	3.362	3.112	2.987	2.802	2.644	2.467
25	3.389	3.135	3.009	2.822	2.663	2.486
26	3.415	3.157	3.029	2.841	2.681	2.502
27	3.440	3.178	3.049	2.859	2.698	2.519
28	3.464	3.199	3.068	2.876	2.714	2.534
29	3.486	3.218	3.085	2.893	2.730	2.549
30	3.507	3.236	3.103	2.908	2.745	2.563
31	3.528	3.253	3.119	2.924	2.759	2.577
32	3.546	3.270	3.135	2.938	2.773	2.591
33	3.565	3.286	3.150	2.952	2.786	2.604
34	3.582	3.301	3.164	2.965	2.799	2.616
35	3.599	3.316	3.178	2.979	2.811	2.628
36	3.616	3.330	3.191	2.991	2.823	2.639
37	3.631	3.343	3.204	3.003	2.835	2.650
38	3.646	3.356	3.216	3.014	2.846	2.661
39	3.660	3.369	3.228	3.025	2.857	2.671
40	3.673	3.381	3.240	3.036	2.866	2.682
41	3.687	3.393	3.251	3.046	2.877	2.692
42	3.700	3.404	3.261	3.057	2.887	2.700
43	3.712	3.415	3.271	3.067	2.896	2.710
44	3.724	3.425	3.282	3.075	2.905	2.719
45	3.736	3.435	3.292	3.085	2.914	2.727

continued

TABLE A.5 *continued*

Number of observations n	Upper 0.1% significance level	Upper 0.5% significance level	Upper 1% significance level	Upper 2.5% significance level	Upper 5% significance level	Upper 10% significance level
46	3.747	3.445	3.302	3.094	2.923	2.736
47	3.757	3.455	3.310	3.103	2.931	2.744
48	3.768	3.464	3.319	3.111	2.940	2.753
49	3.779	3.474	3.329	3.120	2.948	2.760
50	3.789	3.483	3.336	3.128	2.956	2.768
51	3.798	3.491	3.345	3.136	2.964	2.775
52	3.808	3.500	3.353	3.143	2.971	2.783
53	3.816	3.507	3.361	3.151	2.978	2.790
54	3.825	3.516	3.368	3.158	2.986	2.798
55	3.834	3.524	3.376	3.166	2.992	2.804
56	3.842	3.531	3.383	3.172	3.000	2.811
57	3.851	3.539	3.391	3.180	3.006	2.818
58	3.858	3.546	3.397	3.186	3.013	2.824
59	3.867	3.553	3.405	3.193	3.019	2.831
60	3.874	3.560	3.411	3.199	3.025	2.837
61	3.882	3.566	3.418	3.205	3.032	2.842
62	3.889	3.573	3.424	3.212	3.037	2.849
63	3.896	3.579	3.430	3.218	3.044	2.854
64	3.903	3.586	3.437	3.224	3.049	2.860
65	3.910	3.592	3.442	3.230	3.055	2.866
66	3.917	3.598	3.449	3.235	3.061	2.871
67	3.923	3.605	3.454	3.241	3.066	2.877
68	3.930	3.610	3.460	3.246	3.071	2.883
69	3.936	3.617	3.466	3.252	3.076	2.888
70	3.942	3.622	3.471	3.257	3.082	2.893
71	3.948	3.627	3.476	3.262	3.087	2.897
72	3.954	3.633	3.482	3.267	3.092	2.903
73	3.960	3.638	3.487	3.272	3.098	2.908
74	3.965	3.643	3.492	3.278	3.102	2.912
75	3.971	3.648	3.496	3.282	3.107	2.917
76	3.977	3.654	3.502	3.287	3.111	2.922
77	3.982	3.658	3.507	3.291	3.117	2.927
78	3.987	3.663	3.511	3.297	3.121	2.931
79	3.992	3.669	3.516	3.301	3.125	2.935
80	3.998	3.673	3.521	3.305	3.130	2.940
81	4.002	3.677	3.525	3.309	3.134	2.945
82	4.007	3.682	3.529	3.315	3.139	2.949
83	4.012	3.687	3.534	3.319	3.143	2.953
84	4.017	3.691	3.539	3.323	3.147	2.957
85	4.021	3.695	3.543	3.327	3.151	2.961
86	4.026	3.699	3.547	3.331	3.155	2.966
87	4.031	3.704	3.551	3.335	3.160	2.970
88	4.035	3.708	3.555	3.339	3.163	2.973
89	4.039	3.712	3.559	3.343	3.167	2.977
90	4.044	3.716	3.563	3.347	3.171	2.981

TABLE A.5 *continued*

Number of observations n	Upper 0.1% significance level	Upper 0.5% significance level	Upper 1% significance level	Upper 2.5% significance level	Upper 5% significance level	Upper 10% significance level
91	4.049	3.720	3.567	3.350	3.174	2.984
92	4.053	3.725	3.570	3.355	3.179	2.989
93	4.057	3.728	3.575	3.358	3.182	2.993
94	4.060	3.732	3.579	3.362	3.186	2.996
95	4.064	3.736	3.582	3.365	3.189	3.000
96	4.069	3.739	3.586	3.369	3.193	3.003
97	4.073	3.744	3.589	3.372	3.196	3.006
98	4.076	3.747	3.593	3.377	3.201	3.011
99	4.080	3.750	3.597	3.380	3.204	3.014
100	4.084	3.754	3.600	3.383	3.207	3.017
101	4.088	3.757	3.603	3.386	3.210	3.021
102	4.092	3.760	3.607	3.390	3.214	3.024
103	4.095	3.765	3.610	3.393	3.217	3.027
104	4.098	3.768	3.614	3.397	3.220	3.030
105	4.102	3.771	3.617	3.400	3.224	3.033
106	4.105	3.774	3.620	3.403	3.227	3.037
107	4.109	3.777	3.623	3.406	3.230	3.040
108	4.112	3.780	3.626	3.409	3.233	3.043
109	4.116	3.784	3.629	3.412	3.236	3.046
110	4.119	3.787	3.632	3.415	3.239	3.049
111	4.122	3.790	3.636	3.418	3.242	3.052
112	4.125	3.793	3.639	3.422	3.245	3.055
113	4.129	3.796	3.642	3.424	3.248	3.058
114	4.132	3.799	3.645	3.427	3.251	3.061
115	4.135	3.802	3.647	3.430	3.254	3.064
116	4.138	3.805	3.650	3.433	3.257	3.067
117	4.141	3.808	3.653	3.435	3.259	3.070
118	4.144	3.811	3.656	3.438	3.262	3.073
119	4.146	3.814	3.659	3.441	3.265	3.075
120	4.150	3.817	3.662	3.444	3.267	3.078
121	4.153	3.819	3.665	3.447	3.270	3.081
122	4.156	3.822	3.667	3.450	3.274	3.083
123	4.159	3.824	3.670	3.452	3.276	3.086
124	4.161	3.827	3.672	3.455	3.279	3.089
125	4.164	3.831	3.675	3.457	3.281	3.092
126	4.166	3.833	3.677	3.460	3.284	3.095
127	4.169	3.836	3.680	3.462	3.286	3.097
128	4.173	3.838	3.683	3.465	3.289	3.100
129	4.175	3.840	3.686	3.467	3.291	3.102
130	4.178	3.843	3.688	3.470	3.294	3.104
131	4.180	3.845	3.690	3.473	3.296	3.107
132	4.183	3.848	3.693	3.475	3.298	3.109
133	4.185	3.850	3.695	3.478	3.302	3.112
134	4.188	3.853	3.697	3.480	3.304	3.114
135	4.190	3.856	3.700	3.482	3.306	3.116

continued

TABLE A.5 *continued*

Number of observations n	Upper 0.1% significance level	Upper 0.5% significance level	Upper 1% significance level	Upper 2.5% significance level	Upper 5% significance level	Upper 10% significance level
136	4.193	3.858	3.702	3.484	3.309	3.119
137	4.196	3.860	3.704	3.487	3.311	3.122
138	4.198	3.863	3.707	3.489	3.313	3.124
139	4.200	3.865	3.710	3.491	3.315	3.126
140	4.203	3.867	3.712	3.493	3.318	3.129
141	4.205	3.869	3.714	3.497	3.320	3.131
142	4.207	3.871	3.716	3.499	3.322	3.133
143	4.209	3.874	3.719	3.501	3.324	3.135
144	4.212	3.876	3.721	3.503	3.326	3.138
145	4.214	3.879	3.723	3.505	3.328	3.140
146	4.216	3.881	3.725	3.507	3.331	3.142
147	4.219	3.883	3.727	3.509	3.334	3.144

Source: ASTM Designation E178-41, "Standard Recommended Practice for Dealing with Outlying Observations," 1975. Copyright ASTM. Reprinted with permission.

TABLE A.6 Coefficients (a_{N-i+1}) for Shapiro-Wilk W-test of normality

i/n	2	3	4	5	6	7	8	9	10
1	0.7071	0.7071	0.6872	0.6646	0.6431	0.6233	0.6052	0.5888	0.5739
2	—	0.0000	0.1677	0.2413	0.2806	0.3031	0.3164	0.3244	0.3291
3	—	—	—	0.0000	0.0875	0.1401	0.1743	0.1976	0.2141
4	—	—	—	—	—	0.0000	0.0561	0.0947	0.1224
5	—	—	—	—	—	—	—	0.0000	0.0399

i/n	11	12	13	14	15	16	17	18	19	20
1	0.5601	0.5475	0.5359	0.5251	0.5150	0.5056	0.4968	0.4886	0.4808	0.4734
2	0.3315	0.3325	0.3325	0.3318	0.3306	0.3290	0.3273	0.3253	0.3232	0.3211
3	0.2260	0.2347	0.2412	0.2460	0.2495	0.2521	0.2540	0.2553	0.2561	0.2565
4	0.1429	0.1586	0.1707	0.1802	0.1878	0.1939	0.1988	0.2027	0.2059	0.2085
5	0.0695	0.0922	0.1099	0.1240	0.1353	0.1447	0.1524	0.1587	0.1641	0.1686
6	0.0000	0.0303	0.0539	0.0727	0.0880	0.1005	0.1109	0.1197	0.1271	0.1334
7	—	—	0.0000	0.0240	0.0433	0.0593	0.0725	0.0837	0.0932	0.1013
8	—	—	—	—	0.0000	0.0196	0.0359	0.0496	0.0612	0.0711
9	—	—	—	—	—	—	0.0000	0.0163	0.0303	0.0422
10	—	—	—	—	—	—	—	—	0.0000	0.0140

i/n	21	22	23	24	25	26	27	28	29	30
1	0.4643	0.4590	0.4542	0.4493	0.4450	0.4407	0.4366	0.4328	0.4291	0.4254
2	0.3185	0.3156	0.3126	0.3098	0.3069	0.3043	0.3018	0.2992	0.2968	0.2944
3	0.2578	0.2571	0.2563	0.2554	0.2543	0.2533	0.2522	0.2510	0.2499	0.2487
4	0.2119	0.2131	0.2139	0.2145	0.2148	0.2151	0.2152	0.2151	0.2150	0.2148
5	0.1736	0.1764	0.1787	0.1807	0.1822	0.1836	0.1848	0.1857	0.1864	0.1870
6	0.1399	0.1443	0.1480	0.1512	0.1539	0.1563	0.1584	0.1601	0.1616	0.1630
7	0.1092	0.1150	0.1201	0.1245	0.1283	0.1316	0.1346	0.1372	0.1395	0.1415
8	0.0804	0.0878	0.0941	0.0997	0.1046	0.1089	0.1128	0.1162	0.1192	0.1219
9	0.0530	0.0618	0.0696	0.0764	0.0823	0.0876	0.0923	0.0965	0.1002	0.1036
10	0.0263	0.0368	0.0459	0.0539	0.0610	0.0672	0.0728	0.0778	0.0822	0.0862
11	0.0000	0.0122	0.0228	0.0321	0.0403	0.0476	0.0540	0.0598	0.0650	0.0697
12	—	—	0.0000	0.0107	0.0200	0.0284	0.0358	0.0424	0.0483	0.0537
13	—	—	—	—	0.0000	0.0094	0.0178	0.0253	0.0320	0.0381
14	—	—	—	—	—	—	0.0000	0.0084	0.0159	0.0227
15	—	—	—	—	—	—	—	—	0.0000	0.0076

i/n	31	32	33	34	35	36	37	38	39	40
1	0.4220	0.4188	0.4156	0.4127	0.4096	0.4068	0.4040	0.4015	0.3989	0.3964
2	0.2921	0.2898	0.2876	0.2854	0.2834	0.2813	0.2794	0.2774	0.2755	0.2737
3	0.2475	0.2463	0.2451	0.2439	0.2427	0.2415	0.2403	0.2391	0.2380	0.2368
4	0.2145	0.2141	0.2137	0.2132	0.2127	0.2121	0.2116	0.2110	0.2104	0.2098
5	0.1874	0.1878	0.1880	0.1882	0.1883	0.1883	0.1883	0.1881	0.1880	0.1878
6	0.1641	0.1651	0.1660	0.1667	0.1673	0.1678	0.1683	0.1686	0.1689	0.1691
7	0.1433	0.1449	0.1463	0.1475	0.1487	0.1496	0.1503	0.1513	0.1520	0.1526
8	0.1243	0.1265	0.1284	0.1301	0.1317	0.1331	0.1344	0.1356	0.1366	0.1376
9	0.1066	0.1093	0.1118	0.1140	0.1160	0.1179	0.1196	0.1211	0.1225	0.1237
10	0.0899	0.0931	0.0961	0.0988	0.1013	0.1036	0.1056	0.1075	0.1092	0.1108

continued

TABLE A.6 *continued*

i/n	31	32	33	34	35	36	37	38	39	40
11	0.0739	0.0777	0.0812	0.0844	0.0873	0.0900	0.0924	0.0947	0.0967	0.0896
12	0.0585	0.0629	0.0669	0.0706	0.0739	0.0770	0.0798	0.0824	0.0848	0.0870
13	0.0435	0.0485	0.0530	0.0572	0.0610	0.0645	0.0677	0.0706	0.0733	0.0759
14	0.0289	0.0344	0.0395	0.0441	0.0484	0.0523	0.0559	0.0592	0.0622	0.0651
15	0.0144	0.0206	0.0262	0.0314	0.0361	0.0404	0.0444	0.0481	0.0515	0.0546
16	0.0000	0.0068	0.0131	0.0187	0.0239	0.0287	0.0331	0.0372	0.0409	0.0444
17	—	—	0.0000	0.0062	0.0119	0.0172	0.0220	0.0264	0.0305	0.0343
18	—	—	—	—	0.0000	0.0057	0.0110	0.0158	0.0203	0.0244
19	—	—	—	—	—	—	0.0000	0.0053	0.0101	0.0146
20	—	—	—	—	—	—	—	—	0.0000	0.0049

i/n	41	42	43	44	45	46	47	48	49	50
1	0.3940	0.3917	0.3894	0.3872	0.3850	0.3830	0.3808	0.3789	0.3000	0.3751
2	0.2719	0.2701	0.2684	0.2667	0.2651	0.2635	0.2620	0.2604	0.2589	0.2574
3	0.2357	0.2345	0.2334	0.2323	0.2313	0.2302	0.2291	0.2281	0.2271	0.2260
4	0.2091	0.2085	0.2078	0.2072	0.2065	0.2058	0.2052	0.2045	0.2038	0.2032
5	0.1876	0.1874	0.1871	0.1868	0.1865	0.1862	0.1859	0.1855	0.1851	0.1847
6	0.1693	0.1694	0.1695	0.1695	0.1695	0.1695	0.1695	0.1693	0.1692	0.1691
7	0.1531	0.1535	0.1539	0.1542	0.1545	0.1548	0.1550	0.1551	0.1553	0.1554
8	0.1384	0.1392	0.1398	0.1405	0.1410	0.1415	0.1420	0.1423	0.1427	0.1430
9	0.1249	0.1259	0.1269	0.1278	0.1286	0.1293	0.1300	0.1306	0.1312	0.1317
10	0.1123	0.1136	0.1149	0.1160	0.1170	0.1180	0.1189	0.1197	0.1205	0.1212
11	0.1004	0.1020	0.1035	0.1049	0.1062	0.1073	0.1085	0.1095	0.1105	0.1113
12	0.0891	0.0909	0.0927	0.0943	0.0959	0.0972	0.0986	0.0998	0.1010	0.1020
13	0.0782	0.0804	0.0824	0.0842	0.0860	0.0876	0.0892	0.0906	0.0919	0.0932
14	0.0677	0.0701	0.0724	0.0745	0.0775	0.0785	0.0801	0.0817	0.0832	0.0846
15	0.0575	0.0602	0.0628	0.0651	0.0673	0.0694	0.0713	0.0731	0.0748	0.0764
16	0.0476	0.0506	0.0534	0.0560	0.0584	0.0607	0.0628	0.0648	0.0662	0.0685
17	0.0379	0.0411	0.0442	0.0471	0.0497	0.0522	0.0546	0.0568	0.0588	0.0608
18	0.0283	0.0318	0.0352	0.0383	0.0412	0.0439	0.0465	0.0489	0.0511	0.0532
19	0.0188	0.0227	0.0263	0.0296	0.0328	0.0357	0.0385	0.0411	0.0436	0.0459
20	0.0094	0.0316	0.0175	0.0211	0.0245	0.0277	0.0307	0.0335	0.0361	0.0386
21	0.0000	0.0045	0.0087	0.0126	0.0163	0.0197	0.0229	0.0259	0.0288	0.0314
22	—	—	0.0000	0.0042	0.0081	0.0118	0.0153	0.0185	0.0215	0.0244
23	—	—	—	—	0.0000	0.0039	0.0076	0.0111	0.0143	0.0174
24	—	—	—	—	—	—	0.0000	0.0037	0.0071	0.0104
25	—	—	—	—	—	—	—	—	0.0000	0.0035

Source: Shapiro, S. S., and M. B. Wilk, "An Analysis of Variance Test for Normality (Complete Samples)," *Biometrika*, 52, no. 3–4 (1965), pp. 591–611. Reprinted with permission.

TABLE A.7 Percentage points of the Shapiro-Wilk *W*-test

n	0.01	0.05
3	0.753	0.767
4	0.687	0.748
5	0.686	0.762
6	0.713	0.788
7	0.730	0.803
8	0.749	0.818
9	0.764	0.829
10	0.781	0.842
11	0.792	0.850
12	0.805	0.859
13	0.814	0.866
14	0.825	0.874
15	0.835	0.881
16	0.844	0.887
17	0.851	0.892
18	0.858	0.897
19	0.863	0.901
20	0.868	0.905
21	0.873	0.908
22	0.878	0.911
23	0.881	0.914
24	0.884	0.916
25	0.888	0.918
26	0.891	0.920
27	0.894	0.923
28	0.896	0.924
29	0.898	0.926
30	0.900	0.927
31	0.902	0.929
32	0.904	0.930
33	0.906	0.931
34	0.908	0.933
35	0.910	0.934
36	0.912	0.935
37	0.914	0.936
38	0.916	0.938
39	0.917	0.939
40	0.919	0.940
41	0.920	0.941
42	0.922	0.942
43	0.923	0.943
44	0.924	0.944
45	0.926	0.945
46	0.927	0.945
47	0.928	0.946
48	0.929	0.947
49	0.929	0.947
50	0.930	0.947

Source: Shapiro, S. S., and M. B. Wilk, "An Analysis of Variance Test for Normality (Complete Samples)," *Biometrika*, 52, no. 3–4 (1965), pp. 591–611. Reprinted with permission.

TABLE A.8 Percentage points of the Shapiro-Francia W-test for $n \geq 35$

n	0.01	0.05
35	0.919	0.943
50	0.935	0.953
51	0.935	0.954
53	0.938	0.957
55	0.940	0.958
57	0.944	0.961
59	0.945	0.962
61	0.947	0.963
63	0.947	0.964
65	0.948	0.965
67	0.950	0.966
69	0.951	0.966
71	0.953	0.967
73	0.956	0.968
75	0.956	0.969
77	0.957	0.969
79	0.957	0.970
81	0.958	0.970
83	0.960	0.971
85	0.961	0.972
87	0.961	0.972
89	0.961	0.972
91	0.962	0.973
93	0.963	0.973
95	0.965	0.974
97	0.965	0.975
99	0.967	0.976

Source: Shapiro, S. S., and R. S. Francia, "An Approximate Analysis of Variance Test for Normality," *Journal of the American Statistical Association*, 67, no. 337 (1972), pp. 215–216. Reprinted with permission.

TABLE A.9 Percentage points of the normal probability plot correlation coefficient

n	0.01	0.025	0.05
3	0.869	0.872	0.879
4	0.822	0.845	0.868
5	0.822	0.855	0.879
6	0.835	0.868	0.890
7	0.847	0.876	0.899
8	0.859	0.886	0.905
9	0.868	0.893	0.912
10	0.876	0.900	0.917
11	0.883	0.906	0.922
12	0.889	0.912	0.926
13	0.895	0.917	0.931
14	0.901	0.921	0.934
15	0.907	0.925	0.937
16	0.912	0.928	0.940
17	0.912	0.931	0.942
18	0.919	0.934	0.945
19	0.923	0.937	0.947
20	0.925	0.939	0.950
21	0.928	0.942	0.952
22	0.930	0.944	0.954
23	0.933	0.947	0.955
24	0.936	0.949	0.957
25	0.937	0.950	0.958
26	0.939	0.952	0.959
27	0.941	0.953	0.960
28	0.943	0.955	0.962
29	0.945	0.956	0.962
30	0.947	0.957	0.964
31	0.948	0.958	0.965
32	0.949	0.959	0.966
33	0.950	0.960	0.967
34	0.951	0.960	0.967
35	0.952	0.961	0.968
36	0.953	0.962	0.968
37	0.955	0.962	0.969
38	0.956	0.964	0.970
39	0.957	0.965	0.971
40	0.958	0.966	0.972
41	0.958	0.967	0.973
42	0.959	0.967	0.973
43	0.959	0.967	0.973
44	0.960	0.968	0.974
45	0.961	0.969	0.974
46	0.962	0.969	0.974
47	0.963	0.970	0.975
48	0.963	0.970	0.975
49	0.964	0.971	0.977
50	0.965	0.972	0.978

continued

TABLE A.9 *continued*

n	0.01	0.025	0.05
55	0.967	0.974	0.980
60	0.970	0.976	0.981
65	0.972	0.977	0.982
70	0.974	0.978	0.983
75	0.975	0.979	0.984
80	0.976	0.980	0.985
85	0.977	0.981	0.985
90	0.978	0.982	0.985
95	0.979	0.983	0.986
100	0.981	0.984	0.987

Source: Statistical Analysis of Ground-Water Monitoring Data at RCRA Facilities (Addendum to Interim Final Guidance). Office of Solid Waste, U.S. Environmental Protection Agency. July 1992.

TABLE A.10 Critical values of H for the Kruskal-Wallis test

n_1	n_2	n_3	Significance Level			
			0.10	0.05	0.01	0.005
2	1	1				
2	2	1				
2	2	2	4.571			
3	1	1				
3	2	1	4.286			
3	2	2	4.500	4.714	5.357	
3	3	1	4.571	5.143		
3	3	2	4.556	5.361		
3	3	3	4.622	5.600	7.200	7.200
4	1	1				
4	2	1	4.500			
4	2	2	4.056	5.208		
4	3	2	4.511	5.444	6.444	
4	3	3	4.709	5.727	6.746	
4	4	1	4.167	4.967	6.667	
4	4	2	4.555	5.455	7.036	
4	4	3	4.546	5.599	7.144	
4	4	4	4.654	5.692	7.654	
5	1	1				
5	2	1	4.200	5.000		
5	2	2	4.373	5.160	6.533	
5	3	1	4.018	4.960		
5	3	2	4.651	5.251	6.882	
5	3	3	4.533	5.649	7.079	
5	4	1	3.987	4.986	6.955	
5	4	2	4.541	5.268	7.118	
5	4	3	4.549	5.631	7.445	
5	4	4	4.619	5.618	7.760	
5	5	1	4.109	5.127	7.309	
5	5	2	4.508	5.339	7.269	
5	5	3	4.545	5.706	7.543	
5	5	4	4.523	5.643	7.791	
5	5	5	4.560	5.780	7.980	

Source: D. Ebdon, *Statistics in Geography*, 2nd ed. (Cambridge, MA: Blackwell Publishers, 1985), p. 207. Reprinted with permission.

TABLE A.11 Values of lambda (λ) for Cohen's method

	Percentage of Nondetects (*h*)						
γ	**0.01**	**0.05**	**0.10**	**0.15**	**0.25**	**0.40**	**0.50**
0.01	0.0102	0.0530	0.1111	0.1747	0.3205	0.5989	0.8403
0.05	0.0105	0.0547	0.1143	0.1793	0.3279	0.6101	0.8540
0.10	0.0110	0.0566	0.1180	0.1848	0.3366	0.6234	0.8703
0.15	0.0113	0.0584	0.1215	0.1898	0.3448	0.6361	0.8860
0.20	0.0116	0.0600	0.1247	0.1946	0.3525	0.6483	0.9012
0.30	0.0122	0.0630	0.1306	0.2034	0.3670	0.6713	0.9300
0.40	0.0128	0.0657	0.1360	0.2114	0.3803	0.6927	0.9570
0.50	0.0133	0.0681	0.1409	0.2188	0.3928	0.7129	0.9826
0.60	0.0137	0.0704	0.1455	0.2258	0.4045	0.7320	1.0070
0.70	0.0142	0.0726	0.1499	0.2323	0.4156	0.7502	1.0303
0.80	0.0146	0.0747	0.1540	0.2386	0.4261	0.7676	1.0527
0.90	0.0150	0.0766	0.1579	0.2445	0.4362	0.7844	1.0743
1.00	0.0153	0.0785	0.1617	0.2502	0.4459	0.8005	1.0951
1.10	0.0157	0.0803	0.1653	0.2557	0.4553	0.8161	1.1152
1.20	0.0160	0.0820	0.1688	0.2610	0.4643	0.8312	1.1347
1.30	0.0164	0.0836	0.1722	0.2661	0.4730	0.8458	1.1537
1.40	0.0167	0.0853	0.1754	0.2710	0.4815	0.8600	1.1721
1.50	0.0170	0.0868	0.1786	0.2758	0.4897	0.8738	1.1901
1.60	0.0173	0.0883	0.1817	0.2805	0.4977	0.8873	1.2076
1.70	0.0176	0.0898	0.1846	0.2851	0.5055	0.9005	1.2248
1.80	0.0179	0.0913	0.1876	0.2895	0.5132	0.9133	1.2415
1.90	0.0181	0.0927	0.1904	0.2938	0.5206	0.9259	1.2579
2.00	0.0184	0.0940	0.1932	0.2981	0.5279	0.9382	1.2739
2.10	0.0187	0.0954	0.1959	0.3022	0.5350	0.9502	1.2897
2.20	0.0189	0.0967	0.1986	0.3062	0.5420	0.9620	1.3051
2.30	0.0192	0.0980	0.2012	0.3102	0.5488	0.9736	0.3203
2.40	0.0194	0.0992	0.2037	0.3141	0.5555	0.9850	1.3352
2.50	0.0197	0.1005	0.2062	0.3179	0.5621	0.9962	1.3498
2.60	0.0199	0.1017	0.2087	0.3217	0.5686	1.0072	1.3642
2.70	0.0202	0.1029	0.2111	0.3254	0.5750	1.0180	1.3784
2.80	0.0204	0.1040	0.2135	0.3290	0.5812	1.0287	1.3924
2.90	0.0206	0.1052	0.2158	0.3326	0.5874	1.0392	1.4061
3.00	0.0209	0.1063	0.2182	0.3361	0.5935	1.0495	1.4197
3.10	0.0211	0.1074	0.2204	0.3396	0.5995	1.0597	1.4330
3.20	0.0213	0.1085	0.2227	0.3430	0.6054	1.0697	1.4462
3.30	0.0215	0.1096	0.2249	0.3464	0.6112	1.0796	1.4592
3.40	0.0217	0.1107	0.2270	0.3497	0.6169	1.0894	1.4720
3.50	0.0219	0.1118	0.2292	0.3529	0.6226	1.0990	1.4847
3.60	0.0221	0.1128	0.2313	0.3562	0.6282	1.1086	1.4972
3.70	0.0223	0.1138	0.2334	0.3594	0.6337	1.1180	1.5096
3.80	0.0225	0.1148	0.2355	0.3625	0.6391	1.1273	1.5218
3.90	0.0227	0.1158	0.2375	0.3656	0.6445	1.1364	1.5339
4.00	0.0229	0.1168	0.2395	0.3687	0.6498	1.1455	1.5458
4.10	0.0231	0.1178	0.2415	0.3717	0.6551	1.1545	1.5577
4.20	0.0233	0.1188	0.2435	0.3747	0.6603	1.1634	1.5693
4.30	0.0235	0.1197	0.2454	0.3777	0.6654	1.1722	1.5809
4.40	0.0237	0.1207	0.2473	0.3806	0.6705	1.1809	1.5924

TABLE A.11 *continued*

	Percentage of Nondetects (*h*)						
γ	**0.01**	**0.05**	**0.10**	**0.15**	**0.25**	**0.40**	**0.50**
4.50	0.0239	0.1216	0.2492	0.3836	0.6755	1.1895	1.6037
4.60	0.0241	0.1225	0.2511	0.3864	0.6805	1.1980	1.6149
4.70	0.0242	0.1235	0.2530	0.3893	0.6855	1.2064	1.6260
4.80	0.0244	0.1244	0.2548	0.3921	0.6903	1.2148	1.6370
4.90	0.0246	0.1253	0.2567	0.3949	0.6952	1.2230	1.6479
5.00	0.0248	0.1262	0.2585	0.3977	0.7000	1.2312	1.6587
5.10	0.0249	0.1270	0.2603	0.4004	0.7047	1.2394	1.6694
5.20	0.0251	0.1279	0.2621	0.4031	0.7094	1.2474	1.6800
5.30	0.0253	0.1288	0.2638	0.4058	0.7141	1.2554	1.6905
5.40	0.0255	0.1296	0.2656	0.4085	0.7187	1.2633	1.7010
5.50	0.0256	0.1305	0.2673	0.4111	0.7233	1.2711	1.7113
5.60	0.0258	0.1313	0.2690	0.4137	0.7278	1.2789	1.7215
5.70	0.0260	0.1322	0.2707	0.4163	0.7323	1.2866	1.7317
5.80	0.0261	0.1330	0.2724	0.4189	0.7368	1.2943	1.7418
5.90	0.0263	0.1338	0.2741	0.4215	0.7412	1.3019	1.7518
6.00	0.0264	0.1346	0.2757	0.4240	0.7456	1.3094	1.7617

Source: Statistical Analysis of Ground-Water Monitoring Data of RCRA Facilities (Addendum to Interim Final Guidance). Office of Solid Waste, U.S. Environmental Protection Agency. July 1992.

TABLE A.12 Percentiles of the chi-square χ^2 distribution with *df* degrees of freedom, χ^2_{v}, p

df	0.75	0.900	0.950	0.975	0.990	0.995	0.999
1	1.323	2.706	3.841	5.024	6.635	7.879	10.83
2	2.773	4.605	5.991	7.378	9.210	10.60	13.82
3	4.108	6.251	7.815	9.348	11.34	12.84	16.27
4	5.385	7.779	9.488	11.14	13.28	14.86	18.47
5	6.626	9.236	11.07	12.83	15.09	16.75	20.52
6	7.841	10.64	12.59	14.45	16.81	18.55	22.46
7	9.037	12.02	14.07	16.01	18.48	20.28	24.32
8	10.22	13.36	15.51	17.53	20.09	21.96	26.12
9	11.39	14.68	16.92	19.02	21.67	23.59	27.88
10	12.55	15.99	18.31	20.48	23.21	25.19	29.59
11	13.70	17.28	19.68	21.92	24.72	26.76	31.26
12	14.85	18.55	21.03	23.34	26.22	28.30	32.91
13	15.98	19.81	22.36	24.74	27.69	29.82	34.53
14	17.12	21.06	23.68	26.12	29.14	31.32	36.12
15	18.25	22.31	25.00	27.49	30.58	32.80	37.70
16	19.37	23.54	26.30	28.85	32.00	34.27	39.25
17	20.49	24.77	27.59	30.19	33.41	35.72	40.79
18	21.60	25.99	28.87	31.53	34.81	37.16	42.31
19	22.72	27.20	30.14	32.85	36.19	38.58	43.82
20	23.83	28.41	31.41	34.17	37.57	40.00	45.32
21	24.93	29.62	32.67	35.48	38.93	41.40	46.80
22	26.04	30.81	33.92	36.78	40.29	42.80	48.27
23	27.14	32.01	35.17	38.08	41.64	44.18	49.73
24	28.24	33.20	36.42	39.36	42.98	45.56	51.18
25	29.34	34.38	37.65	40.65	44.31	46.93	52.62
26	30.43	35.56	38.89	41.92	45.64	48.29	54.05
27	31.53	36.74	40.11	43.19	46.96	49.64	55.48
28	32.62	37.92	41.34	44.46	48.28	50.99	56.89
29	33.71	39.09	42.56	45.72	49.59	52.34	58.30
30	34.80	40.26	43.77	46.98	50.89	53.67	59.70
40	45.62	51.80	55.76	59.34	63.69	66.77	73.40
50	56.33	63.17	67.50	71.42	76.15	79.49	86.66
60	66.98	74.40	79.08	83.30	88.38	91.95	99.61
70	77.58	85.53	90.53	95.02	100.4	104.2	112.3
80	88.13	96.58	101.9	106.6	112.3	116.3	124.8
90	98.65	107.6	113.1	118.1	124.1	128.3	137.2
100	109.1	118.5	124.3	129.6	135.8	140.2	149.4

Source: Adapted from Pearson, E.S., and H.O. Hartley, eds., *Biometrika Tables for Statisticians, volume II* (Cambridge, UK: Cambridge University Press, 1972), pp. 160–169. Reprinted with permission.

TABLE A.13 Critical values of D for the Kolmogorov-Smirnov goodness-of-fit test

Degrees of freedom	Significance level				
	0.20	**0.15**	**0.10**	**0.05**	**0.01**
1	0.900	0.925	0.950	0.975	0.9950
2	0.684	0.726	0.776	0.842	0.929
3	0.565	0.597	0.642	0.708	0.829
4	0.494	0.525	0.564	0.624	0.734
5	0.446	0.474	0.510	0.563	0.669
6	0.410	0.436	0.470	0.521	0.618
7	0.381	0.405	0.438	0.486	0.577
8	0.358	0.381	0.411	0.457	0.543
9	0.339	0.360	0.388	0.432	0.514
10	0.322	0.342	0.368	0.409	0.486
11	0.307	0.326	0.352	0.391	0.468
12	0.295	0.313	0.338	0.375	0.450
13	0.284	0.302	0.325	0.361	0.433
14	0.274	0.292	0.314	0.349	0.418
15	0.266	0.283	0.304	0.338	0.404
16	0.258	0.274	0.295	0.328	0.391
17	0.250	0.266	0.286	0.318	0.380
18	0.244	0.259	0.278	0.309	0.370
19	0.237	0.252	0.272	0.301	0.361
20	0.231	0.246	0.264	0.294	0.352
25	0.21	0.22	0.242	0.264	0.32
30	0.19	0.20	0.22	0.242	0.29
35	0.18	0.19	0.21	0.23	0.27
40	0.17	0.18	0.19	0.21	0.25
50	0.15	0.16	0.17	0.19	0.23
60	0.14	0.15	0.16	0.17	0.21
70	0.13	0.14	0.15	0.16	0.19
80	0.12	0.13	0.14	0.15	0.18
90	0.11	0.12	0.13	0.14	0.17
100	0.11	0.11	0.12	0.14	0.16

Source: D. Ebdon, *Statistics in Geography*, 2nd ed. (Cambridge, MA: Blackwell Publishers, 1985), p. 199. Reprinted with permission.

TABLE A.14 *k*-values for Pearson type III distribution

	Recurrence interval in years						
	1.01	**1.11**	**1.25**	**5**	**25**	**50**	**100**
Skew coefficient C_S	**Percent chance**						
	99	**90**	**80**	**20**	**4**	**2**	**1**
	Positive skew						
3.0	–0.667	–0.660	–0.636	0.420	2.278	3.152	4.051
2.5	–0.799	–0.771	–0.711	0.518	2.262	3.048	3.845
2.0	–0.990	–0.895	–0.777	0.609	2.219	2.912	3.605
1.5	–1.256	–1.018	–0.825	0.690	2.146	2.743	3.330
1.0	–1.588	–1.128	–0.852	0.758	2.043	2.542	3.022
0.5	–1.955	–1.216	–0.856	0.808	1.910	2.311	2.686
0.0	–2.326	–1.282	–0.842	0.842	1.751	2.054	2.326
–0.1	–2.400	–1.292	–0.836	0.846	1.716	2.000	2.252
–0.5	–2.686	–1.323	–0.808	0.856	1.567	1.777	1.955
–1.0	–3.022	–1.340	–0.758	0.852	1.366	1.492	1.588
–1.5	–3.330	–1.333	–0.690	0.825	1.157	1.217	1.256
–2.0	–3.605	–1.302	–0.609	0.777	0.959	0.980	0.990
–2.5	–3.845	–1.250	–0.518	0.711	0.793	0.798	0.799
–3.0	–4.051	–1.180	–0.420	0.636	0.666	0.666	0.667

Source: After Water Resources Council, Bulletin no. 15, December 1967.

TABLE A.15 95th percentiles of the Bonferroni *t*-statistics, $t\,(df, \alpha/m)$

	m				
	1	2	3	4	5
	α/m				
df	**0.05**	**0.025**	**0.0167**	**0.0125**	**0.01**
4	2.13	2.78	3.20	3.51	3.75
5	2.02	2.57	2.90	3.17	3.37
6	1.94	2.45	2.74	2.97	3.14
7	1.90	2.37	2.63	2.83	3.00
8	1.86	2.31	2.55	2.74	2.90
9	1.83	2.26	2.50	2.67	2.82
10	1.01	2.23	2.45	2.61	2.76
15	1.75	2.13	2.32	2.47	2.60
20	1.73	2.09	2.27	2.40	2.53
30	1.70	2.04	2.21	2.34	2.46
—	1.65	1.96	2.13	2.24	2.33

Where df = degrees of freedom associated with the mean square error
m = number of comparisons
α = 0.05, the experimentwise error level

Source: Statistical Analysis of Ground-Water Monitoring Data at TLCRA Facilities, Interim Final Guidance. Washington, DC: Office of Solid Waste, U.S. Environmental Protection Agency, April 1989.

TABLE A.16 Significance levels of Dixon's criteria for testing extreme observations in a single sample

n	Criterion	Level 5%	Level 1%
3	$\frac{X_n - X_{n-1}}{X_n - X_1}$ *or* $\frac{X_2 - X_1}{X_n - X_1}$	0.941	0.988
4		0.765	0.889
5		0.642	0.780
6		0.560	0.698
7		0.507	0.637
8	$\frac{X_n - X_{n-1}}{X_n - X_2}$ *or* $\frac{X_2 - X_1}{X_{n-1} - X_1}$	0.554	0.683
9		0.512	0.635
10		0.477	0.597
11	$\frac{X_n - X_{n-2}}{X_n - X_2}$ *or* $\frac{X_3 - X_1}{X_{n-1} - X_1}$	0.576	0.679
12		0.546	0.642
13		0.521	0.615
14	$\frac{X_n - X_{n-2}}{X_n - X_3}$ *or* $\frac{X_3 - X_1}{X_{n-2} - X_1}$	0.546	0.641
15		0.525	0.616
16		0.507	0.595
17		0.490	0.577
18		0.475	0.561
19		0.462	0.547
20		0.450	0.535
21		0.440	0.524
23		0.421	0.505
24		0.413	0.497
25		0.406	0.489

Source: Dixon, W.J. "Processing Data for Outliers," *Biometrics*, 9 (1953), 74–89. Reprinted with permission.

TABLE A.17 Supplementary data for exercises attached in diskette file "NORM.XLS"

Sample number	Sampling location									
	A	B	C	D	E	F	G	H	I	J
1	37.2	35.6	21.8	26.7	26.0	35.4	44.6	28.9	37.2	55.9
2	24.6	38.4	51.6	50.9	34.0	42.4	23.8	35.7	36.3	39.7
3	36.1	36.8	44.1	45.4	36.6	24.2	18.8	30.7	25.9	31.4
4	29.2	24.8	33.5	38.5	23.0	32.6	29.8	46.4	31.1	29.7
5	30.0	36.2	43.4	35.1	25.2	36.9	40.9	29.5	53.3	21.8
6	34.9	30.0	49.4	27.6	46.8	43.9	11.1	31.6	25.5	44.1
7	31.1	18.3	24.6	39.1	37.6	33.1	42.5	26.4	20.0	43.6
8	27.9	34.8	17.2	32.7	33.0	39.8	27.4	43.4	36.1	24.6
9	50.6	28.3	43.9	25.3	51.8	35.9	34.0	40.1	38.5	45.8
10	53.3	29.2	37.9	39.1	21.0	30.8	34.2	34.4	34.8	37.7
11	20.8	32.8	35.7	21.5	41.0	42.6	52.1	36.7	32.8	38.0
12	25.5	49.7	40.4	34.8	31.9	41.9	34.2	45.6	31.3	41.6
13	32.7	44.2	36.0	15.8	30.9	29.0	52.1	35.2	55.9	52.8
14	63.0	42.3	21.9	18.9	34.2	25.3	47.6	48.9	51.6	33.2
15	45.1	31.3	39.6	21.4	32.6	30.6	32.7	27.4	44.1	34.3
16	19.0	40.1	32.9	34.0	43.6	38.1	15.0	25.0	49.4	52.5
17	31.7	44.6	52.8	36.9	30.2	26.9	38.2	33.3	43.9	28.7
18	36.1	55.9	37.6	53.5	48.2	44.8	49.9	65.5	40.4	47.4
19	44.2	46.1	31.2	31.6	43.3	51.3	30.6	40.8	39.5	37.1
20	36.8	39.7	28.6	34.9	38.8	25.3	32.1	31.2	37.6	29.7
21	29.7	24.6	32.3	45.3	25.1	39.1	38.9	51.6	32.3	34.7
22	48.6	31.7	37.8	46.9	23.7	37.0	32.8	24.5	50.5	39.0
23	33.4	36.1	23.1	33.9	34.4	32.1	51.9	30.7	50.8	36.8
24	33.9	29.7	50.5	32.4	29.4	43.1	55.2	46.8	38.5	20.9
25	46.5	38.3	43.0	29.1	43.9	39.2	5.4	21.7	5.7	36.0

continued

TABLE A.17 *continued*

Sample	Sampling location									
number	K	L	M	N	O	P	Q	R	S	T
1	23.3	33.2	31.1	39.5	47.4	37.3	46.5	41.6	43.2	40.9
2	43.1	33.6	35.4	25.5	38.2	39.0	22.9	46.9	41.3	31.6
3	41.9	45.2	56.9	28.2	40.3	34.9	46.3	38.8	31.4	40.3
4	41.2	40.2	28.4	29.3	36.0	25.8	20.8	34.3	43.3	37.8
5	38.0	35.4	38.9	37.0	39.2	54.8	25.8	38.9	39.8	40.5
6	31.4	38.2	26.7	17.1	27.9	31.9	29.5	33.4	23.5	38.7
7	28.6	24.4	29.6	32.9	25.0	37.3	22.3	43.3	26.7	53.8
8	43.3	46.3	30.9	16.1	25.2	32.7	18.4	44.4	42.1	30.9
9	67.5	27.4	38.8	22.9	30.3	26.2	17.5	46.1	45.4	22.1
10	39.8	43.0	25.7	30.4	34.3	30.6	32.5	39.5	32.9	40.5
11	41.0	36.1	35.9	29.3	29.0	58.9	31.4	29.0	37.2	41.5
12	23.5	34.2	30.0	44.3	49.3	28.9	32.5	74.6	38.5	28.2
13	35.3	32.2	38.5	27.0	26.5	29.7	29.7	34.5	33.2	37.0
14	26.6	22.6	35.4	30.8	24.3	33.1	42.0	28.0	45.2	34.9
15	36.7	36.1	36.7	38.3	28.9	17.9	25.8	29.2	38.2	32.4
16	42.1	40.1	29.2	43.5	41.9	24.1	31.2	27.7	46.3	29.0
17	39.9	37.9	51.8	25.9	34.0	29.8	54.8	52.2	43.0	40.3
18	45.4	18.6	35.2	29.5	36.2	45.4	25.6	28.6	34.2	31.5
19	18.8	44.6	29.0	50.8	30.7	33.5	28.5	34.7	36.1	23.0
20	32.9	28.2	35.8	21.0	26.8	32.0	23.3	34.6	38.0	33.5
21	35.4	24.9	35.0	54.9	44.9	29.4	27.8	32.2	28.2	38.9
22	37.2	40.3	20.1	31.5	18.2	29.2	38.7	35.9	40.3	49.4
23	41.3	36.7	34.9	34.9	57.7	27.1	26.1	24.7	30.0	42.2
24	38.5	30.1	38.5	36.9	54.3	40.1	29.7	21.9	31.1	41.2
25	34.1	46.8	31.7	19.3	46.0	62.3	43.2	32.5	56.9	27.0

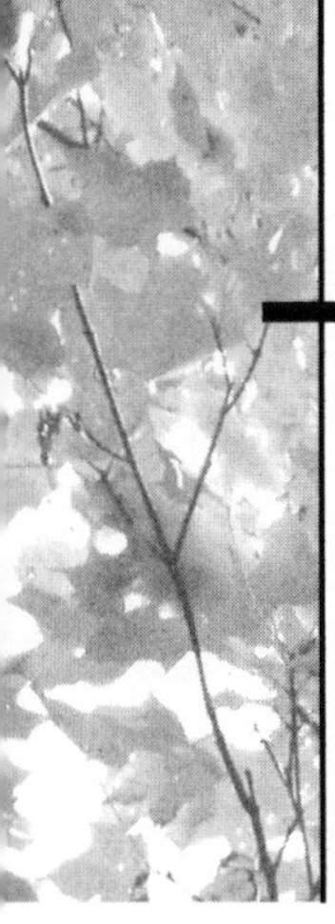

INDEX

A

B

C

D

E

F

G

H

I

K

L

M

N

O

P

Q

R

S

T

U

V

W